T0140666

Natural Computing Series

More information about this series at http://www.springer.com/series/4190

Mike Preuss · Michael G. Epitropakis ·
Xiaodong Li · Jonathan E. Fieldsend
Editors

Metaheuristics for Finding Multiple Solutions

Springer

Editors
Mike Preuss
LIACS
Universiteit Leiden
Leiden, The Netherlands

Xiaodong Li
Computer Science and Software
Engineering
RMIT University
Melbourne, VIC, Australia

Michael G. Epitropakis
The Signal Group
Athens, Greece

Jonathan E. Fieldsend
Department of Computer Science
University of Exeter
Exeter, UK

ISSN 1619-7127
Natural Computing Series
ISBN 978-3-030-79555-9 ISBN 978-3-030-79553-5 (eBook)
https://doi.org/10.1007/978-3-030-79553-5

This Springer imprint is published by the registered company Springer Nature Switzerland AG
The registered company address is: Gewerbestrasse 11, 6330 Cham, Switzerland

Foreword

My first encounter with the notion of multimodal optimization came in the early 1970s from discussions with my father, an acoustical engineer, regarding his approach to optimizing the acoustical properties of auditoriums used for concerts, religious services, congress halls, etc. The computational tools of his generation of engineers were slide rules and desk calculators, and so design optimization involved the manual evaluation of a relatively few, carefully chosen points in design space. Our conversations were motivated by the increasing availability of computers and their use in science and engineering and the possibility of significantly improving his time-intensive design process. It also happened to be at the time that I was doing my thesis and deeply involved in understanding how Holland's adaptive plans could be instantiated as algorithms useful for solving optimization problems.

As these initial (genetic) algorithms were developed, early notions of "premature convergence" emerged out of observations that most optimization problems of interest had multiple modes (optima) or equivalently, from a landscape perspective, multiple peaks and valleys and that these simple evolutionary algorithms were quite adept at quickly converging to a peak, but not necessarily the highest one. That lead to a variety of early ideas about algorithmic modifications such as "crowding", "fitness sharing", and "niching" to slow down the rate of convergence, maintain more population diversity, and improve problem-solving performance.

So, here we are, 50 years later, during which time we have seen the development of a wide variety of evolutionary algorithms far more complex and sophisticated than those simple genetic algorithms. At the same time, we continue to be confronted with multimodal optimization problems of increasing size and complexity. The result is the need for continuing the development and advancement of algorithmic techniques for solving them. This book fills an important gap and provides a valuable service with its clear introduction to the area, a precise summary of the issues, and a well-rounded survey of the current state of the art in multimodal optimization. It should be recommended reading for anyone interested in research in this area or involved in solving difficult application problems.

P.S. Those early discussions with my father led to the development of a simple evolutionary algorithm written in Basic, running on a family TRS-80, and allowed him to explore his acoustical design spaces in much more detail.

Kenneth De Jong
Professor Emeritus
George Mason University
Fairfax, USA

Preface

The goal of maintaining a diverse set of solutions in population-based stochastic optimization algorithms has been pursued for the past several decades. This has been driven by the need to effectively distribute their search over the design space, which in practice often has complicated mappings under an optimized quality function that can lead to premature convergence and the delivery of sub-optimal solutions if diversity is not maintained. Techniques designed specifically for achieving this goal are commonly referred to as *niching techniques*. Iconic niching techniques such as *crowding* and *fitness sharing* were developed as early as the 70s and 80s in the previous century by Kenneth De Jong and David Goldberg, respectively. At first, obtaining several distinct optimal solutions in one optimization run was not the primary motivation, but rather improving the chances to detect the sought single global optimum. These two goals are still intertwined in today's meta-heuristic algorithms, as it is not possible to know in practice if a solution is globally optimal without knowing other optima that may exist. Furthermore, in many application settings, it is also important to know alternative structurally distinct optimal solutions, e.g., design optimization. This is the main aim of what we see as *multimodal optimization* nowadays, also framed as *multi-solution optimization* by some researchers.

It has been common to subsume these types of algorithms by the term *niching* algorithms in recent decades, taking inspiration from biological niching mechanisms. However, modern evolutionary biology rather thinks of niches as the co-creation of organisms as opposed to a static setting that just has to be found. Interestingly, this perspective on population behavior and dynamics is the main motivation for *quality diversity*, which is nowadays a flourishing research topic area with close links to multimodal optimization. The term *niching* now often refers to the niching effects induced inside a standard optimization algorithm, which aims to ensure the found solutions are well distributed.

This edited book is for interested researchers who wish to learn more about the latest trends and developments in multimodal optimization and niching. The included chapters touch on algorithmic improvements and developments, representation and visualization issues, as well as new research directions, namely the addition of the multimodal viewpoint to multi-objective optimization, and quality diversity.

To our knowledge, this edited book is a first of this kind specifically on the topic of niching techniques. The chapter authors invited are active researchers who have made important contributions to niching in recent years.

In the following, we give a short summary of each of the 12 chapters of this book, to allow readers to have a quick glimpse and find the chapters they are most interested in. The order of presentation we chose is driven by the content of the chapters, but this does not mean that one builds on top of the other, and it shall not be necessary to have read previous chapters in order to understand subsequent ones. However, the novice reader who has not had any exposure to multimodal optimization and niching previously is encouraged to read the chapter "Multimodal Optimization: Formulation, Heuristics, and a Decade of Advances" as a primer.

The chapter "Multimodal Optimization: Formulation, Heuristics, and a Decade of Advances" provides an overview of the current state of aims, algorithms, measurements, and recent competition results in the field of multimodal optimization. It may serve as an introduction especially for readers not familiar with the area. Additionally, the results of the niching competitions of recent years are aggregated and discussed for the first time.

The chapter "Representation, Resolution and Visualization in Multimodal Optimization" describes how the properties and complexity of continuous multimodal problems can change with the *resolution* they are viewed under—including the curious situations where the number of modes can decrease with particular resolution increases. The chapter also presents the visualization of multi-dimensional multimodal landscapes through the use of *local optima networks*. As part of this, the chapter author provides network visualizations of the IEEE CEC test problems discussed in the chapter "Multimodal Optimization: Formulation, Heuristics, and a Decade of Advances" and shows how their complexities change with resolution.

The chapter "Finding Representative Solutions in Multimodal Optimization for Enhanced Decision-Making" proposes a novel niching algorithm, namely the Suppression-Radius-based Niching (SRN) algorithm, based on the principle of suppression radius (from the field of image processing). SRN is able to obtain a user-defined number of representative solutions. The notion of *representativeness* is examined to ensure the obtained solutions are sufficiently different and representative. SRN is especially appealing when incorporating a decision maker's knowledge and saving computational costs are paramount.

The chapter "Lifting the Multimodality-Fog in Continuous Multi-objective Optimization" transfers the multimodal perspective to multi-objective problems and introduces a visualization technique based on local gradients which helps to understand the search dynamics on these problems. This in turn enables building new multi-objective optimization algorithms that use the discovered dynamics to their advantage: in multi-objective optimization, multimodality may actually be helpful, not obstructive.

The chapter "Towards Basin Identification Methods with Robustness Against Outliers" investigates basin identification methods that are employed to steer the search process in global and multimodal optimization according to findings of the problem landscape. It reviews existing methods and shows how to parametrize them in order to make them less sensitive regarding outliers.

The chapter “Deflection and Stretching Techniques for Detection of Multiple Minimizers in Multimodal Optimization Problems” presents a comprehensive overview of the deflection and stretching techniques with simple examples, to address multimodal optimization problems. The main idea behind these techniques is to apply proper transformations to the objective function in order to facilitate any optimization algorithm to effectively search the space and detect multiple optima. The chapter also presents a concise review of the latest developments and applications of the techniques over the past two decades.

The chapter “Multimodal Optimization by Evolution Strategies with Repelling Subpopulations” describes a niching method based on the concept of repelling subpopulations. The method employs several existing ideas such as the taboo points in Tabu search, the normalized Mahalanobis distance, and the Hill-Valley function. The key consideration for adopting these techniques is to overcome the assumptions on distribution, size, and shape of the basins in MMO, which are often regarded as impractical in many real-world problems.

The chapter “Two-Phase Real-Valued Multimodal Optimization with the Hill–Valley Evolutionary Algorithm” describes the properties of the HillVallEA multimodal optimizer, currently the state of the art in terms of performance, as judged in the annual competition (described in the chapter “Multimodal Optimization: Formulation, Heuristics, and a Decade of Advances”). It investigates the properties of the two-phase clustering/optimization the method uses and contrasts it with other recent clustering-based multimodal optimizers, demonstrating its continued state-of-the-art performance on the CEC benchmark suit.

The chapter “Probabilistic Multimodal Optimization” introduces a number of probabilistic-based approaches to the multimodal optimization task, including locality-sensitive hashing to attack the computational cost associated with traditional niching approaches. The final optimization framework is also validated on common benchmark problems and also a practical problem from computer vision—automated pedestrian detection.

The chapter “Reduced Models of Gene Regulatory Networks: Visualising Multi--modal Landscapes” introduces multimodal optimization problems from the system biology domain, where mathematical models need to be calibrated to experimental data. In this case, tuning the kinetic parameters of a reduced-order gene regulatory network model. The complexity and highly multimodal task in this application domain are shown, which include using some of the visualization methods described in the chapter “Representation, Resolution and Visualization in Multimodal Optimization”, along with describing other measures on such landscapes.

The chapter “Grammar-Based Multi-objective Genetic Programming with Token Competition and Its Applications in Financial Fraud Detection” describes an application of the niching concept called *token competition* to Financial Fraud Detection (FFD) Problems. The proposed approach adopts several techniques including Grammar-based Genetic Programming (GBGP), token competition, multi-objective optimization, and ensemble learning. The method compares competitively with widely used data mining methods applied to the FFD problems in the literature.

The chapter "Phenotypic Niching Using Quality Diversity Algorithms" deals with the quality diversity approach and its similarities to multimodal optimization. Both apply niching methods, but quality diversity rather deals with behavioral systems than with static problem landscapes. However, the similarities are striking and the fields can certainly learn from one another.

Each of these chapters has been carefully selected and reviewed by the guest editors, who are either the current or former chairs/vice-chairs of the IEEE CIS Task-force on Multi-modal Optimization (MMO).[1] It is our hope that this book will be not only beneficial to beginners but also to more savvy niching researchers. Considering niching's broader applicability to a wide range of practical applications, we hope the book will help attract more researchers and practitioners to work on niching and multimodal optimization in the years to come.

Leiden, The Netherlands — Mike Preuss
Athens, Greece — Michael G. Epitropakis
Melbourne, Australia — Xiaodong Li
Exeter, UK — Jonathan E. Fieldsend
December 2020

Acknowledgments We would like to take this opportunity to thank all the chapter authors for their contributions to this book. It would not be possible without their dedication and patience for the past few years. We would also like to thank Springer, in particular Ronan Nugent, for his unreserved support during this difficult time due to the COVID-19 pandemic.

---

[1] http://www.epitropakis.co.uk/ieee-mmo/.

Contents

Multimodal Optimization: Formulation, Heuristics, and a Decade of Advances

Mike Preuss, Michael Epitropakis, Xiaodong Li, and Jonathan E. Fieldsend

Abstract Multimodal optimization is a relatively young term for the aim of finding several solutions of a complex objective function simultaneously. This has been attempted under the denomination 'niching' since the 1970s, transferring ideas from biological evolution in a very loose fashion. In this chapter we more formally define it, and then highlight its most important perspectives: how do we measure what is good? On what problems do we measure it? Which type of algorithms may be effectively employed for multimodal optimization? How do they relate to each other? Competitions at two major evolutionary computation conferences have driven algorithm development in recent years. We therefore report, in a concise fashion, what we have learned from competition results and give an outlook on interesting future developments.

1 Introduction

What is the essence of *multimodal optimization* (MMO) when lots of mathematical optimization methods have been used to deal with multimodal problems since the 1970s and even before? Experience shows that real-world problems are very often too complex to be treated successfully with classical optimization methods. This is

M. Preuss (✉)
Universiteit Leiden, Leiden, Netherlands
e-mail: m.preuss@liacs.leidenuniv.nl

M. Epitropakis
The Signal Group, Athens, Greece

X. Li
RMIT University, Melbourne, VIC, Australia

J. E. Fieldsend
University of Exeter, Exeter, UK

M. Preuss et al. (eds.), *Metaheuristics for Finding Multiple Solutions*,
Natural Computing Series,
https://doi.org/10.1007/978-3-030-79553-5_1

certainly a biased view, as simple problems are already solvable by means of standard optimization methods, and only the 'leftovers' are difficult enough to require some reasoning on how to treat them. These non-trivial real-world problems are almost always multimodal, either by nature of the underlying function or enforced by the islands of feasibility that are induced by constraints, or both. Although it is definitively as interesting as it is important to also treat constrained multimodal optimization, we abstain from doing so as this area is not that well explored yet and thus much more speculative.

Here, we provide an introduction and overview of the state-of-the-art of (unconstrained) multimodal optimization, primarily, but not exclusively, from an evolutionary computation (EC) perspective. Interestingly, many of the ideas and techniques that emerged for multimodal optimization in the EC field in recent years have predecessors in global optimization. Some of these have been forgotten for some decades and only recently been brought to light again (see, e.g., [1, 2]). We may therefore speak of *convergent evolution* in this case, as obviously the need to overcome a specific set of difficulties induced similar ideas in several research areas.

Let us first attempt to non-formally approach the meaning of the term 'multimodal optimization'. In our view, it is an umbrella term that contains several types of applications of specialized optimization methods to multimodal functions, but always with the focus on eventually delivering multiple solutions. As previously stated, very many application problems are multimodal in nature. However, they are often treated by simpler methods of convex optimization [3] that make strong assumptions on the geometrical shape of the function, or by methods from global optimization [2, 4] that are designed to find *one* global optimum of multimodal optimization problems. Why shall we be interested in determining several optima, and possibly some that are not even globally optimal? Li et al. [5] discussed this issue nearly 20 years ago and gave two main reasons:

- to increase the chance of actually locating the global optimum, and
- to provide insights into the problem and suggest alternative innovative solutions.

One may argue that the latter argument is actually two arguments, as understanding optimization problems in order to better adapt algorithms to them, is of value even if no alternative solutions are desired. The other half of the argument reflects practitioners' difficulties to accurately model a problem. The provided global optima may actually be infeasible (see [6] for an example), or cannot be realized exactly (i.e., when only a discrete set of shapes/parts/sizes is available, or machining is limited to a particular fidelity). The model itself may be in error, globally, or locally, meaning a range of disparate (but predicted high quality) solutions is of considerable value to mitigate the risk when the realizing the design(s) in practice. Besides, we may also encounter the situation that the decision-maker rejects specific solutions for reasons that have never been seen as part of the model, either because they have not been expected to be relevant, or because they are very hard to be expressed mathematically and are rather a matter of taste (so-called soft constraints). An example that is somewhere between both cases is the sound of a car combustion engine that is optimized for fuel efficiency but should still sound somewhat typical (not too different from the expectation of the customer).

Once we accept that searching for multiple optima at once is sometimes required, the next question is then how to define the properties of the *particular* optima we are interested in discovering. This becomes increasingly important if there are a large number of potential optima in an optimization problem. Even if we insist on returning global optima in our solution set, we still have at least three choices [1]:

1. find all 'best' solutions, that is all points in the search space that are globally optimal, or
2. obtain all optima, global or not, possibly better than a specified threshold, or
3. find at least one globally optimal solution.

Taking into account that for most practical black-box problems, we do not know the value of the global optimum and the problem to decide if a point is a global optimum is thus as hard as solving the optimization problem itself, we may further relax the requirements and just ask for some 'good' optima. However, this does not help much, because there is no clear criterion on which ones to look for if there are many, how to balance diversity and quality. The only thing that is clear is that the sought optima shall be distant enough from each other to actually make a difference in their practical implementation. This requires what has been very early termed as 'diversity maintenance', albeit also a fuzzy term.

We have come a long way already with niching and closely related similar approaches from global optimization, but the reshaped and unified field of *multimodal optimization* is still forming. We aim to consolidate it here and also want to point to interesting developments as well as open problems. However, this work is not a survey. For that, you may consider consulting one of [1, 7, 8]. Here, we want to point to the important lines of scientific development, not to every single attempt, and we largely do so from the perspective of the multimodal optimization competitions of the last years.

The chapter now continues with a definition of the optimization task before we turn to the available performance measures in Sect. 3. Section 4 then deals with test problems and generators, and Sect. 5 provides an overview over common algorithmic approaches. We report on the outcome of recent years competitions in Sect. 6, and then conclude the chapter with an outlook to possible future research directions.

2 Definitions

We now outline the multimodal optimization task more formally, starting by disambiguating it from the standard optimization task.

2.1 The General Optimisation Problem

The general optimization task is to find the 'best-performing' design, $\mathbf{x}^\star$, from some search space (or *domain*) $\mathcal{X}$, $\mathbf{x}^\star \in \mathcal{X}$, given some cost function, f. Without loss of generality, maximization problems can be converted to minimization variants through the simple multiplication of -1. $\mathcal{X}$ may be in a continuous space, a discrete space, a permutation space, mixed integer space, etc. For the use of our illustrations here, $\mathcal{X}$ is a boxed-constrained continuous space, i.e., $\mathcal{X} = [\boldsymbol{\ell}, \mathbf{u}] \subset \mathbb{R}^n$. $n \in \mathbb{N}$ is the fixed (in this case) number of decision variables. The vectors $\boldsymbol{\ell} = (\ell_1, \dots, \ell_n)^\top$ and $\mathbf{u} = (u_1, \dots, u_n)^\top$ are called the lower and upper bounds of $\mathcal{X}$, respectively.

There may be other constraints for the problem, which may be expressed as equality or inequality functions of $\mathbf{x}$, $g_i(\mathbf{x})$, however, in this chapter we shall focus exclusively on continuous box-constrained problems—which reflects the properties of the test problems in the competition results we discuss in later sections.

2.2 The Multimodal Optimization Problem

The combination of $\mathcal{X}$, f and $\mathbf{g}$ together are sufficient to describe the general optimization task, however in the *multimodal* optimization task we are also concerned with a *neighborhood-function*, which is used to characterize the local search landscape features.

For optimization problems in the continuous space, a common neighborhood-function is $N(\mathbf{y}) = \{\mathbf{x} \in \mathcal{X} \mid d(\mathbf{x}, \mathbf{y}) \le \epsilon\}$ which describes a ball-shaped neighborhood of a point $\mathbf{y} \in \mathcal{X}$. Using neighborhood-function N, and an $\epsilon > 0$, we can say $\mathbf{x}'$ is a local minimum at radius ϵ if $\nexists \mathbf{x} \in N(\mathbf{x}') : f(\mathbf{x}) \le f(\mathbf{x}')$. From the definition provided in [9, p. 6], the multimodal optimization task may be formulated explicitly as:

Definition 1 (*Multimodal minimization problem*) Let there be ν local minima $\mathbf{x}_1^*, \dots, \mathbf{x}_\nu^*$ of f in $\mathcal{X}$. If the ordering of these optima is $f(\mathbf{x}_1^*) \le \dots \le f(\mathbf{x}_l^*) < h \le \dots \le f(\mathbf{x}_\nu^*)$, a multimodal minimization problem is given as the task to approximate the set $\bigcup_{i=1}^{l} \{\mathbf{x}_i^*\}$.

The variable h in this definition is simply a threshold to potentially exclude some of the lower quality optima. Conversely, $h = \infty$ indicates we are interested in discovering *all* local optima of a problem. If $h = f(\mathbf{x}_1^*)$, we are only interested in approximating (all) the global optima.

Let P be the obtained approximation set. Additional constraints may be applied to this set, to obtain more specific problem definitions. For example, the cardinality of P could be restricted by requiring $|P| \le k$. If $k = 1$, we have the conventional global optimization problem, where typically only one solution is sought. Another issue relates to the diversity requirements, which could be formulated by demanding $\forall \mathbf{x}, \mathbf{y} \in P, \mathbf{x} \ne \mathbf{y} : d(\mathbf{x}, \mathbf{y}) > \kappa$, i.e., the distance between any two solutions may not

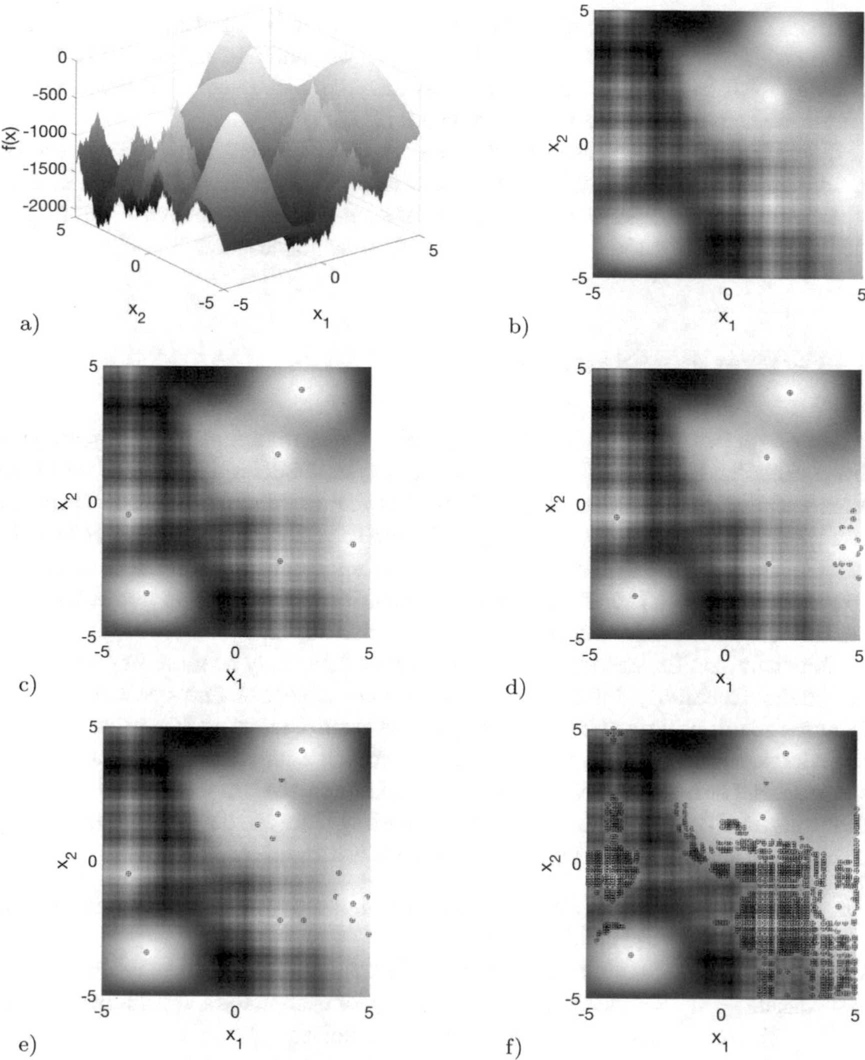

Fig. 1 Illustration of different targets under the multimodal optimization task. Composite Function 1 used from [10] with $n = 2$. Panels **a** and **b** illustrate the problem in three dimensions and the plane. Panel **c** has the desired solutions marked for $h = 0$, $\epsilon = 0.02$ and $\kappa = 0$ (the global optima), with $\kappa \gtrapprox 9.56$ a single global optima is satisfactory. Panel **d** has the desired solutions for $h = -200$, $\epsilon = 0.02$ and $\kappa = 0.2$ (the global optima and some high quality local optima). Panel **e** has the desired solutions for $h = -500$, $\epsilon = 0.02$ and $\kappa = 0.5$ (the global optima and some high quality local optima which are distant from them). Panel **f** has the desired solutions for $h = -1000$, $\epsilon = 0.02$ and $\kappa = 0.1$ (all global optima, and many local optima of moderate performance and better)

be smaller than some threshold κ. Figure 1 illustrates the desired set for various h, ϵ and κ on a multimodal test problem from [10]. Alternatively, a more sophisticated diversity measure on the set P could be calculated, which may, for instance, lead to a multi-objective formulation of the problem [11].

Given the above, one can see there is a range of different ways the multimodal optimization task may be defined, which unsurprisingly means there are a number of different (aligned) ways that the degree of *success* in solving such a task may be measured. We now outline some of these performance measures.

3 Performance Measures

Measuring performance in multimodal optimization has additional challenges beyond those exhibited by standard global optimization, as the definition of goodness is far more complex. The first striking difference is that as in multi-objective optimization, measurements are always computed from sets, not from single solutions [11, 12]. Furthermore, it is necessary to know where the sought optima are, or at least how good they are. The three principal dimensions of measuring performance in multimodal optimization are quality, diversity, and speed.

Depending on the application context, any of these may be more important than the others. Generally, this means that we cannot expect to find one measure that captures all our criteria, the measuring process is ambiguous by design and we have to make decisions regarding what we consider as the most important features of a measure, and indeed their relative importance. Leaving the temporal aspect (when are solutions found) aside for the moment, and starting from all search points (solutions) an algorithm has seen doing one run, this leads to a multi-step process:

subset selection remove all solutions that do not meet the minimum performance threshold

attribution decide which optima each solution belongs to

assignment provide performance values for each of the single known optima/ basins and their associated solutions

aggregation compute an overall performance value on the basis of the single values

Thus, apart from the detection of multiple local or global optima, also an explicit identification or subset selection of solutions that are needed for measuring within the possibly large set of explored solutions is necessary (see Fig. 2). Note that for global optimization, this is usually implicitly done by just selecting the best evaluated solution.

Ideally, the optimization algorithms will already perform this task. However, not all algorithms provide this explicit identification. Sometimes, it is easy to add, e.g., when the algorithm relies on several convergent local searches. Then, we can simply collect the best solutions of the local searches, and remove the duplicates. For some

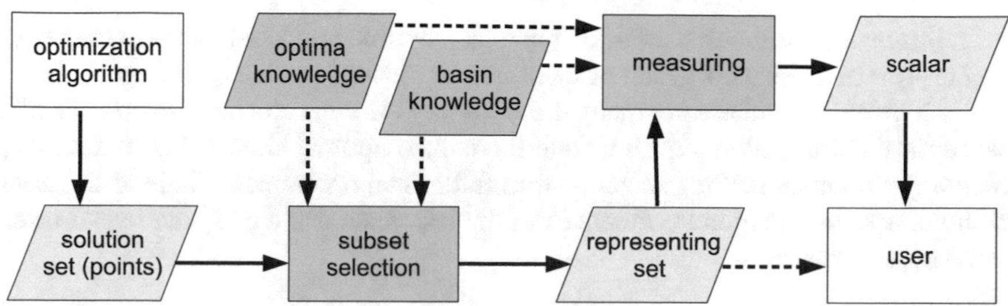

Fig. 2 How to evaluate performance of a multimodal optimization algorithm, general view (from [11])

population-based approaches, however, the whole history of evaluations may have to be searched for suitable solutions. This might require quadratic-time algorithms for a reasonable identification, which can get computationally expensive. Also, depending on the concrete measures, it may be necessary to use knowledge on the location of optima and/or basins for the selection process. However, without subset selection, multimodal optimization algorithms are not very useful in practice as they emit huge solution sets the decision-maker then has to choose from.

In the following, we list some possible measures and discuss briefly their advantages and disadvantages, mostly based on [11]. The first group of indicators is purely statistical and does not need any prior knowledge of a problem. It corresponds to the diversity goal only. In order to measure diversity of a solution set, one may use the sum of (all) distances, or the sum of nearest neighbor distances. The latter may be better at avoiding clustering of solutions. Solow-Polasky diversity is a biologically motivated diversity measure that has some advantages but is computationally expensive and thus suitable for rather smaller solution sets. Additionally, it uses a critical parameter, which is not trivial to specify. For more details on these measures please see [9].

Given the prior knowledge of the optima, i.e., the global optima are known in advance, we can measure the peak ratio (PR) by checking for each desired optimum if there is a point in the solution set that approximates it closely enough. This performance measure also requires a parameter in order to define how 'close' it is from an known optimum. By measuring the peak distance, that is the average distance from each optimum to the nearest point in the solution set, we can avoid that parameter. This measure is related to the inverted generational distance well known in multi-objective optimization. We can generalize it even further and define an averaged Hausdorff distance, which is especially suitable for comparisons of different solution set sizes as it penalizes unnecessary points. Whereas these three measures only account for the search space distances to the sought optima, we can also use the objective space distance, that is the average error per peak, in the peak inaccuracy (originally designed as peak accuracy) measure. Similarly, with knowledge of the basins of attraction, we can also define basin ratio and basin inaccuracy. However, as basin information is much harder to obtain than optima information, these measures are hardly ever used. This group of measures corresponds to the quality goal.

We have seen that there are many different ways to measure the performance of a multimodal optimization algorithm and there is no optimal choice. For that reason, the competitions on niching methods for multimodal optimization held at different evolutionary computation conferences in the last years use a combination of three performance criteria:

Peak Ratio (PR): This measure is kept for comparability to earlier results and is used as described in [10]. It measures the average percentage of the identified known global optima by an algorithm over multiple execution runs. In detail, given the fixed number of maximum function evaluations (MaxFEs) and the desired level of accuracy (ϵ), it can be calculated according to the following equation:

$$PR = \frac{\sum_{i=1}^{NR} GO_{\alpha}^{i}}{GO * NR}, \tag{1}$$

where GO_{α}^{i} denotes the number of global optima found at the end of the ith execution run by algorithm α, GO the number of known global optima, and NR the number of runs. Note that we do not do subset selection for this measure, the whole archive of sampled solutions in one run can be submitted and is searched for points that are near to the known (global) optima. This is also the recall (sensitivity) value for the other two criteria.

Static F_1: Here we are not only interested in the ability of the algorithm to find all sought optima, but also to do proper subset selection and deliver only the important portion of the solution set. In information retrieval, this is called precision (the fraction of useful information), and its product with the recall (the fraction of the total amount of relevant information that was actually retrieved) results in the F_1 measure.

More specifically, given an algorithm α, and a problem instance, let GO be the number of all global optima for that problem. Let $S\alpha$ be the set of potential solutions that have been obtained by algorithm α and let GO_{S_α} be the number of distinct global optima that exist in $S\alpha$. The precision metric denotes the fraction of retrieved data that are useful, where under the MMO formulation this can be translated as the fraction of the number of identified global optima by algorithm α over the number of solutions identified by α, i.e., $precision = GO_{S_\alpha}/|S_\alpha|$, where $|\cdot|$ denotes the cardinality of a set. Similarly, recall denotes the fraction of the total amount of relevant data that were actually retrieved, which in our context means the fraction of the number of global optima an algorithm α identifies over the number of existing global optima for a given problem $recall = GO_{S_\alpha}/GO$. It can be clearly observed that the recall measure essentially defines the peak ratio (PR) of an algorithm α. The static F_1 measure can be calculated according to the following equation:

$$F_1 = 2 \times \frac{precision \times recall}{precision + recall} = \frac{2 \times GO_{S_\alpha}}{GO + |S_\alpha|} \tag{2}$$

Just as the simple PR measure, the F_1 value is bounded by 1. It resembles the harmonic mean of precision and recall and takes its best value if all sought optima are found, and only these. Both measures correspond to the goals quality and diversity, but not to speed.

Dynamic F_1: The efficiency of an algorithm in terms of speed (how fast it identifies the desired global optima) cannot be captured by the above-mentioned metrics. Therefore, we introduce a dynamic F_1 measure in order to separate slower from faster algorithms. This can be achieved if the point in time (evaluation number under a fixed budget) is known when each solution that ends up in the final solution set $|S_\alpha|$ has been obtained. We simply compute the static F_1 value for each point in time when the solution set changes, and then integrate over this value (over the number of evaluations allowed in one run). Thus, the dynamic F_1 is the area under the curve of the static version F_1 over time and is also bounded by 1.
It is computed as follows: let the maximum budget of function evaluations of a problem instance be $MaxFEs$. Let GO_{S_α} be the solution set identified by algorithm α, ordered by the number of function evaluations f_i when solution i was added to the set, where $i \in [1, |GO_{S_\alpha}|]$. Then we can denote the subset of GO_{S_α} from solution i up to, including, solution j, where $i < j$ and $i, j \in [1, |GO_{S_\alpha}|]$, as $GO_{S_\alpha}(i, j)$. The dynamic F_1 measure ($dynF_1$) can be defined according to the following equation:

$$dynF_1 = \sum_{i=2}^{|GO_{S_\alpha}|} \left(\frac{f_i - f_{i-1}}{MaxFEs_p} F_1(GO_{S_\alpha}(1, i-1)) \right) + \frac{MaxFEs_p - f_{|GO_{S_\alpha}|}}{MaxFEs_p} F_1(GO_{S_\alpha}). \tag{3}$$

Intuitively the first term of the equation calculates the area under the curve of the static F_1 metric for a given solution set, i.e., $GO_{S_\alpha}(1, i-1)$, while the second term performs the last step of the integral to reach the maximum available budget. Larger values of the metric indicate faster conversion rates to the desired global optima set.

Please note that all these three measures are based on the peak ratio (PR) and therefore need information on the sought optima in order to be computed. Whereas diversity-based measures discussed above do not need that, if used alone, but they can only provide a very limited view onto the multimodal optimization performance. Thus, it is much harder to reasonably measure performance for real-world test problem when optima information is not known and we very much depend on benchmarks in order to develop better algorithms.

4 Benchmark Suites and Problem Generators

The capability of a niching method in locating multiple optimal (or near optimal) solutions within a single optimization run can be assessed by using a series of test functions constructed explicitly with multimodality in mind, i.e., the fitness landscape as computed by the objective function needs to be multimodal. In such a scenario, a search algorithm equipped with a niching method is challenged to find as many as possible optimal (or near optimal) solutions in an optimization run. This is clearly more demanding than just finding a single global optimal solution, which is typical for a standard metaheuristic algorithm. The earliest work on designing benchmark multimodal test functions was done by Deb in his 1989 master thesis [13]. Deb's test suite includes several simple 1 and 2-dimensional test functions, with multiple peaks of varying heights and distances between them. Subsequent to that, a more challenging test function with millions of local optima and 32 global optima was proposed by Goldberg et al. [14].

Test function generators have also been developed to construct multimodal test functions with more versatile and sophisticated sets of properties, also with the goal of being able to benchmark on a larger set of instances and not single problems. Gallagher and Yuan proposed such a generator based on the superposition of Gaussians [15]. Rönkkönen et al. [16] developed a more general multimodal test function generator using tunable function families, such as the *cosine* and *quadratic* function families. These test functions have more controllable numbers of global and local optima, scalable to higher dimensions. The functions can be rotated to a random angle, and each dimension of a function can be stretched independently using Bezier curves. Similarly, Singh and Deb proposed to use the *hump* function family to generate a series of multimodal test functions with different levels of complexity, e.g., several peaks can be generated at random locations with different shapes and sizes [17].

More complex multimodal fitness landscapes (i.e., with multiple global optima) can be constructed by superposition of several basic functions each with a single global optimum, as proposed by Qu and Suganthan [18]. Several such composition functions were included in the technical report 'Benchmark Functions for CEC'2013 Special Session and Competition on Niching Methods for Multimodal Function Optimization' [10]. Twenty multimodal test functions ranging from simple, separable, symmetric, and non-scalable to more challenging, non-separable, non-symmetric, and scalable were included in this benchmark suite, which has been used in the GECCO/CEC niching competition series running since 2013. Summarized results and rankings on many state-of-the-art niching methods are made publicly available (see details in a subsequent section on niching competition results). It has been widely adopted as the standard benchmark in evaluating niching methods in recent literature. The niching competition series aims to provide a common platform to facilitate meaningful evaluation and fair comparisons of different niching methods across a range of problem characteristics and difficulty levels.

Wessing [9] suggested the highly versatile polynomial based *Multiple Peaks Model 2* (MPM2) generator (in more detail described in [19]) that is based on former works by Preuss and Lasarczyk [20]. The most recent attempt in improving benchmark functions was carried out by Ahrari and Deb [21], where a general procedure for generating scalable multimodal test functions was proposed. This new procedure constructs a composite function through an application of two levels of basic functions, with the first level made up of two basic functions g_I and g_{II}, whose outputs in turn are fed into another basic function g_{III} in the 2nd level. Desirable properties such as ill-conditioning and variable interaction can be designed through manipulation of the adopted basic functions within the general framework. The composite functions generated are scalable to dimensions, with controllable number of global optima, variable shapes, and sizes of basins. In addition, constrained multimodal test functions can be also constructed through introducing constraints in the basic function g_I. Note that the first attempt in designing constrained multimodal test functions was made by Deb and Saha in [22].

To evaluate niching methods in dealing with uncertainty, multimodal test functions with consideration of robustness were also developed by Alyahya et al. [23]. Some of these robustness test functions were built upon the widely used standard multimodal benchmark, i.e., CEC'2013 niching competition technical report [10].

5 Popular Algorithmic Approaches and History of the Field

Studies on niching methods can be traced back to the early days of development of genetic algorithms, initially as attempts to promote population diversity [24–26] in order to improve global search capabilities. However, as a side-effect, it was found that these diversity maintenance techniques could be also used to locate multiple optimal solutions. One of the earliest studies attempting to induce niching behavior in a Genetic Algorithm (GA) is probably Cavicchio's dissertation [24], where several schemes were proposed in which offspring directly replace the parents that produce them. These so-called pre-selection schemes were later generalized by De Jong in a scheme called *crowding* [25]. Some years later, Goldberg and Richardson proposed *fitness sharing* [27] as alternative niching scheme, and both methods have been quite influential on later developments.

Whereas crowding attempts to restrict recombination or rather replacement to individuals that are similar, fitness sharing explicitly penalizes candidate solutions that are near to each other in terms of 'similarity' and reduces their fitness accordingly. Both methods suggest to use a search space distance function as similarity measure. The overall idea is of course to enable the survival of 'fitter as well as different' individuals occupying different promising regions of the search space. Related niching methods developed in this time include restricted tournament selection [28] and clearing [29].

However, it has been shown that the parameters introduced by these methods (i.e., niche radius and scaling factor) are rather difficult to determine [14] without any prior knowledge of a problem. Nevertheless, the notion of a niche radius and the matching metaphorical idea of a sub-population or species has dominated the multimodal optimization field for quite some time. From the large number of proponents of niche radius-based approaches, we name two examples: species conservation [5] and niching in evolution strategies with its dynamic peak identification [30].

At some point, it became clear that niche radii, even if adapted at runtime, limits the performance of multimodal optimization algorithms, simply because they cannot be expected to correctly characterize basins of attraction. Furthermore, evidence accumulated that it was unrealistic to try to employ a single population to manage a set of basins for a prolonged time [31]. At the same time, reasoning about the nature of niching and multimodal optimization led to the belief that researchers were actually try to handle a process that consisted of two distinct parts, and their interplay [32]:

- detecting and characterizing single basins of attraction, and
- performing local optimization in the most interesting of these basins.

Optimization algorithms may perform these steps explicitly or implicitly and in a continuous or alternating fashion, as summarized in [1, 9]. Modern multimodal optimization algorithms, therefore, diverge in these two major dimensions:

- how are basins identified (niching method, exploration), and
- how their local optimization is steered (base algorithm, exploitation).

Next to the previously described implicit basin identification, e.g., via restricted tournament selection, and looking beyond simple niche radius-based schemes, a number of interesting approaches have been suggested. In global optimization, simple clustering [2] evolved into more sophisticated specialized methods such as topographical selection [33]. The new geometrically motivated approaches consequently also use the objective value as an additional dimension. Another related method that does so is the hill-valley test of Ursem [34] that uses additional samples between suspected cluster centers. If used economically, it can be highly beneficial for multimodal optimization methods to spend a few additional evaluations, as [35] and, more recently, [36] show. Nearest-better clustering (NBC) [1] is a similar, geometrically motivated heuristic to detect basins that attempt the same goal, but without additional sample points. Wessing et al. [37] provide a thorough comparison of some of these methods and show that ensembles may improve performance.

The second important aspect of multimodal optimization algorithms is the layout of the base algorithm, how are the different optima pursued? As already described by Mahfoud, there are both sequential and parallel approaches to this [38]. In particular, the parallel methods are mostly population-based metaheuristic methods including Evolutionary Algorithms. They use different means such as, e.g., communication topologies [39] between sub-populations in order to prevent loss of information that would lead to fixation (the search space shrinks to one point). New algorithms have also been developed leveraging the unique characteristics of other meta-heuristics, e.g., Particle Swarm Optimization [40] and Differential Evolution [41]. Interested readers can find detailed information on these methods in a latest survey on niching methods [8].

Sequential algorithms tend to run repeatedly, each time saving a different optimum, and they do not necessarily possess a population. However, they need to make sure that search is started at different locations. Beasley has shown that this approach can work well [42], and other algorithms as the NEA+ [1] also followed this path.

6 Niching Competition Result Analysis

This section presents a summary of the results and analysis of participating niching algorithms in the competition series on 'Niching Methods for Multimodal Optimization' that took place between 2016 and 2019. These participating algorithms and presented results provide a glimpse of the latest trends in developing effective niching algorithms and the state-of-the-art in the field.

The competition series aims to provide a common platform allowing for easy and fair comparisons across different niching algorithms. This niching competition series started its first competition in 2013, at CEC'2013, based on the niching benchmark suite 'Benchmark Functions for CEC'2013 Special Session and Competition on Niching Methods for Multimodal Function Optimization' [10]. The competition has been organized by the IEEE CIS Task-force on MMO.[1] It has been held at both GECCCO and CEC conferences almost every year since 2013. The competition series has been playing a key role in engaging with the niching research community and has received numerous entries, among which several represents the state-of-the-art niching algorithms in the field [8]. The competition series has adopted widely used performance measures, i.e., the Average Peak Ratio and Success Rate [10], for evaluating the performance of niching methods. Most importantly, the competition provides a common platform to facilitate niching researchers to evaluate their algorithms' performance on a diverse set of test functions with different characteristics with increasing complexity. Different solution accuracy levels have been adopted. Intuitively, the higher the accuracy level, the more challenging the benchmark instance becomes for locating and maintaining all the global optima. Such setup first appeared in [41], while up until the CEC 2013 competition, the community only considered arbitrary accuracy levels that might lead to inaccurate assessment of the performance of niching approaches, since a successful located solution measured by a low accuracy level (e.g., 0.1) could be still far away from the true global optimum. In contrast, higher accuracy levels require effective search capabilities from the applied algorithm to accurately locate the basin of attraction of the global optimum.

In 2016, the competition series adopted two new performance measures, to better reflect the ability and efficiency of a niching algorithm to accurately identify the set of the known global optima, i.e., how comparable is the provided set of potential solutions with the set of known global optima, and how fast the global optima are identified by the algorithm. As a result, from 2016 onwards, the competition series included the following three new scenarios to rank participating niching algorithms

[1] http://www.epitropakis.co.uk/ieee-mmo/.

based on both existing and newly proposed performance measures (see Sect. 3 for details on how to compute these):

- **Scenario I:** This scenario adopts the CEC 2013 competition ranking procedure [10], which is based on the average Peak Ratio measure value across all benchmark instances and accuracy measures, to facilitate straightforward comparison with all previous competition entries. This measure is essentially the same as the *recall* measure commonly used in pattern recognition and information retrieval, where *recall* is defined as the fraction of the total amount of relevant instances that were actually retrieved.
- **Scenario II:** This scenario adopts the (static) F_1 measure (as described in Eq. 2, Sect. 3) to take into consideration the performance characteristics of the corresponding algorithm in terms of the *recall* and *precision* of its final solution set. The ranking procedure, in this case, is performed by the average F_1 measure value across the different benchmark instances and accuracy levels. The F_1 measure is usually understood as the product of *precision* and *recall*. Here, *precision* is defined as the fraction of relevant instances among the retrieved instances. In the case of niching (assuming maximization), *precision* is the fraction of detected peaks among all the retrieved solutions. Ideally, if all sought peaks are found and only these in the submitted entry, it would result in an F_1 value of 1 (as both recall and precision would be equal to 1).
- **Scenario III:** This scenario adopts the (dynamic) F_1 measure integral ($dynF_1$ as described in Eq. 3, Sect. 3) over the entire run time of an algorithm, for a benchmark instance, to take into account the computational efficiency of the submitted algorithm. The integral of the F_1 measure over time (or $dynF_1$) computes the area-under-the-curve (AUC) of F_1, divided by the maximum number of function evaluations that are allowed for that specific problem instance. It is, therefore, also bounded by the value of 1. This measure takes into account the time (in terms of the number of function evaluations) at which the peaks are detected.

It is worth noting that to accurately measure the performance of the algorithms on the three scenarios, the setup includes the option for competitors to perform actions on the submitted solution set as the budget is consumed. This is an essential feature to correctly measure the behavior of algorithms on the third scenario since an algorithm should be able to adapt its decision on which solutions are characterized as potential global optima through time. As such, three actions have been defined that give the option to add and delete solutions from the submitted solution set as well as to reset it.

The competition result analysis starts with a presentation of the final rankings of the submitted entries and some initial statistical analysis on the reported results. It then proceeds with a short analysis per scenario. Table 1 summarizes the competition results on all three scenarios for all competition entries submitted to GECCO/CEC competitions from 2016 to 2019. For each entry, the name of the algorithm (Algorithm), the mean Peak Ratio value (μ_{PR}), the mean F_1 (μ_{F_1}) and the mean $dynF_1$ (μ_{dynF_1}) are reported along with their corresponding ranking per scenario. The last two columns denote the mean rank per algorithm for all three scenarios and the

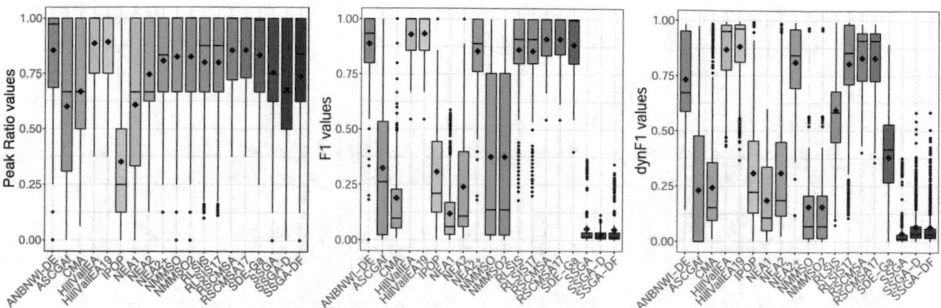

Fig. 3 Boxplots of the performance measure values across all accuracy levels and all benchmark instances. The first boxplot represents Scenario I (Peak Ratio values), the second boxplot corresponds to Scenario II (static F_1 values) and the third to Scenario III ($dynF_1$ values)

final rank across all algorithms and all scenarios.[2] Figure 3 provides the boxplots which show the variations on these performance measures for each algorithm across all accuracy levels and all benchmark instances. Note that the mean value of each distribution is marked by a (black) diamond in the plots. Some of these entries are variants from their original implementation, e.g., HillVallEA19 from the originally proposed HillVallEA. It is worth pointing out that, in most cases, differences between these variants are small, and the new versions still retain the original key ideas and principles. The top three ranking algorithms are highlighted in bold. Clearly, the best-performing entries across all three scenarios are the HillVallEA19, HillVallEA, and RSCMSA17. It is interesting to note that ranks in Scenario I and Scenario II are identical for the top six entries, showing remarkable consistency. Scenario III ranks are the same only for the top two entries. Scenario III ranks for the 3rd and 4th places are the 4th and 3rd places (i.e., swapping places) according to Scenario I and II. This suggests the amount of computational effort in detecting the sought peaks starts to differ. For Scenario III, it is worth noting the performance of NEA+ and RLSIS17 algorithms that exhibit much better performance in Scenario III comparing with the first two scenarios. This is justified from the fact that both approaches incorporate strong local search algorithms (e.g., CMA-ES) which can quickly refine potential solutions, and the submitted sets of potential solutions throughout the execution run is always small and includes mostly the best-estimated solutions without adding any noise.

The graphical illustration of the performance distributions (in Fig. 3) suggests a more clear view of the best-performing algorithms across different scenarios. In Scenario I, where the peak ratio measure is used, the best-performing algorithms can be clearly identified and are on par with the reported rankings in Table 1. However, rankings become a bit difficult to distinguish between the algorithms with mean and median performance values around 0.75. The performance differences of such entries become more evident in Scenario II, where the static F_1 metric considers both preci-

[2] Note that, due to space limitation on graphs, the SSGA-DMRTS-DDC and SSGA-DMRTS-DDC-F entries are denoted as SSGA-D and SSGA-DF accordingly.

Table 1 Result summary of niching competition results for entries submitted to GECCO/CEC from 2016 to 2019

Algorithm	Scenario I		Scenario II		Scenario III		Mean	Final
	μ_{PR}	Rank	μ_{F_1}	Rank	μ_{dynF_1}	Rank	Rank	Rank
ANBNWI-DE	0.8544	5	0.8872	5	0.7268	7	17/3 = 5.66	5
ASCGAf	0.6000	18	0.3244	12	0.2139	13	43/3 = 14.33	12
CMA	0.6680	16	0.1891	15	0.2410	12	43/3 = 14.33	12
HillVallEA	**0.8851**	**2**	**0.9297**	**2**	**0.8689**	**2**	6/3 = 2.00	**2**
HillVallEA19	**0.8916**	**1**	**0.9335**	**1**	**0.8827**	**1**	3 /3 = 1.00	**1**
IPOP	0.3522	19	0.3083	13	0.3033	11	43/3 = 14.33	12
NEA1	0.6088	17	0.1187	16	0.1693	14	47/3 = 15.66	13
NEA2	0.7458	13	0.2408	14	0.3052	10	37/3 = 12.33	11
NEA2+	0.8071	9	0.8548	9	0.8064	5	23/3 = 7.66	7
NMMSO	0.8254	8	0.3763	11	0.1549	16	35/3 = 11.66	10
NMMSO2	0.8254	7	0.3763	10	0.1549	15	32/3 = 10.66	9
RLSIS	0.8001	10	0.8616	7	0.5903	8	25/3 = 8.33	8
RLSIS17	0.7995	11	0.8553	8	0.8007	6	25/3 = 8.33	8
RSCMSA	0.8548	4	0.9103	4	**0.8299**	**3**	11/3 = 3.66	4
RSCMSA17	**0.8557**	**3**	**0.9106**	**3**	0.8290	4	10/3 = 3.33	**3**
SDE-Ga	0.8329	6	0.8835	6	0.3764	9	21/3 = 7.00	6
SSGA	0.7552	12	0.0519	17	0.0293	19	48/3 = 16.00	14
SSGA-D	0.6793	15	0.0268	19	0.0565	17	51/3 = 17.00	16
SSGA-DF	0.7380	14	0.0482	18	0.0529	18	50/3 = 16.66	15

sion and recall of each algorithm. In this scenario, the top performers can be clearly separated from the rest ones. The top performers are ANBNWI-DE, HillVallEA, HillVallEA19, RLSIS, RLSIS17, RSCMSA, and RSCMSA17. In Scenario III, the top-performing group of algorithms includes HillVallEA and HillVallEA19, which are followed by the RSCMSA variants (RSCMSA and RSCMSA17). The next group includes RLSIS17, NEA2+, and ANBNWI-DE.

To statistically validate the observed differences among the algorithm across the three scenarios we perform statistical analysis on the reported performance values. We first employ the Friedman rank-sum test [43] to assess whether at least two algorithms exhibit significant differences in the observed performance values. The null hypothesis of the Friedman rank-sum test states that the performance distributions of all samples are the same, while the alternative hypothesis states that at least two samples differ. The statistical analysis proceeds with a posthoc analysis to determine which pairs of algorithms exhibit significant differences in performance (per scenario). In this step, pairwise Wilcoxon-signed rank tests are performed on the performance samples of each pair of algorithms. In addition, to alleviate the issue of having Type I errors given multiple comparisons, the Bonferroni correction method is applied.

For all three scenarios, the applied Friedman rank-sum tests reveal statistically significant differences in performances among the submissions across all considered problem instances and levels of accuracy, with all p-values being $\lll 10^{-16}$. As such, for all scenarios, we proceed to the posthoc analysis to investigate the pairwise performance differences among the algorithms. Figures 4, 5, and 6 present tile-plots to illustrate all pairwise differences in the observed performance samples across the different scenarios, with the outcomes of the pairwise tests at the 5% level of significance. More specifically, the outcomes of the pairwise Wilcoxon-signed rank tests, without and with the application of the Bonferroni correction method, are provided on the left and right-hand side of the figures respectively. Each tile corresponds to a pairwise significance test between the algorithms of the corresponding row and column. The color of the tile indicates if the observed performance differences were enough to reject the null hypothesis at the 5% significance level (p-value < 0.05). As such, gray tiles indicate significant differences between the pair of algorithms, while black tiles indicate that the observed performance differences did not support the rejection of the null hypothesis, i.e., no significant differences were observed.

Both sets of significance tests suggest that the majority of the previously mentioned observations on the performance differences are statistically significant. The differences between the algorithms in the top-performing group are significant for the majority of the cases across different scenarios. There are only a few pairs of algorithms that exhibit similar behaviors, for example, NMMSO, RLSIS, and RSCMSA variants against some specific cases.

The competition series checks five different accuracy levels with complexity increases substantially as the accuracy level increases. The highest accuracy level being checked is $\epsilon = 10^{-5}$, where an algorithm (apart from identifying the regions of different global optimal solutions) has to also identify corresponding global optima with the highest accuracy. Figure 7 shows the boxplot with median and the average

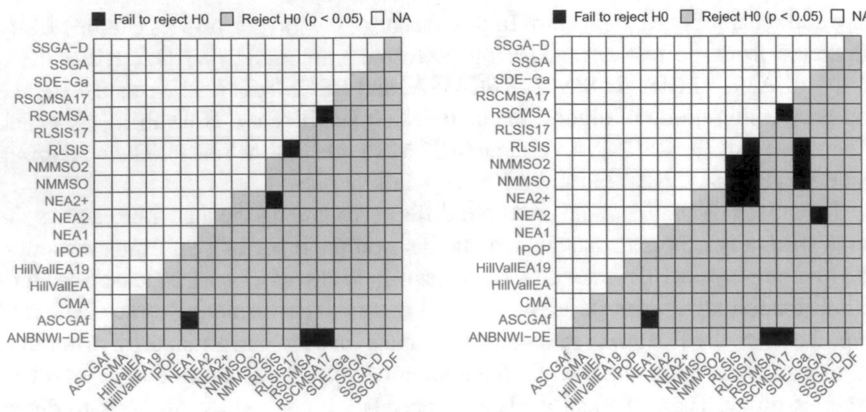

Fig. 4 Pairwise Wilcoxon statistical tests (left) with Bonferroni posthoc analysis (right) on results across all accuracy levels and all benchmark instances for Scenario I

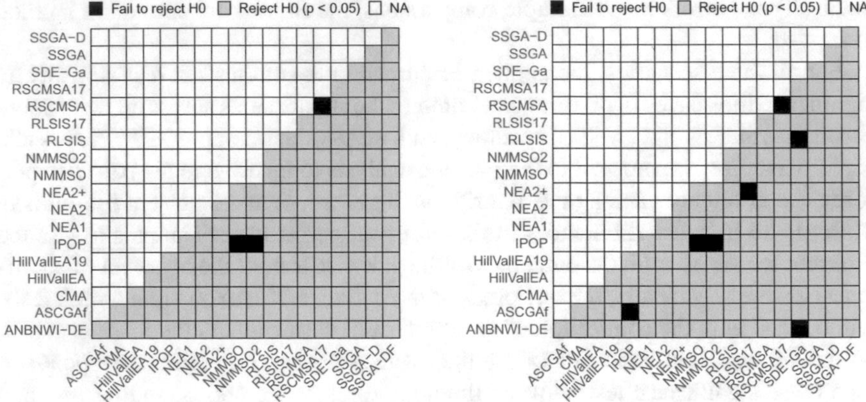

Fig. 5 Pairwise Wilcoxon statistical tests (left) with Bonferroni posthoc analysis (right) on results across all accuracy levels and all benchmark instances for Scenario II

peak ratio (i.e., Scenario I) of all participants across all benchmark problem instances for the $\epsilon = 10^{-5}$ level of accuracy. It is noticeable that for most algorithms, their average performance is lower than their median performance. It is worth noting that the group of best-performing algorithms usually incorporate effective local search algorithms, such as CMA-ES, which enables them to accurately search and identify the global optima when approaching their basins of attraction. As a result, the top-performing algorithms can still maintain their ranks even at such a challenging accuracy level.

Similarly, for Scenario II, Fig. 8 illustrates the performance distributions on precision, recall, and the F_1 measure of the algorithms across all benchmark problem instances at the highest accuracy level ($\epsilon = 10^{-5}$). It can be observed that most of the

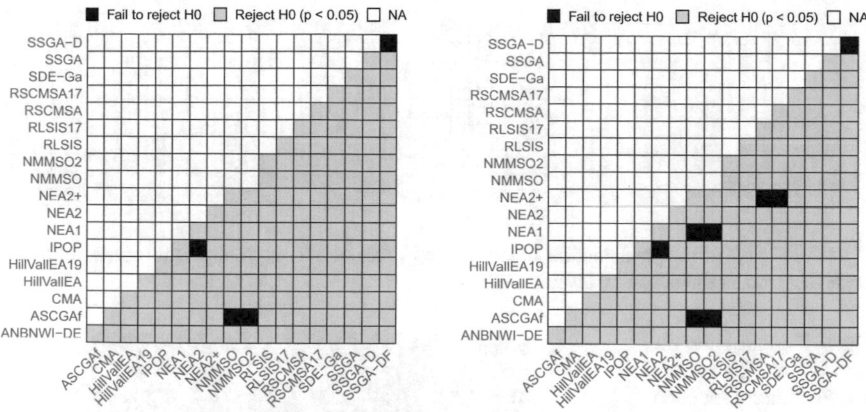

Fig. 6 Pairwise Wilcoxon statistical tests (left) with Bonferroni posthoc analysis (right) on results across all accuracy levels and all benchmark instances for Scenario III

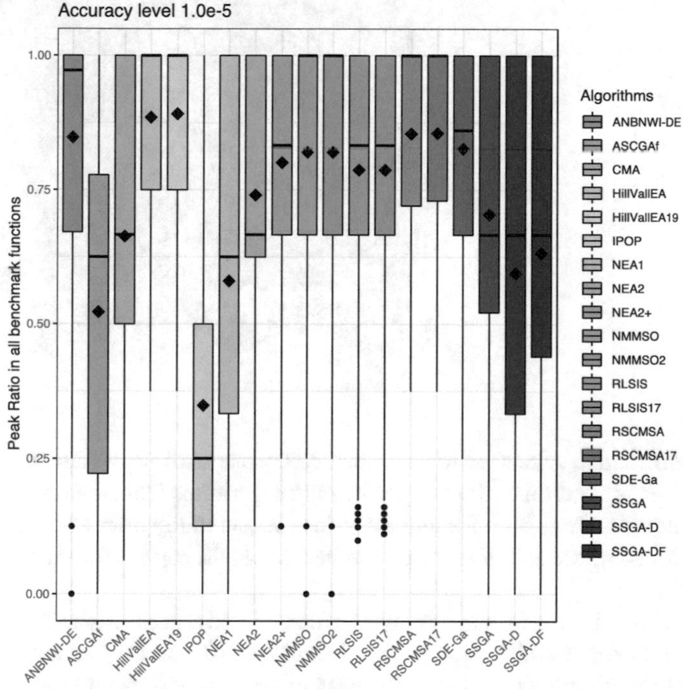

Fig. 7 Boxplot of median and the average peak ratio values for each algorithm across all benchmark instances at the accuracy level of 10^{-5}. Note that the average peak ratio value is denoted by a black diamond

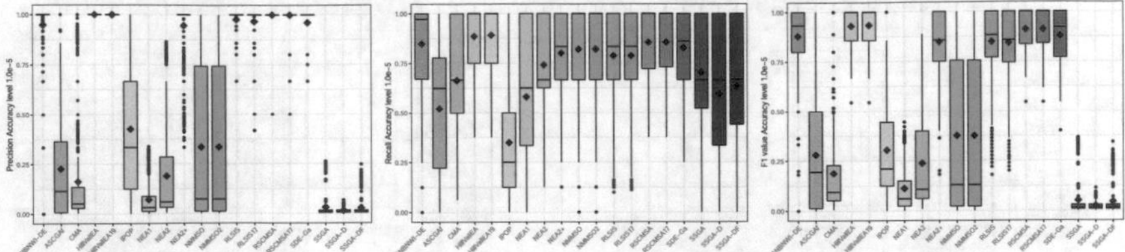

Fig. 8 Boxplots of the precision, recall, and static F_1 measure for each algorithm across all benchmark instances at the accuracy level of 10^{-5}

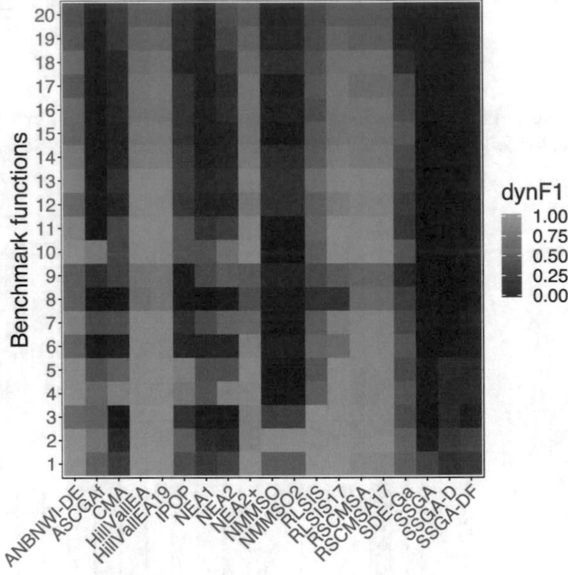

Fig. 9 Heatmap plot that demonstrates the average $dynF_1$ measure for each algorithm per benchmark instance at the accuracy level of 10^{-5}

median performance values are lower than their corresponding average values, while there is a clear separation of the top-performing algorithms from the remaining ones across all metrics. Here, high PR and SR values substantially affect the performance of an algorithm, where effective local searchers boost the top-performing algorithms substantially.

Figure 9 shows the performance pattern of considered algorithms across the benchmark instances for the highest accuracy level ($\epsilon = 10^{-5}$) in a heatmap plot. It is evident that the efficiency and effectiveness of the top-performing algorithms can be observed across the benchmark instances. The group of the top-performing algorithms includes HillVallEA19 and HillVallEA, RSCMSA and RSCMSA17, RLSIS and RLSIS17, the NEA+, and ANBNWI-DE, respectively, which constitute the most robust algorithms across the different scenarios.

Here we will focus on providing some analysis on the group of top-performing algorithms. Note that the Scenario I ranking is fairly consistent with the final ranks, as shown in Table 1. The ranks are the same for the first 6 algorithms and only differ slightly after that.

The 1st and 2nd places belong to HillVallEA19 and HillVallEA respectively. The HillVallEA variants [36] employ a Hill-Valley Clustering (HVC) mechanism that combines the Hill-Valley test [34] with the notion of the nearest better tree. For the Hill-Valley test, the number of sample points to be checked is adaptively chosen. A mechanism is also in place to check if a solution belongs to the niches of its nearest better neighbors, then a new cluster is created. This process is repeated for all individuals during the selection. A restart strategy with an increased population size (starting with very small size) is adopted to allow elite individuals to continue to survive into the future iterations. The population size parameters for both the HVC and the core algorithm have been made adaptable with different problem sizes. The HillVallEA algorithm (which adopts AMaLGaM-Univariate as the core algorithm) has been shown to outperform the state-of-the-art niching algorithms of the niching competition series if more computational budget is given [36].

At the 3rd and 4th places are RSCMSA and RSCMSA17 [44]. These RSCMSA variants are developed with an effective repelling sub-population (RS) scheme in conjunction with CMA-ES. The RS is a hybrid method built on several existing ideas such as the taboo points in Tabu search [45], normalized Mahalanobis distance [46], and the Hill-Valley function [34]. The rationale behind the choices of these techniques is to overcome the often made assumptions on distribution, size, and shape of the basins in MMO, which are impractical in many real-world problems. RSCMSA allows several sub-populations to search in parallel. The previously identified basins are marked as taboo points to avoid revisiting these basins in the future. A normalized Mahalanobis distance metric (instead of the usual Euclidean distance) is adopted to better handle non-spherical basin shape. The size of each basin is also adaptively defined by some parameters such that different sizes of basins are considered individually (instead of a uniform parameter defining all basin sizes). The Hill-Valley function is used to determine if a solution belongs to a new basin. The core algorithm adopts CMSA-ES (covariance matrix self-adaptation evolution strategy) [47]. RSCMSA builds on the strength of CMSA-ES so that it can handle ill-conditioned problems efficiently. To save computational cost, RSCMSA also employs a restart strategy with an increasing population size.

At the 7th and 8th places are NMMSO and NMMSO2 (according to Scenario I). These algorithms employ subswarms, with an exploitative particle swarm optimizer directed at each basin of attraction currently identified. Hill-Valley detection is used to distinguish these basins, and the number of swarms is unbounded. This means a priori information regarding the number of basins to attain is not required, but the search can be biased toward those basins of attractions that appear to be of the higher quality. New swarms are generated to look for additional basins through random seeding and splitting off exiting swarm members where Hill-Valley detection suggests they are on a different peak than the swarm they are a member of.

At the 9th place is NEA2+ (according to Scenario I), an enhanced version of NEA (Nearest-better EA) [48]. NEA builds on the idea of Nearest Better Clustering (NBC) that constructs a spanning tree to connect each solution to its nearest solution that has better fitness. NBC assumes that the best solutions are often located at different basins apart from each other and that the distances between them are usually larger than the average distance between all solutions and their nearest better neighbors. Therefore, by removing the longest edges (of the spanning tree) that are larger than the average distance between all solutions and their nearest better neighbor, clusters of solutions (or niches) can be identified. For each identified niche, NEA2+ employs CMA-ES to search its basin for an optimal solution.

7 Conclusion

In this chapter we have provided a glimpse of multimodal optimization (MMO), including its background, MMO formulation, performance measures, its historical and recent development. We also provided a result summary on the niching competition series between 2016 and 2019, hoping that such analysis can help capture the latest trends and identify techniques that have been harnessed to induce effective niching behaviors. Clearly there are still many open research questions, and we have only witnessed the early stage of development of some new breeds of niching methods.

From the previous section on niching competition results, there are clearly emerging trends in designing more powerful and robust niching algorithms, e.g., the use of more adaptive schemes for parameter specification, mechanisms to avoid revisiting the same area of the search space, restart strategies with incremental population sizes, and the adoption of individually adapted niche sizes.

Beyond developing more competent niching algorithms for just unconstrained and continuous optimization problems, e.g., as those shown in the CEC 2013 benchmark suite for MMO [10], there are abundance of research opportunities for further algorithmic development and applications of niching methods, including the following:

- Most real-world problems are highly constrained and combinatorial in nature, however, existing niching methods are mostly developed only for handling problems of unconstrained and continuous nature.
- In today's big data era, many large-scale (high dimensional) problems exist. However, most existing work on MMO have been confined to studies on low dimensional problems, and those which are (relatively) quick to evaluate. The scalability of these methods, be it to problem size and/or expensiveness, is relatively underexplored.
- Niching methods are often associated with many user-specified parameters, which could make them difficult to use in practice. Many efforts have been attempted to make niching parameters adaptive. However, there seems to be little study on how we can make use of the preference information (which can be easily supplied by a decision-maker) for niching. The benefit of this is that it will break down the barrier and allow niching methods to be more readily accepted by practitioners.

- Existing niching methods are mostly evaluated using benchmark test functions with known global optima. However, in a real-world situation, it is common that we do not have prior knowledge of the global optima (neither number nor best objective values). In such cases, how can we still measure the performance of a niching method in a meaningful manner?
- Since a decision-maker may have a fatigue issue [49] (considering the cognitive load), it may be unnecessary to locate and present too many global optimal solutions. How many solutions are adequate, and how do we determine an appropriate number? There could be also some mechanism for controlling the number of solutions obtained. If a small number is chosen, then the distribution of the solutions in the set can be made more sparse (to allow a better variety of selections by thc decision-maker).
- Since many design problems, e.g., engineering structural design problems are typically multimodal [50], what would be the best approach to hybridize engineering optimization methods with niching methods without too much computational cost?
- Often real-world optimization problems exhibit uncertainty or noise. There has been very little work on how optimizers developed for MMO cope on such problems.

Niching methods, which are specifically designed to locate multiple solutions in a single optimization run, are a common task that transcends the disciplinary boundaries [8]. In many real-world applications, we can see the prevalent need of discovering more than one optimal solutions, which can provide alternative solutions and additional information (of the problem under study) to a decision-maker. This is in sharp contrast to the conventional view of single-optimum seeking behavior of an optimization algorithm. As we face ever more challenging and complex problems in the big data era, there will be increasing demands on designing more powerful niching algorithms for handling ever more challenging real-world MMO problems. Niching as a research topic is currently experiencing some new found rejuvenation. It is our hope that this chapter will draw more attention to the significance and benefits of doing niching, and attract even more researchers to contribute to this classic and yet fascinating area of research.

References

1. Preuss, M.: Multimodal Optimization by Means of Evolutionary Algorithms. Springer, Berlin (2015)
2. Törn, A., Žilinskas, A.: Global Optimization. Springer, Berlin (1989)
3. Boyd, S., Boyd, S.P., Vandenberghe, L.: Convex Optimization. Cambridge university press, Cambridge (2004)
4. Locatelli, M., Schoen, F.: Global Optimization: Theory, Algorithms, and Applications, Vol. 15. Siam (2013)
5. Li, J.-P., Balazs, M.E., Parks, G.T., John Clarkson, P.: A species conserving genetic algorithm for multimodal function optimization. Evol. Comput. **10**(3), 207–234 (2002)

6. Preuss, M., Wessing, S., Rudolph, G., Sadowski, G.: Solving phase equilibrium problems by means of avoidance-based multiobjectivization. In: Kacprzyk, J., Pedrycz, W. (eds.) Springer Handbook of Computational Intelligence, Springer Handbooks, pp. 1159–1171. Springer, Berlin (2015)
7. Das, S., Maity, S., Qu, B.-Y., Suganthan, P.N.: Real-parameter evolutionary multimodal optimization - a survey of the state-of-the-art. Swarm Evol. Comput. **1**(2), 71–88 (2011)
8. Li, X., Epitropakis, M.G., Deb, K., Engelbrecht, A.P.: Seeking multiple solutions: an updated survey on niching methods and their applications. IEEE Trans. Evol. Comput. **21**(4), 518–538 (2017)
9. Wessing, S.: Two-stage methods for multimodal optimization. Ph.D. thesis, Technische Universität Dortmund (2015)
10. Li, X., Engelbrecht, A., Epitropakis, M.G.: Benchmark functions for CEC'2013 special session and competition on niching methods for multimodal function optimization. Technical report, RMIT University, Evolutionary Computation and Machine Learning Group, Australia (2013)
11. Preuss, M., Wessing, S.: Measuring multimodal optimization solution sets with a view to multiobjective techniques. In: Emmerich, M., Deutz, A., Schütze, O., Bäck, T., Tantar, E., Tantar, A.-A., Moral, P.D., Legrand, P., Bouvry, P., Coello, C.A. (eds.) EVOLVE – A Bridge between Probability, Set Oriented Numerics, and Evolutionary Computation IV, volume 227 of Advances in Intelligent Systems and Computing, pp. 123–137. Springer (2013)
12. Kerschke, P., Wang, H., Preuss, M., Grimme, C., Deutz, A.H., Trautmann, H., Emmerich, M.T.M.: Search dynamics on multimodal multiobjective problems. Evol. Comput. **27**(4), 577–609 (2019)
13. Deb, K.: Genetic Algorithms in multimodal function optimization (Master thesis and TCGA Report No. 89002). Ph.D. thesis, Tuscaloosa: University of Alabama, The Clearinghouse for Genetic Algorithms (1989)
14. Goldberg, D.E., Deb, K., Horn, J.: Massive multimodality, deception, and genetic algorithms. In: Männer, R., Manderick, B. (eds.) PPSN 2, Amsterdam (1992). Elsevier Science Publishers, B. V
15. Gallagher, M., Yuan, B.: A general-purpose tunable landscape generator. IEEE Trans. Evol. Comput. **10**(5), 590–603 (2006)
16. Rönkkönen, J., Li, X., Kyrki, V., Lampinen, J.: A framework for generating tunable test functions for multimodal optimization. Soft. Comput. **15**(9), 1689–1706 (2011)
17. Singh, G., Deb, K.: Comparisons of multi-modal optimization algorithms based on evolutionary algorithms. In: Proceedings of the Genetic and Evolutionary Computation Conference 2006 (GECCO'06), pp. 1305–1312, Washington, USA (2006)
18. Qu, B.-Y., Suganthan, P.N.: Novel multimodal problems and differential evolution with ensemble of restricted tournament selection. In: 2010 IEEE Congress on Evolutionary Computation (CEC), pp. 1–7 (2010)
19. Wessing, S.: The multiple peaks model 2. Algorithm Engineering Report TR15-2-001. Technische Universität Dortmund (2015). https://ls11-www.cs.uni-dortmund.de/_media/techreports/tr15-01.pdf
20. Preuss, M., Lasarczyk, C.: On the importance of information speed in structured populations. In: Parallel Problem Solving from Nature - PPSN VIII. Lecture Notes in Computer Science, vol. 3242, pp. 91–100. Springer (2004)
21. Ahrari, A., Deb, K.: A novel class of test problems for performance evaluation of niching methods. IEEE Trans. Evol. Comput. **22**(6), 909–919 (2018)
22. Deb, K., Saha, A.: Multimodal optimization using a bi-objective evolutionary algorithm. Evol. Comput. **20**(1), 27–62 (2012)
23. Alyahya, K., Doherty, K., Akman, Q., Fieldsend, J.: Robust multi-modal optimisation. In: Proceedings of the 2018 Conference on Genetic and Evolutionary Computation (GECCO'18), pp. 1783–1790 (2018)
24. Cavicchio, D.J.: Adapting Search Using Simulated Evolution. Ph.D. Thesis, University of Michigan, Ann Arbor, Michigan (1970)

25. Jong, K.A.De.: An analysis of the behavior of a class of genetic adaptive systems. Ph.D. thesis, University of Michigan (1975)
26. Holland, J.H.: Adaptation in Natural and Artificial Systems. University of Michigan Press, Ann Arbor, Michigan (1975)
27. Goldberg, D.E., Richardson, J.: Genetic algorithms with sharing for multimodal function optimization. In: Proceedings of the Second International Conference on Genetic algorithms and their application, pp. 41–49. Lawrence Erlbaum Associates, Inc. (1987)
28. Harik, G.R.: Finding multimodal solutions using restricted tournament selection. In: Eshelman, L. (ed.) Proceedings of the Sixth International Conference on Genetic Algorithms, pp. 24–31. Morgan Kaufmann, San Francisco, CA (1995)
29. Pétrowski, A.: A clearing procedure as a niching method for genetic algorithms. In: Proceedings of the 3rd IEEE International Conference on Evolutionary Computation, pp. 798–803 (1996)
30. Shir, O.M.: Niching in evolution strategies. In: Beyer, H.-G. (ed.) GECCO '05: Proceedings of the 2005 conference on Genetic and Evolutionary Computation, pp. 865–872, New York, NY, USA. ACM Press (2005)
31. Preuss, M., Schönemann, L., Emmerich, M.: Counteracting genetic drift and disruptive recombination in $(\mu +/, \lambda)$-EA on multimodal fitness landscapes. In: Proceedings of the 2005 Conference on Genetic and Evolutionary Computation, GECCO '05, pp. 865–872. ACM (2005)
32. Preuss, M.: Niching prospects. Bioinspired Optimization Methods and their Applications, pp. 25–34 (2006)
33. Törn, A., Viitanen, S.: Topographical global optimization. In: Floudas, C.A., Pardalos, P.M. (eds.) Recent Advances in Global Optimization, Princeton Series in Computer Sciences, pp. 384–398. Princeton University Press (1992)
34. Ursem, R.K.: Multinational evolutionary algorithms. In: Angeline, P.J. (ed.) Proceedings of the Congress of Evolutionary Computation (CEC 99), vol. 3, pp. 1633–1640. IEEE Press (1999)
35. Stoean, C., Preuss, M., Stoean, R., Dumitrescu, D.: Multimodal optimization by means of a topological species conservation algorithm. IEEE Trans. Evol. Comput. **14**(6), 842–864 (2010)
36. Maree, S.C., Alderliesten, T., Thierens, D., Bosman, P.A.N.: Real-valued evolutionary multimodal optimization driven by hill-valley clustering. In: Aguirre, H.E., Takadama, K. (ed.) Proceedings of the Genetic and Evolutionary Computation Conference, GECCO 2018, Kyoto, Japan, July 15–19, pp. 857–864. ACM (2018)
37. Wessing, S., Rudolph, G., Preuss, M.: Assessing basin identification methods for locating multiple optima. In: Advances in Stochastic and Deterministic Global Optimization, pp. 53–70. Springer, Berlin (2016)
38. Mahfoud, S.W.: A comparison of parallel and sequential niching methods. In: Proceedings of the 6th International Conference on Genetic Algorithms, pp. 136–143, San Francisco, CA, USA (1995). Morgan Kaufmann Publishers Inc
39. Li, X.: Niching without niching parameters: particle swarm optimization using a ring topology. IEEE Trans. Evol. Comput. **14**(1), 150–169 (2010)
40. Li, X.: Developing niching algorithms in particle swarm optimization. In: Panigrahi, B.K., Shi, Y., Lim, M.-H. (eds.) Handbook of Swarm Intelligence. Adaptation, Learning, and Optimization, vol. 8, pp. 67–88. Springer, Berlin (2011)
41. Epitropakis, M.G., Plagianakos, V.P., Vrahatis, M.N.: Finding multiple global optima exploiting differential evolution's niching capability. In: IEEE Symposium on Differential Evolution (SDE) (2011)
42. Beasley, D., Bull, D.R., Martin, R.R.: A sequential niche technique for multimodal function optimization. Evol. Comput. **1**(2), 101–125 (1993). June
43. Hollander, M., Wolfe, D.A., Chicken, E.: Nonparametric Statistical Methods, 3rd edn. Wiley, New York (2013)
44. Ahrari, A., Deb, K., Preuss, M.: Multimodal optimization by covariance matrix self-adaptation evolution strategy with repelling subpopulations. Evol. Comput. **25**(3), 439–471 (2017)
45. Glover, F.: Future paths for integer programming and links to artificial intelligence. Comput. Oper. Res. **13**(5), 533–549 (1986)

46. Shir, O.M., Emmerich, M., Bäck, T.: Adaptive niche radii and niche shapes approaches for niching with the cma-es. Evol. Comput. **18**(1), 97–126 (2010)
47. Beyer, H.-G., Sendhoff, B.: Covariance matrix adaptation revisited – the CMSA evolution strategy. In: Rudolph, G., Jansen, T., Beume, N., Lucas, S., Poloni, C. (eds.) Parallel Problem Solving from Nature – PPSN X, pp. 123–132. Springer, Berlin (2008)
48. Preuss, M.: Improved topological niching for real-valued global optimization. In: Applications of Evolutionary Computation. Lecture Notes in Computer Science, vol. 7248, pp. 386–395. Springer (2012)
49. Schwartz, B.: The Paradox of Choice: Why More Is Less. Harper Perennial (2004)
50. Islam, Md.J., Li, X., Deb, K.: Multimodal truss structure design using bilevel and niching based evolutionary algorithms. In: Proceedings of the Genetic and Evolutionary Computation Conference, GECCO '17, pp. 274–281, New York, NY, USA, Association for Computing Machinery (2017)

Representation, Resolution, and Visualization in Multimodal Optimization

Jonathan E. Fieldsend

Abstract This chapter examines the fundamental limits placed on algorithm capabilities due to the internal solution representation used by a search algorithm, and its implications for mode discovery and problem tractability. It presents various approaches to visualizing the problem landscape, based on both exhaustive enumeration and sampling—which may be of use to both practitioners and problem owners. It also provides the first comprehensive examination of the widely used IEEE CEC 2013 benchmark multimodal problems using local optima networks. These visualize the fitness landscape as a directed graph, and convey such information as local optima quality, basin size, and how easy it is to traverse the fitness landscape from one local optimum to another.

1 Multimodal Optimization: The What and the Why

If you are reading this book, then there is a good chance you have an optimization problem with multiple modes that you wish to solve. Two obvious questions this poses are

1. "*What do you mean by a mode?*" And
2. "*Why might you want to search across and return solutions on multiple modes?*"

To start answering these questions, we first define what a mode is and discuss some different interpretations, and their implications. We also discuss why it may be useful to return a set of modal design solutions to an optimization problem. Later in the chapter, we will explore how representation affects the search landscape itself, alongside how to visualize the search landscape in high dimensions. Following this, we discuss the visualization of the search population, and end with a discussion of the different quality measures you may want to employ to determine if an optimizer is

---

J. E. Fieldsend (✉)
Computer Science, University of Exeter, North Park Road, Exeter EX4 4QF, UK
e-mail: J.E.Fieldsend@exeter.ac.uk

M. Preuss et al. (eds.), *Metaheuristics for Finding Multiple Solutions*,
Natural Computing Series,
https://doi.org/10.1007/978-3-030-79553-5_2

effective (and their issues). Throughout, we will strive for a balance between formal definitions and informal examples.

First, without loss of generality, let us consider the optimization problem

$$\arg\min_{\mathbf{x}\in\mathscr{X}} f(\mathbf{x}).$$

That is, we seek a *design* $\mathbf{x}$, from the domain of legal designs $\mathscr{X}$ (i.e. those designs which meet any and all equality and inequality constraints of the design problem) which has the minimum response under the cost function $f(\cdot)$. Commonly, a feasible design which minimizes $f(\cdot)$ is denoted by $\mathbf{x}^*$. On the one hand, such problems may be solved with a linear program if the function form and constraints are linear, or at the other extreme the problem may be tackled with a search heuristic (for example, a genetic algorithm) if for instance, the $f(\cdot)$ is a 'black box' (an oracle, which may be queried to return the quality of a putative $\mathbf{x}$, but whose internal processes are not exposed). Either way, we are seeking *an* optimal solution to $f(\cdot)$—though in reality we may only at best get some *approximation* of this.

1.1 What Is a Mode?

Here, we shall provide a general notation for a mode, and also highlight how, depending on the definition used, the number of modes exhibited by a particular problem may be observed to vary.

Optimization problems can reside in a range of search spaces. For instance, each element of the design vector $\mathbf{x}$ may contain one of a set of symbols (a combinatorial optimization problem), and the vector may represent a permutation of symbols, discrete values, continuous values, or a mixture (e.g. mixed integer problems). The *fitness landscape* of a multimodal problem is a function of the triple $(\mathscr{X}, f(\cdot), N(\cdot))$. That is, the design space $\mathscr{X}$, the cost function $f(\cdot)$, and a *neighborhood* function $N(\cdot)$, where $N(\mathbf{x})$ is the set of neighbors of $\mathbf{x}$. In the case of minimization, $\mathbf{x}$ is commonly categorized as a *local* mode if

$$f(\mathbf{x}) \leq f(\mathbf{u}), \quad \forall \mathbf{u} \in N(\mathbf{x}), \tag{1}$$

that is, if it is no worse than any of its neighbors. $\mathbf{x}$ is commonly categorized as a *global* mode if

$$f(\mathbf{x}) \leq f(\mathbf{u}), \quad \forall \mathbf{u} \in \mathscr{X}.$$

Note, the equality check in (1) can lead to vast numbers of designs categorized as modes in landscapes where there are areas of neutrality—i.e. where adjacent solutions in design space give the same response under $f(\cdot)$. Conversely, if

$$f(\mathbf{x}) < f(\mathbf{u}), \quad \forall \mathbf{u} \in N(\mathbf{x})$$

is used, then no member of a neutral region will be identified as a mode, however, it can also mean a global mode is *not* categorized as a local mode (if the global solutions lie in neutral regions). Stating how many modes exist in a landscape, even if $\mathscr{X}$ may be completely enumerated and queried under $f(\cdot)$, therefore depends on the mode categorization used, and also crucially the neighborhood itself.

Let us consider first a combinatorial problem, where each element of a design variable x_i holds a symbol from the binary alphabet $\{0, 1\}$. If $|\mathbf{x}| = m$, then, given no additional constraints, the feasible search space size is 2^m. A typical neighborhood function would be all those designs which have a hamming distance of 1 from a design (a single bit-flip away). In this case, every design would have m neighbors. This is illustrated in Fig. 1 using an undirected graph, where vertices are designs, and an edge between vertices indicates that they are neighbors. The grayscale shading of a vertex denotes quality under $f(\cdot)$ (darker indicating better performance). In the illustration, there are three global modes (0001, 0010, 0100), and one local mode (1111).

Varying the neighborhood size, however, will change the landscape. For instance, increasing to a hamming distance of 2 will merge the three separate global modes into connected designs sitting on a flat (neutral) 'ridge'. This illustrates a general property: increasing the neighborhood size in the mode definition will always either leave the number of modes unchanged or reduce them (as long as the lower-sized neighborhood set is a subset of the larger neighborhood). At the limit, if the neighborhood was defined as the hamming distance of m, then all designs are immediate neighbors of each other, and there is just a single mode in the fitness landscape, containing either a single optimal solution, or multiple optimal solutions on a plateau. Naively, this may appear to make the problem 'easy', simply increase the neighborhood size so that the global optima are the immediate neighbors of all designs. However, if we consider a search algorithm employing this neighborhood, which proposes a new

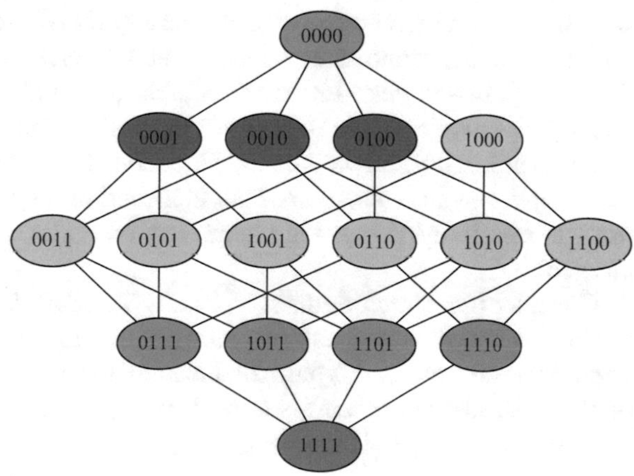

Fig. 1 Neighborhood graph induced on a four-bit search domain using a hamming distance of 1. Darker vertices indicate better quality. Edges between vertices denote neighbors

Table 1 Incrementally, defining basin size by shifting mass to the best neighbor

x	$f(\mathbf{x})$	Basin size estimate at t		
		$t = 1$	$t = 2$	$t = 3$
0000	0.5	1	1	0
0001	0.7	1	$3\frac{1}{3}$	$3\frac{2}{3}$
0010	0.7	1	$3\frac{1}{3}$	$3\frac{2}{3}$
0011	0.3	1	0	0
0100	0.7	1	$3\frac{1}{3}$	$3\frac{2}{3}$
0101	0.3	1	0	0
0110	0.3	1	0	0
0111	0.5	1	0	0
1000	0.3	1	0	0
1001	0.3	1	0	0
1010	0.3	1	0	0
1011	0.5	1	0	0
1100	0.3	1	0	0
1101	0.5	1	0	0
1110	0.5	1	0	0
1111	0.6	1	5	5

design by selecting a neighbor at random, then the probability of actually making the single neighborhood move necessarily to jump from a suboptimal design to an optimal design is $\frac{k}{|\mathscr{X}|-1}$—where k is the number of globally optimal designs.

Another important concept when dealing with modes is that of *basins*. These are regions in design space that, due to the neighborhood structure, lead to a particular mode (be it a single design or set of designs on a plateau) when making moves across the landscape. A mode that is supported by a large basin will tend to be easier to find than one in a small basin. There are various ways to define a basin, but here we will illustrate an incremental algorithmic approach based on a *greedy* neighborhood search. If we distribute an equal mass to each vertex, and then incrementally move the mass from a vertex to its best neighbor (or divide it equally between the set of equally best neighbors, in the case of neutrality), then progressively this mass will end up being deposited on the mode location. The process terminates when either a time step is reached where no mass is moved or a cyclical state is achieved where mass is merely cycling among neighbors on a plateau (which may effectively be merged).

Consider the search domain in Fig. 1. The data in Table 1 illustrates how this incremental basin size calculation progresses, with a final state achieved after 3 time steps. Interestingly, on this particular example, the single mode with the largest basin ($\frac{5}{16}$th of the search domain) is a local mode, though collectively the global modes have basins covering $\frac{11}{16}$th of the domain.

1.2 Why Optimize Multiple Modes?

In multimodal optimization, although the cost function remains *uni-objective*, we are concerned with obtaining a *set* of solutions which are of high quality but are *distinct* in that they represent different modes in the cost landscape. The natural question is—“*Why is this modification of use?*” There are a number of reasons why returning a set of solutions would be preferable over the return of a single design:

(i) It can *add value* to the problem owner.
(ii) It can *derisk* errors in (or missing) constraints.
(iii) It can *derisk* any error in $f(\cdot)$.
(iv) It may *aid* the optimization process itself.

Let us expand on these points. It is often of significant interest to a problem owner if they have distinct design parameterizations which result in identical or similar performance. This is because it can often add insight into the problem itself, and there may be additionally ‘unexposed’ preference criteria the problem owner would apply to select between such equivalent solutions, meaning the final choice would be more satisfactory than an arbitrary $\mathbf{x}^*$ returned by an optimizer from this set. For instance, given the choice the problem owner may well prefer a high-quality design further away from a constraint boundary than another, or ‘closer’ to an existing manufactured design.

It may be the case that the constraints used during the optimization run are actually a subset of those practically experienced. Some design configurations that appear to be of high quality under $f(\cdot)$ may require factory reconfigurations that weren’t anticipated and effectively invalidate them, or may require materials/inputs whose availability/price may be at higher risk of change—meaning that having one or more alternative solutions readily at hand is of great advantage if a design needs to be switched at short notice.

In many optimization situations, $f(\cdot)$ is some computational model of a physical system. This is because of the punitive cost in time or money (or both!) of manufacturing and evaluating each putative design proposed by an optimizer (often referred to as *embodied optimization* [23]). Instead, the system is *emulated* on a computer, and the effect of a design parametrization is evaluated under this emulation. If the model is in error, then an optimized design may well not behave as predicted when manufactured. Therefore, having a set of designs, which all appear high quality under the model, provides an opportunity to discard those which do not perform as well as expected in practice and still (potentially) have *some* which meet requirements. Relatedly, there is a body of work in the expensive optimization literature whose focus is the effective updating of $f(\cdot)$ *during* an optimization run through periodically querying a higher fidelity model, or a real implementation of some carefully selected designs [11].

Finally, and more problem-dependent, is that the act of optimizing a problem using a multimodal optimization framework may speed the optimization process, even if only a single best $\mathbf{x}^*$ is desired. By their nature, multimodal optimizers attempt to maintain and search in diverse regions in design space on modes they identify, and so for complicated landscapes with variable basin sizes, this added diversity can be beneficial over the course of an optimization, separate from the goal of returning a single design or a set of solutions.

2 Representation, Resolution, and Basic Visualizations Plots

Although neighborhoods are more easily conceptualized in combinatorial problems (as illustrated above), multimodal problems in continuous domains have increasingly become the focus of the multimodal optimization community. Indeed the international Competition on Niching Methods for Multimodal Optimization, run since 2013 in the IEEE Congress on Evolutionary Computation (CEC), and more recently also in the Genetic and Evolutionary Computation Conference (GECCO), has been solely focused on algorithm performance on continuous problems (with 20 benchmark problems used) [16].

Although a problem may be continuous, it is worth reemphasizing here that current digital computing technology represents such designs to a finite floating-point precision. One popular meta-heuristic for optimizing such problems, the *genetic algorithm* often represents floating-point numbers at a much coarser precision [17]. Furthermore, tackling a problem progressively at increasing resolutions (also called multi-scale optimization) has been a fruitful avenue in various applications (see, e.g. [3–5, 7, 13, 14, 18, 19, 22, 24, 25]). Work has also focused on changing or evolving the resolution used *during* an optimization run [2, 4, 5, 10, 13, 15, 24]. A natural question that follows is: "*What are the properties of the multimodal fitness landscape at different precision resolutions?*". As observed by, e.g. [13], imposing a resolution through a bit-string to represent a floating point value also limits the ability to converge to the optima (in the real-valued space) due to the granularity of the resolution used. We now explore this here, and its implications, using some of the competition test problems from [16].

The 20 problems from [16] span a range of properties (mixtures of mode types and shapes and distributions, spanning from only global modes, to those with vast numbers of local modes). Problem sizes vary from 1 up to 20 design dimensions. Some characteristics are detailed in Table 2.

Figure 2 shows how the fitness of the best design in the search domain at a given resolution differs from the best achievable under the continuous representation. A common quality measure employed in the assessment of a multimodal optimizer on test problems, the peak ratio, is the proportion of modes found within a given ε of the best possible value achievable. Commonly, ε values from 10^{-1} to 10^{-5}

Table 2 "Competition on Niching Methods for Multimodal Optimization" problem properties

Func.	Peak height	# Glob. opt.	Loc. Opt?	# dim.	Max Func Evals.
1	200.0	2	✓	1	50000
2	1.0	5		1	50000
3	1.0	1	✓	1	50000
4	200.0	4		2	50000
5	1.031628453	2	✓	2	50000
6	186.7309088	18	✓	2	200000
7	1.0	36		2	200000
8	2709.093505	81	✓	3	400000
9	1.0	216		3	400000
10	−2.0	12		2	200000
11	0	6	✓	2	200000
12	0	8	✓	2	200000
13	0	6	✓	2	200000
14	0	6	✓	3	400000
15	0	8	✓	3	400000
16	0	6	✓	5	400000
17	0	8	✓	5	400000
18	0	6	✓	10	400000
19	0	8	✓	10	400000
20	0	8	✓	20	400000

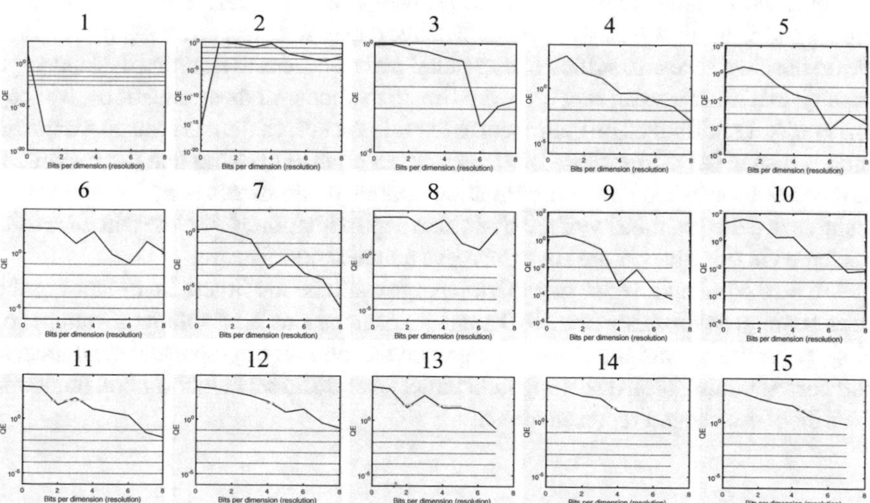

Fig. 2 Difference between the best $f(\cdot)$ value possible under the continuous representation, and that achievable under different resolutions for the CEC'13 multimodal test problems. Dashed horizontal lines highlight $\varepsilon = \{10^{-1}, 10^{-2}, 10^{-3}, 10^{-4}, 10^{-5}\}$

are used, which are also highlighted on the panels in Fig. 2.[1] A few interesting properties are immediately apparent. Firstly, increasing the resolution (the search space size) does not always lead to better designs being discoverable. Indeed in all the problems looked at, there are times when increasing the resolution *decreases* the best quality obtainable. This is because not all mappings available at resolution 2^n will be available at 2^{n+1}. A second observation is that for many problems, designs may be found only slightly worse than those obtainable in continuous space, when using very coarse resolutions. This gives an indication as to why tackling industrial problems at varying resolutions has proved popular, however, varying the resolution can also affect the *landscape* of the problem being tackled, and also the success of algorithms such as DIRECT, which progressively increase design resolution in areas of high quality observed at lower resolutions [12]. We consider this now in some standard plots, and examine it in the next section using local optima networks.

When the search domain is continuous rather than combinatorial, the neighborhood of a design is typically defined as those set of solutions within some radius of the design (within $\mathscr{X}$). Usually, this distance is calculated as the Euclidean distance, though it may be on a normalized transformation of the design vectors (if, for instance, different design variables are on substantially different ranges). Although in a continuous space there are an infinite number of neighbors within any non-zero radius of a design, in reality on current computer systems there is a finite precision (typically 64 or 32 bits) used to represent a floating-point value, and therefore we may compute exactly the number of numerically *distinct* neighbors of a design. In reality, enumerating the response of these under $f(\cdot)$ is likely to be prohibitive, so a grid or Monte Carlo sampling may be employed instead of approximating whether a design may be categorized as a mode.

For small discrete or combinatorial problems, we may calculate the number of modes and basin sizes exactly via an exhaustive process. However, for many problems we do not have access to sufficient computational resources to examine the landscape exactly, and must instead rely on approximate methods. In these situations, we may use randomly initialized hill-climbers to explore the search domain and approximate such statistics [21]. Algorithm 1 details such an algorithm. Note that the maximum number of modes such an approximation approach can detect is n; the number of restarts, and the granularity of the basin size approximation is $\frac{1}{n}$. Note that all results presented in this chapter are for exhaustive enumeration.

We now conduct a series of experiments to evaluate the fitness landscapes of the continuous problems from the 2013 benchmark at various bit resolutions (number of bits), r, per design variable. We also compare the number of global and local modes, and corresponding basin sizes apparent under three different neighborhood functions over these discretized representations:

---

[1] Note that such a visualization is limited to those problems where the optimal quality value is known a priori, and is therefore typically limited to test problems rather than real-world applications. However, it is of course possible to plot the *raw* quality values achievable as resolution increases.

```
Data: Number of restart locations, n
Result: Approximated modes, M, and basin sizes, B (indexed by mode location)
counter := 0;
M := ∅;
while counter < n do
    counter := counter + 1;
    x := random_draw(𝒳);
    converged := false;
    while converged = false do
        x' := arg max_{x'∈N(x)} f(x);
        if f(x) < f(x') then
            x := x'
        else
            converged := true;
        end
    end
    if {x} ∉ M then
        M := M ∪ {x};
        B_x := 1;
    else
        B_x := B_x + 1;
    end
end
B := B/n;
```

Algorithm 1: Approximating mode number and basin size using greedy local hill-climbers

1. A single-axis parallel shift using the single increment and decrement for the integer value corresponding to locations at that resolution for each of the design variables in turn (meaning each non-boundary design has $2 \times m$ neighbors).
2. A single bit-flip of the binary representation of the design (all solutions have exactly $m \times r$ neighbors).
3. A single bit-flip of the binary representation of the design when Gray encoding is used; in this case, the *binary reflected Gray code* as in [6] (all solutions have exactly $m \times r$ neighbors).

Gray encoding is popular due to its property that single bit-flip neighbors include adjacent integer neighbors when mapped back to the represented values (see Table 3 for an illustration of this for the three-bit representation case). Furthermore, the general view is that a Gray encoding makes a problem easier to solve with a genetic algorithm when it results in fewer modes than the standard binary case, though exhaustive work in this area for all possible $r = 3$ landscapes shows that this is not *always* the case [6]. The integer value, v_i represented by a bit-string, is converted to the corresponding floating-point value for a problem by

$$x_i = \frac{v_i}{2^r - 1}(u_i - l_i) + l_i \tag{2}$$

Table 3 Neighbors under different encodings (integer increment and decrement or single bit-flip), r = 3, search space size is 2^3

Integer	Binary	Gray
0, $N(\mathbf{x}) = \{1\}$	000, $N(\mathbf{x}) = \{1, 2, 4\}$	000, $N(\mathbf{x}) = \{1, 2, 4\}$
1, $N(\mathbf{x}) = \{0, 1\}$	001, $N(\mathbf{x}) = \{0, 3, 5\}$	001, $N(\mathbf{x}) = \{0, 2, 6\}$
2, $N(\mathbf{x}) = \{1, 3\}$	010, $N(\mathbf{x}) = \{0, 3, 6\}$	011, $N(\mathbf{x}) = \{1, 3, 5\}$
3, $N(\mathbf{x}) = \{2, 4\}$	011, $N(\mathbf{x}) = \{1, 2, 7\}$	010, $N(\mathbf{x}) = \{0, 2, 4\}$
4, $N(\mathbf{x}) = \{3, 5\}$	100, $N(\mathbf{x}) = \{0, 5, 6\}$	110, $N(\mathbf{x}) = \{4, 5, 7\}$
5, $N(\mathbf{x}) = \{4, 6\}$	101, $N(\mathbf{x}) = \{1, 4, 7\}$	111, $N(\mathbf{x}) = \{2, 4, 6\}$
6, $N(\mathbf{x}) = \{5, 7\}$	110, $N(\mathbf{x}) = \{2, 4, 7\}$	101, $N(\mathbf{x}) = \{2, 5, 7\}$
7, $N(\mathbf{x}) = \{6\}$	111, $N(\mathbf{x}) = \{3, 5, 6\}$	100, $N(\mathbf{x}) = \{0, 4, 6\}$

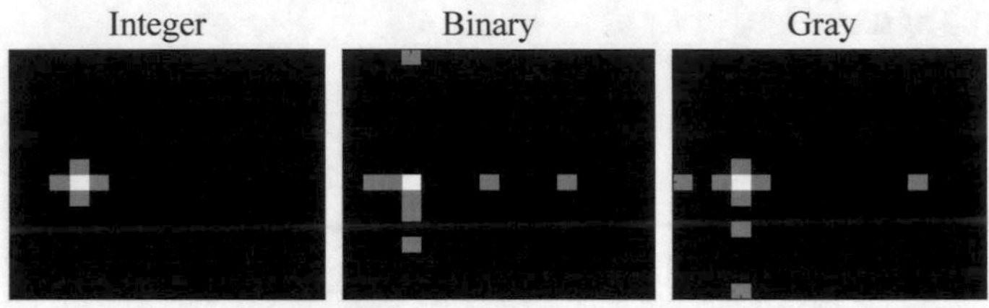

Fig. 3 Neighborhoods induced using different representations at 2^4 resolution per dimension. Left: direct integer representation. Middle: standard binary encoding. Right: Gray encoding

where u_i is the upper boundary value for the ith design variable, and l_i is the lower boundary for the ith design variable.

Figure 3 illustrates in 2D the neighbors (gray cells) of an arbitrary location (white cell), using the step size imposed by a resolution of 2^4, for the three different encodings (integer, binary, and Gray). As can be seen, binary and Gray one-hamming distance neighborhoods are the same size, but distributed differently. The one increment neighbors of the integer representation are fewer, and a subset of the neighbors in the Gray encoding.

Figure 4 shows how the total number of modes in the landscape varies according to resolution and neighborhood type for each problem. This usefully highlights that for the same search domain, on the same problem, vastly different number of modes can be present depending on the neighborhood function used. Furthermore, the representation with the highest number of modes is not the same between problems, and indeed the relative performance is not even consistent for the case on the same problem when different design resolutions are used. For instance, as can be seen in the upper right panel (corresponding to test problem 1), at $r = \{0, 1, 2, 3, 4, 5\}$ the binary representation has the fewest modes on the induced landscape, however, for $r = \{6, 7, 8\}$ it has the highest. In some problems, there are vastly different numbers of modes induced, up to two orders of magnitude difference in the examples shown in Fig. 4 (see, e.g. the panel for test problem 11).

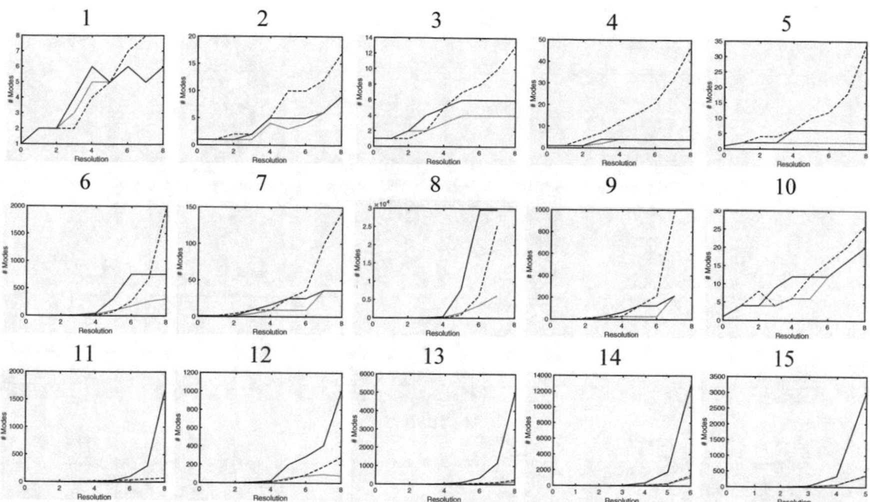

Fig. 4 Number of modes at each resolution for each problem. Solid line is for integer single value neighborhood, dashed line for binary single hamming distance neighborhood, and dotted line for Gray single hamming distance neighborhood

Figures 5, 6, and 7 show scatter plots of basin quality versus basin size for problems 6–14 at various resolutions for the three different encodings. Immediately apparent is that this distribution changes between problems, between representation, and between resolutions on the same problem. Note, for some problems, not all panels are provided as due to the exhaustive nature of the plots, there is a computational limit on their calculation (for instance, when $d = 3$, $(2^8)^d = 16,777,216$).

Although these plots give us some useful raw information (Gray encoding tends to have fewer modes, and a general trend in most of the problems for fitter modes to have larger basins), they do not convey how easy or difficult it may be to move between basins and search the landscapes. We explore this now through the use of local optima networks.

3 Visualizing the Multimodal Landscape: Local Optima Networks

In recent years, an increasing amount of effort has been directed toward the *visualization* of problems by the optimization community, to aid both those optimizing a problem to have a sense of the landscape they are dealing with, and to inform problem owners. A simple visualization derived from the analysis above is the scatter plots of mode fitness versus basin size. A much richer visualization, however, is provided by the local optima network (LON) graph [8, 9, 20, 26–28]. A LON can

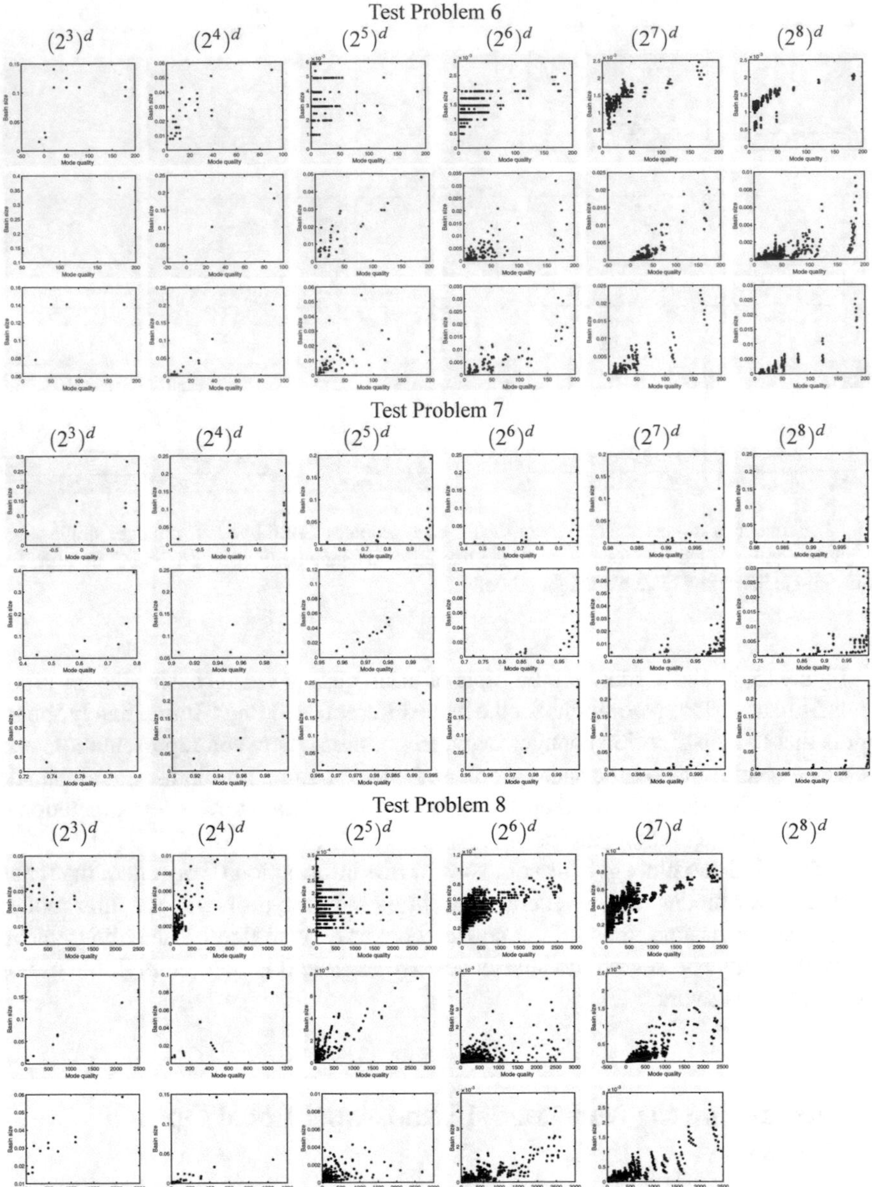

Fig. 5 Mode quality versus basin size (as a proportion of total design space) (top row) for test problems 6–8. Using integer encoding (top row), binary encoding (middle row) and Gray encoding (bottom row)

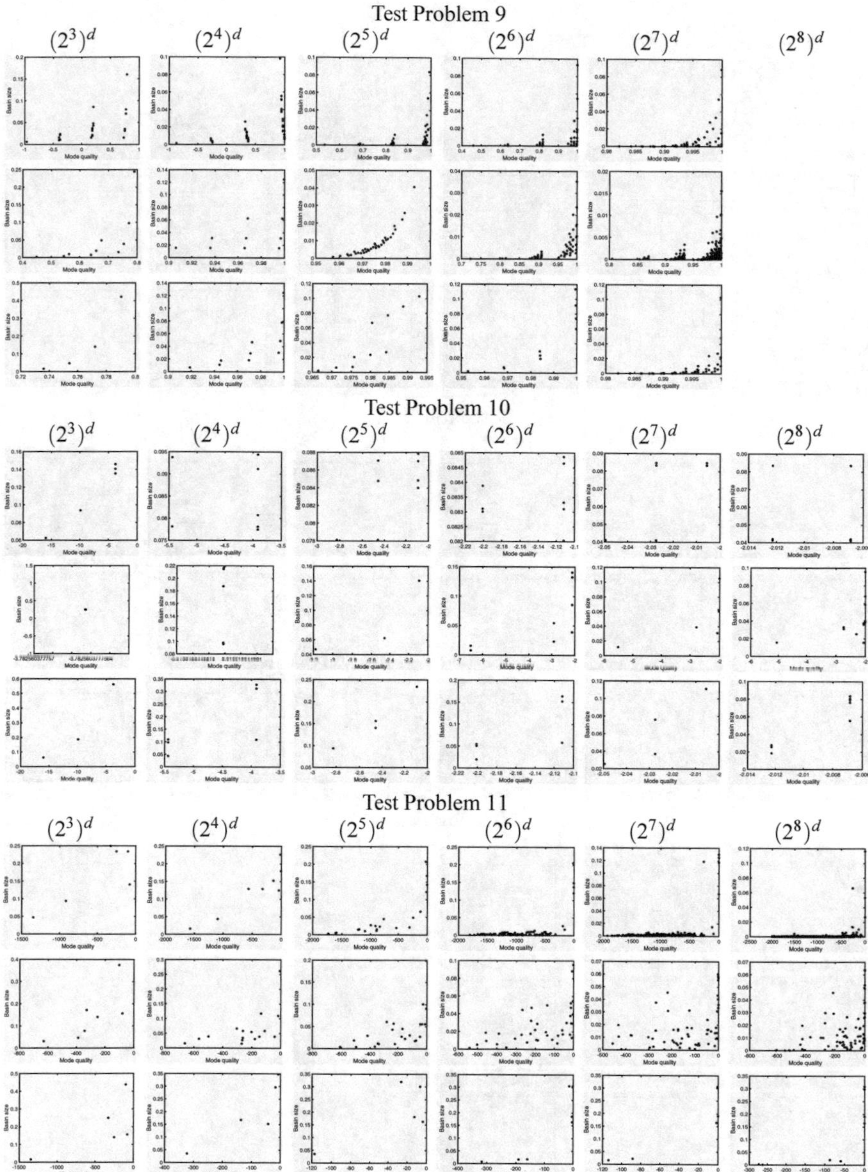

Fig. 6 Mode quality versus basin size (as a proportion of total design space) (top row) for test problems 9–11. Using integer encoding (top row), binary encoding (middle row), and Gray encoding (bottom row)

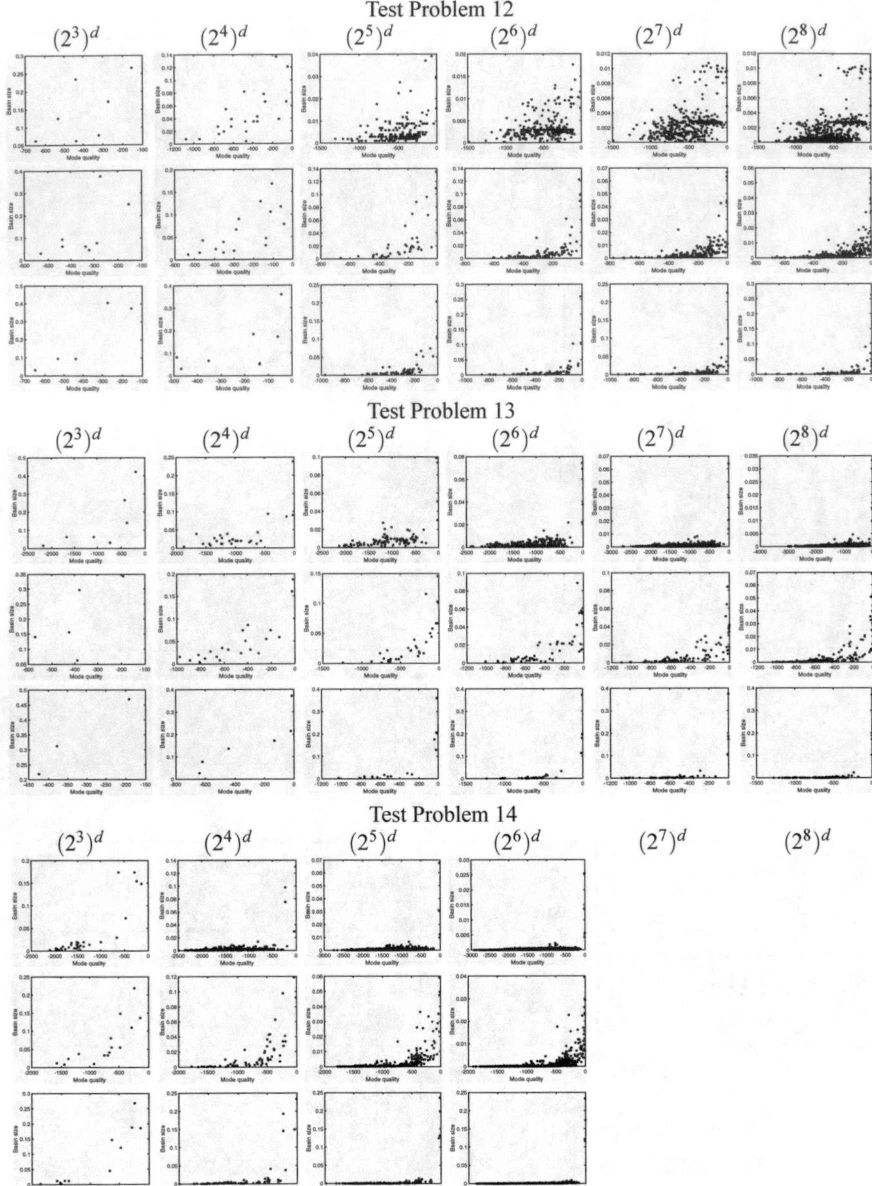

Fig. 7 Mode quality versus basin size (as a proportion of total design space) (top row) for test problems 12–14. Using integer encoding (top row), binary encoding (middle row), and Gray encoding (bottom row)

Fig. 8 Local optima network (LON) of example in Fig. 1 using basin-transition. Darker vertices indicate local optima with higher quality. Note that in this case, the largest basin corresponds to a local optima

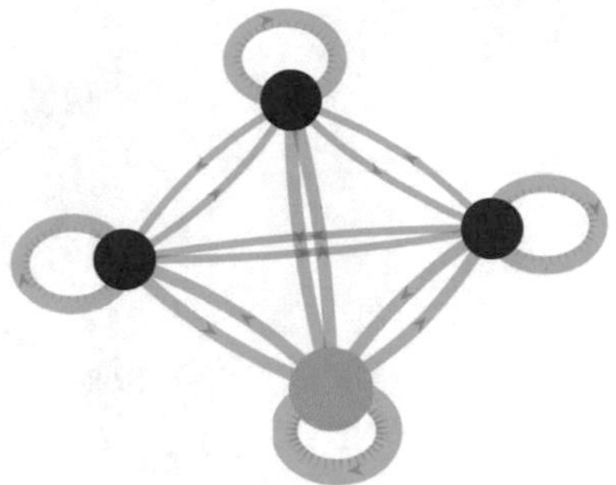

convey basin size, mode quality, and search transition likelihood between modes in one compact form. A LON is a graph $G = (V, E)$, which is made up of vertices V denoting modes, and directed edges E denoting connectivity between mode regions. Typically, the weight of a directed edge between vertex i and vertex j is proportional to how easy it is to move from mode/basin i to mode/basin j. The size of a vertex indicates how large the corresponding basin for that mode is, and the color of a vertex and the quality of the corresponding mode location under $f(\cdot)$.

We have already discussed how to calculate the size of a basin, and the quality of a mode, however, what does ease of movement between modes/basins mean? Two proposed approaches for this calculation are *basin-transition* and *edge escape* [21]. For basin-transition, we assume a uniformly random selection of a neighbor for a design. We may then calculate the probability of moving from $\mathbf{x}$ to $\mathbf{x}'$, $p(\mathbf{x} \rightarrow \mathbf{x}')$, in terms of $N(\mathbf{x})$ as $\frac{1}{|N(\mathbf{x})|}$ if $\mathbf{x}' \in N(\mathbf{x})$, and 0 otherwise. Using this calculation, we may calculate the total probability of transiting from basin k (B_k) to basin m (B_m) as

$$p(B_k \rightarrow B_m) = \frac{1}{L} \sum_{\mathbf{x} \in B_k} \sum_{\mathbf{x}' \in B_m} p(\mathbf{x} \rightarrow \mathbf{x}')p(\mathbf{x} \rightarrow B_k)p(\mathbf{x} \rightarrow B_m) \qquad (3)$$

where there are L basins in the landscape. $p(\mathbf{x} \rightarrow B_k)$ is the probability that $\mathbf{x}$ is in basin k—where there is neutrality in a landscape this is not necessarily 1, as although $\mathbf{x} \in B_k$, it may also be a member of another basin also, if there is more than one neighbor of $\mathbf{x}$ which would satisfy the greedy move criteria (and who may lead to a different mode).

Figure 8 shows the LON for the example in Fig. 1, constructed using the basin-transition probabilities. Note that besides the directed edges between basins, there are also edges from a basin back to itself, indicating the likelihood of moves between members of the same basin.

In the case of escape edges, these are parametrized by a distance d, which is the minimum number of moves (e.g. bit-flips) between a solution in B_k to the *local optima* position associated with basin m, $\mathbf{o}_m$, in order to weight the edge between basin k and m. More formally, the edge weight between basin k and m is

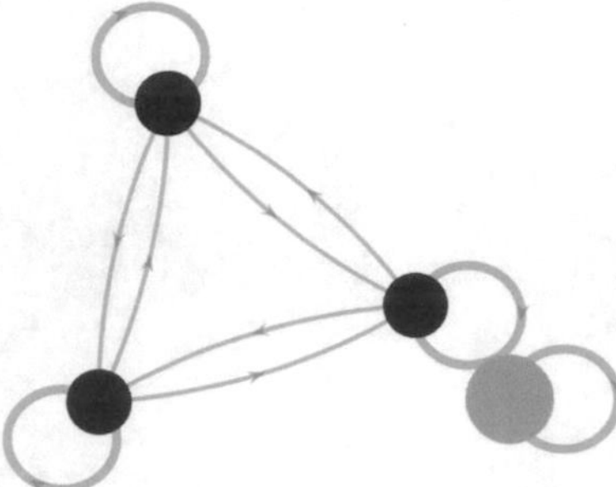

Fig. 9 Local optima network (LON) of example in Fig. 1 using escape edges for hamming distance ≤ 1. Note that basin sizes are unchanged—however, the connectivity and edge widths have changed. The former change is due to the change in edge definition. The latter is due to fewer greedy paths being considered: with basin-transition, the combined flow out from a basin matches the basin volume, whereas in escape edges this is fixed by the number of neighbors of a mode

$$w_{k,m} = \sum_{\mathbf{x} \in B_k} \delta(moves(\mathbf{x}, \mathbf{o}_m) \leq d) p(\mathbf{x} \rightarrow B_k), \tag{4}$$

$\delta(\cdot)$ returns 1 if its argument evaluates to true, 0 otherwise. $moves(\mathbf{x}, o_m)$ returns the number of moves from $\mathbf{x}$ to $\mathbf{o}_m$. Although not in [28], we include the weighting term $p(\mathbf{x} \rightarrow B_k)$ to allow for neutrality. Figure 9 shows the LON with escape edges for $d = 1$ for the previous example.

We now examine the LONs induced for resolutions 2^3 through to 2^8 per dimension (up to the resolution for higher d problems where the memory requirements remain below 8 GB). Figures 10, 11, 12, 13, and 14 shows these exact LONs using integer neighbors, standard binary representation neighbors, and Gray encoding neighbors using basin transition edges. Figures 15, 16, 17, 18, and 19 shows the LONs using standard escape edges.

As previously indicated in the fitness/basin size plots, often the binary encoding tends to induce vastly more modes compared to Gray encoding for these problems, with the number growing as the resolution increases. Furthermore, although growing in lower resolutions, the number of modes for Gray encoding stabilizes for problem 3 but oscillates for the other two 1D problems, as does the edge structure (note, for instance, in problem 1 how the LON at 2^7 is distinct from both 2^6 and 2^8, which are in turn similar to one another). As earlier work introducing escape edges has shown [28], this approach tends to result in sparser LONs than those created using basin transition, and this is also apparent in the plots shown here. It should be noted that the number of modes (and basin sizes) are not altered by the edge generation method—unlike when the neighborhood function $N(\cdot)$ is changed.

Simple integer neighbors (which in this context mean small fixed-step movements imposed on a continuous space) increasingly lead to entirely *disconnected* basin escape edges as the resolution increases (and the step size imposed by the integer movement on the continuous space mapping reduces). This illustrates how a hill

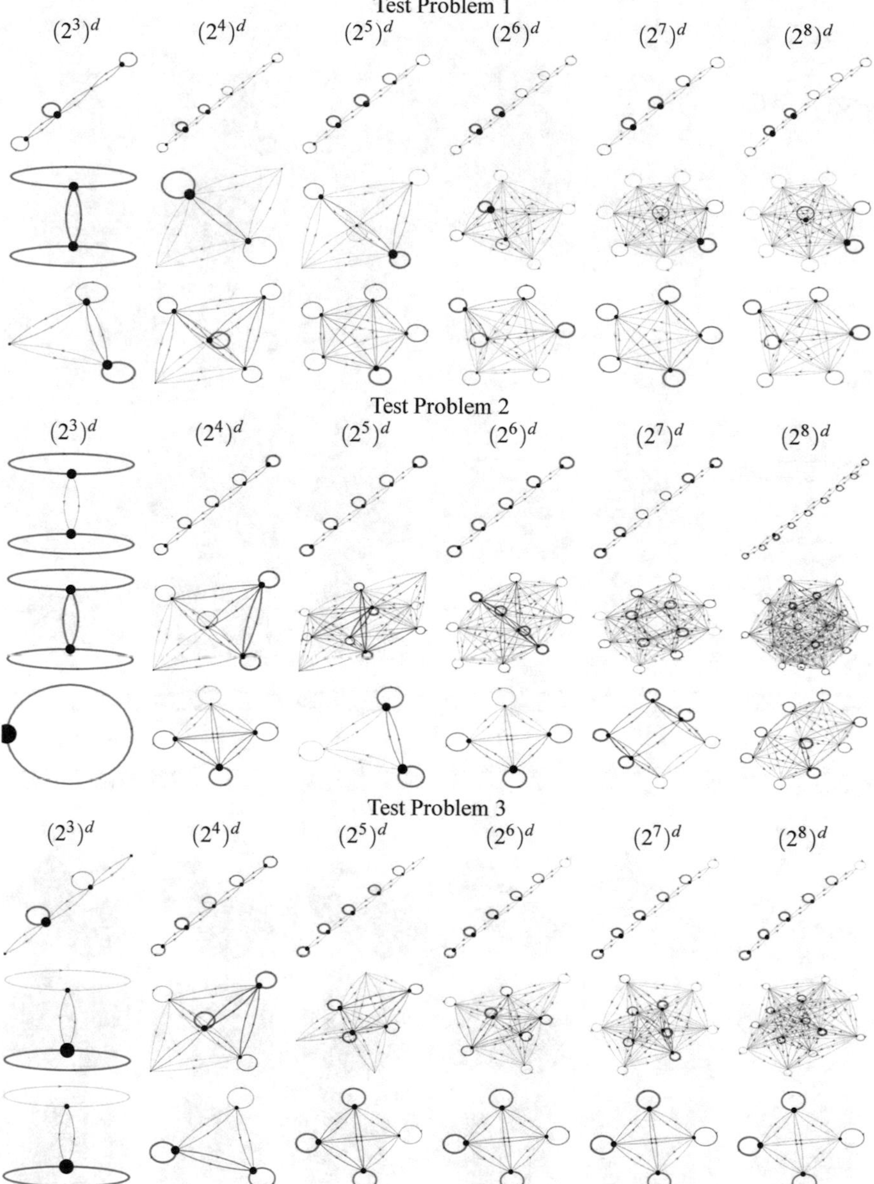

Fig. 10 Exact LON of test problems 1–3 using integer encoding (top row), binary encoding (middle row), and Gray encoding (bottom row). Basin transition for edges weights

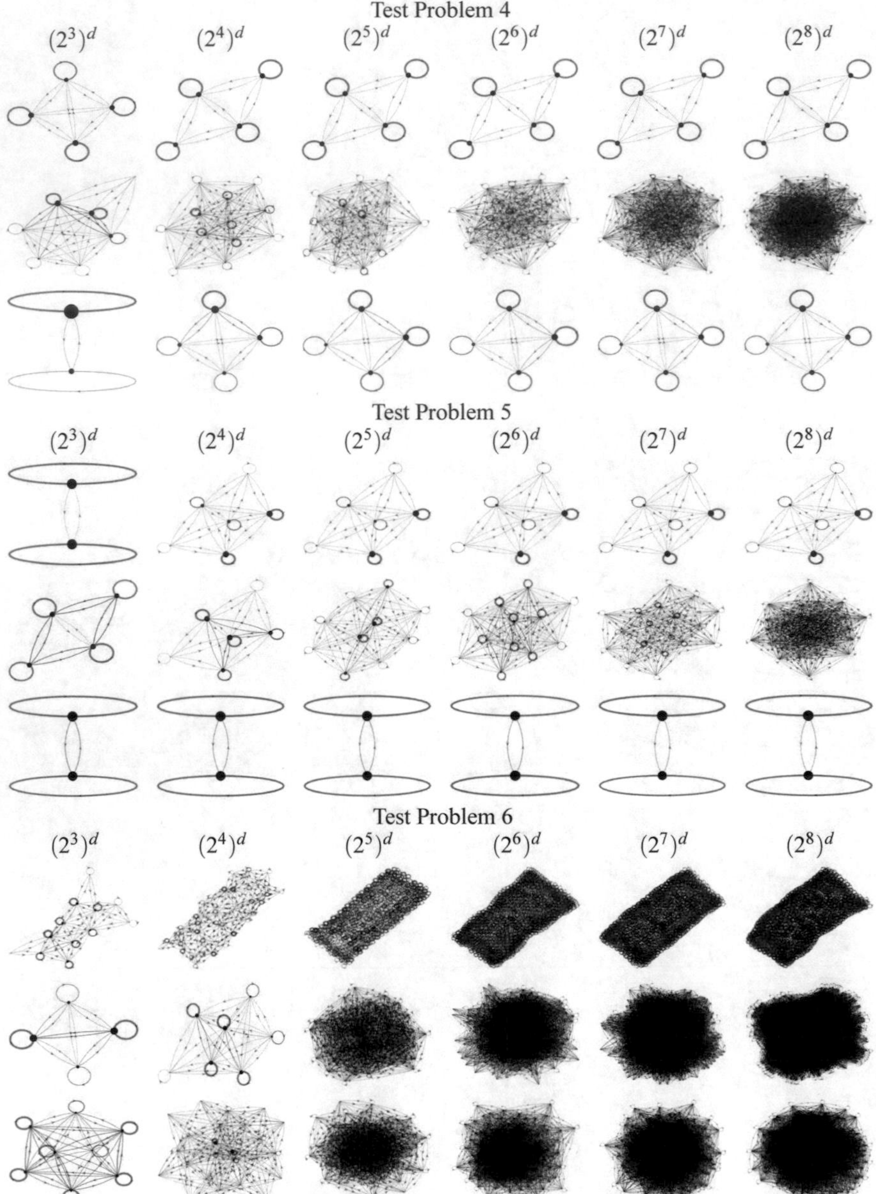

Fig. 11 Exact LON of test problems 4–6 using integer encoding (top row), binary encoding (middle row), and Gray encoding (bottom row). Basin transition for edges weights

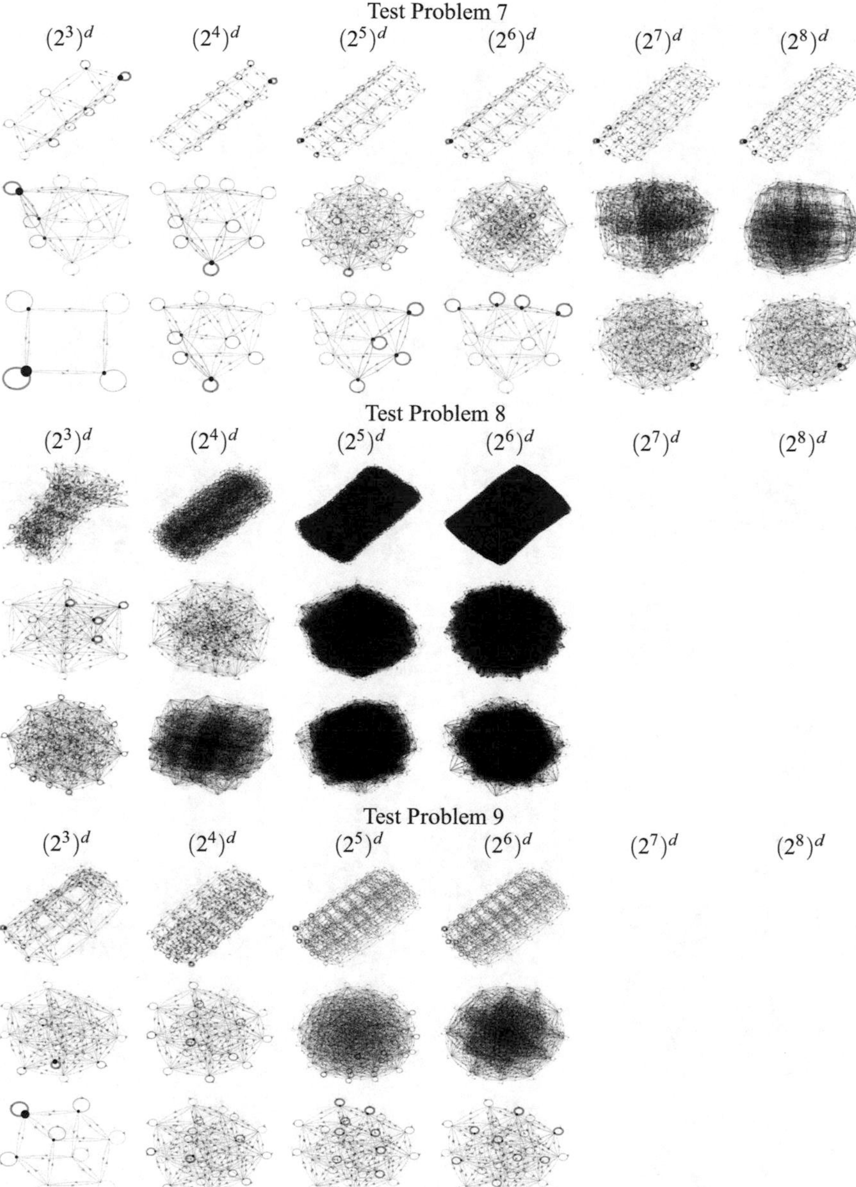

Fig. 12 Exact LON of test problem 7–9 using integer encoding (top row), binary encoding (middle row), and Gray encoding (bottom row). Basin transition for edges weights

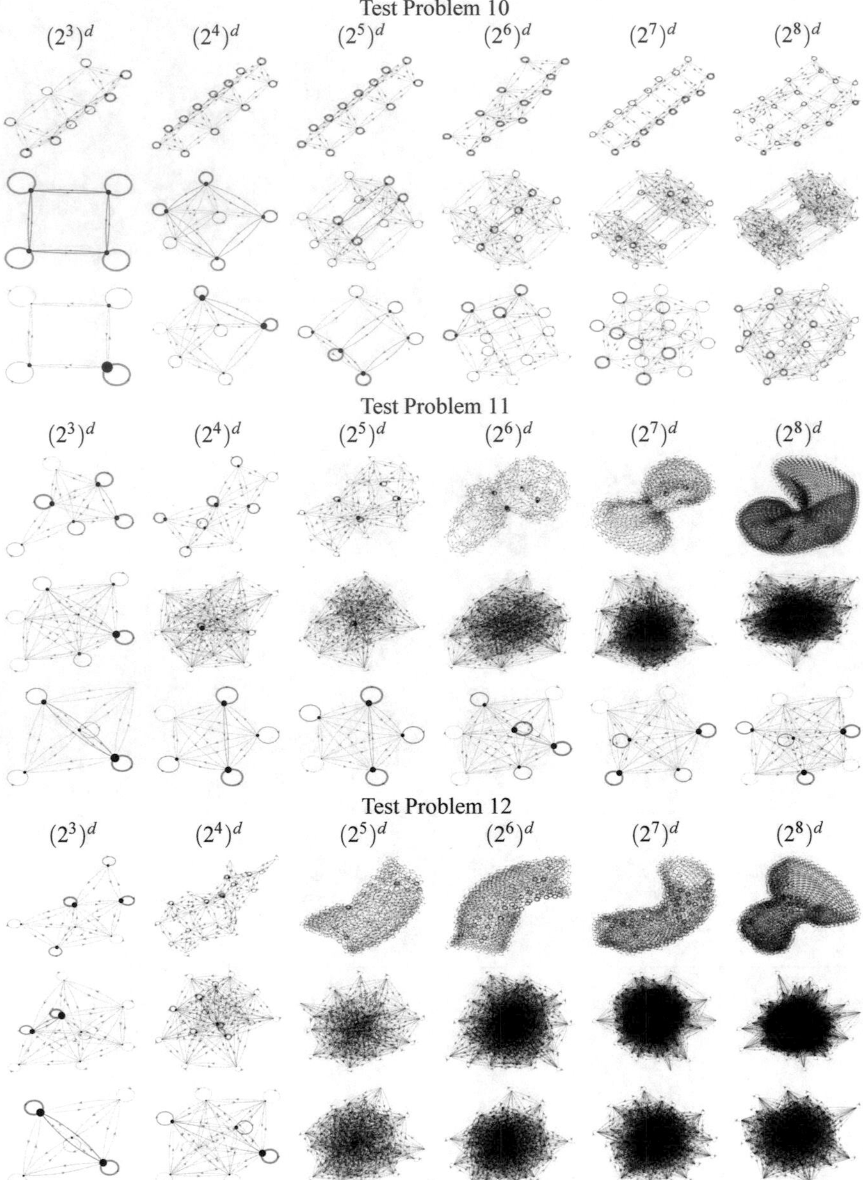

Fig. 13 Exact LON of test problems 10–12 using integer encoding (top row), binary encoding (middle row), and Gray encoding (bottom row). Basin transition for edges weights

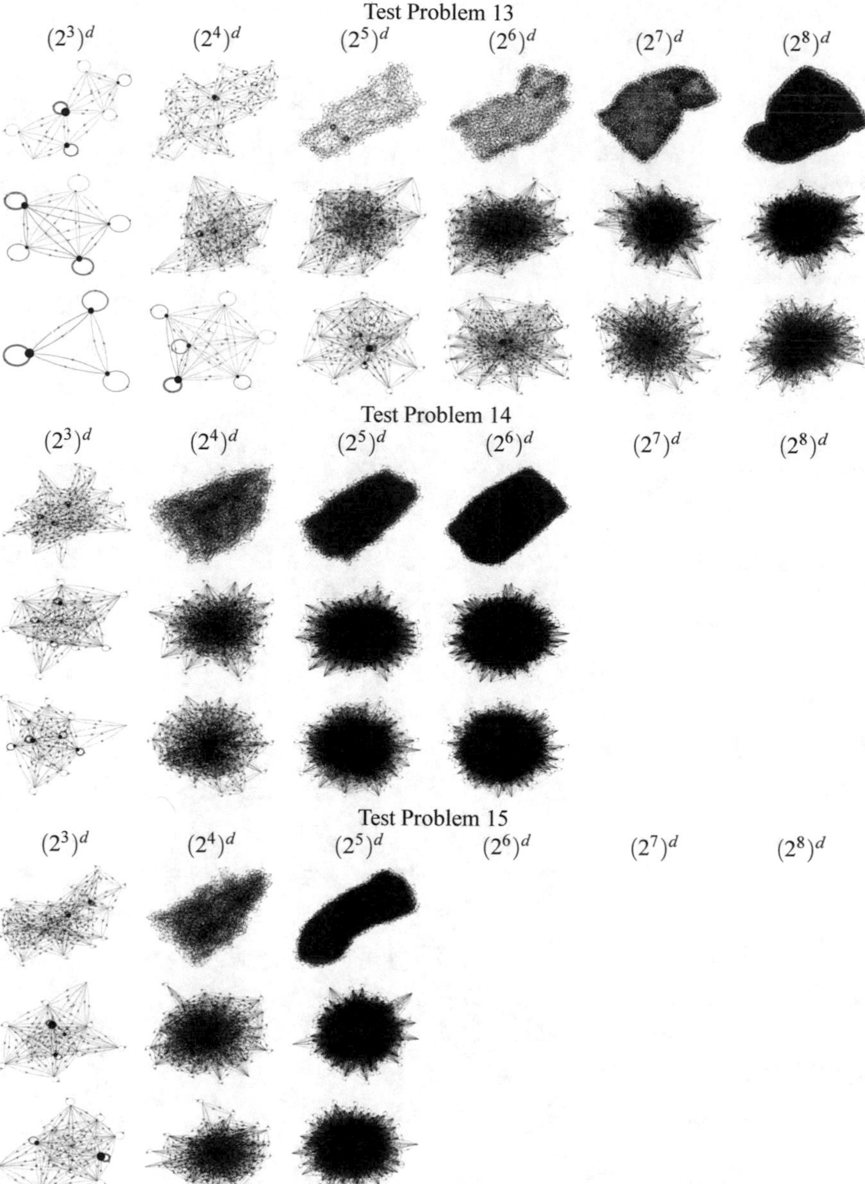

Fig. 14 Exact LON of test problem 13–15 using integer encoding (top row), binary encoding (middle row), and Gray encoding (bottom row). Basin transition for edges weights

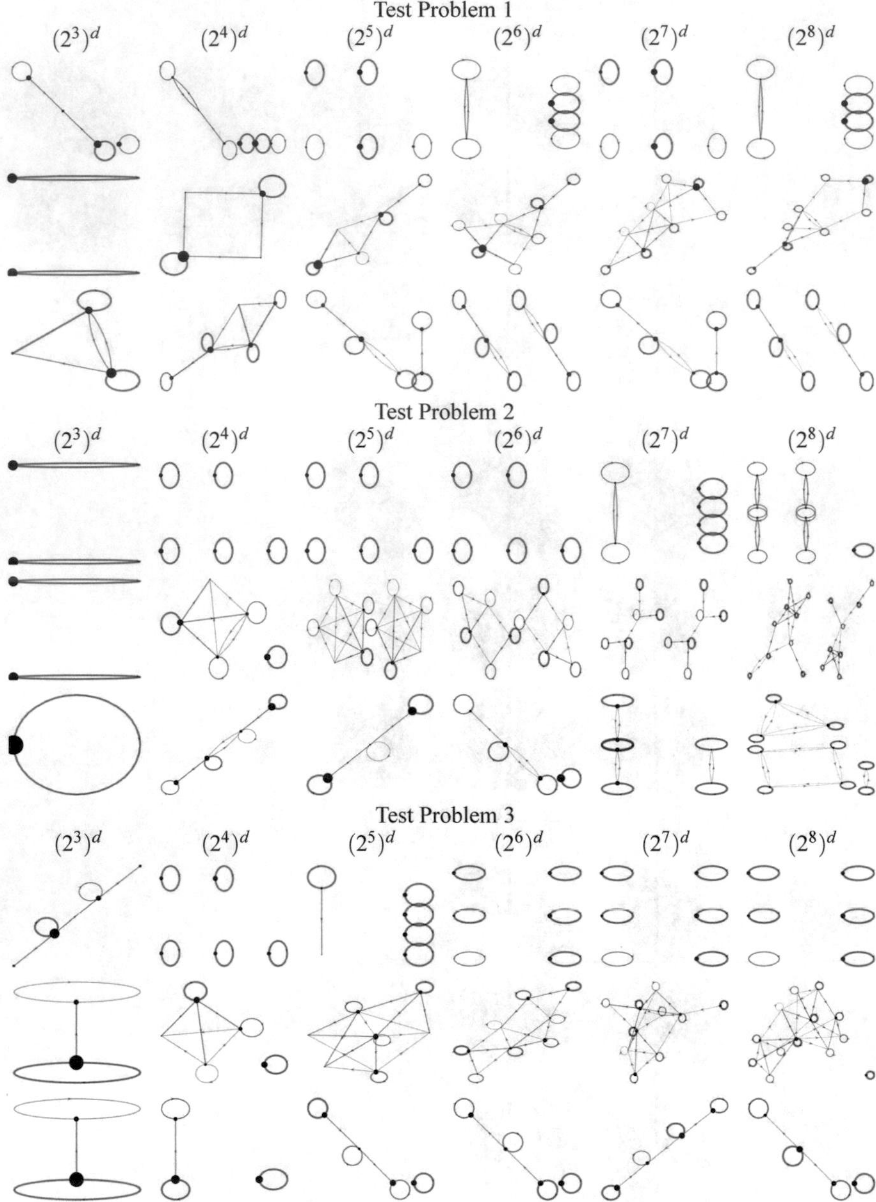

Fig. 15 Exact LON of test problems 1–3 using integer encoding (top row), binary encoding (middle row), and Gray encoding (bottom row). Escape edges for hamming distance ≤ 1 for edges weights

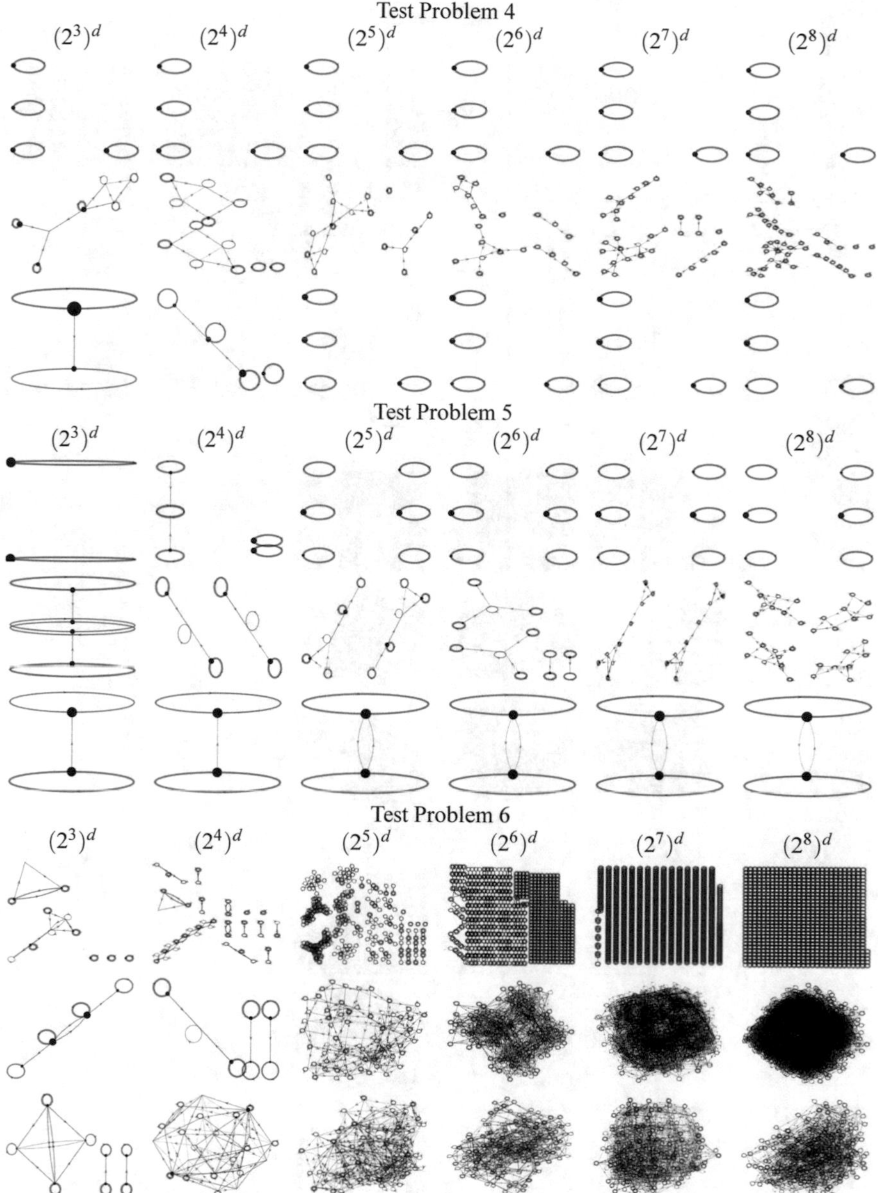

Fig. 16 Exact LON of test problems 4–6 using integer encoding (top row), binary encoding (middle row), and Gray encoding (bottom row). Escape edges for hamming distance ≤ 1 for edges weights

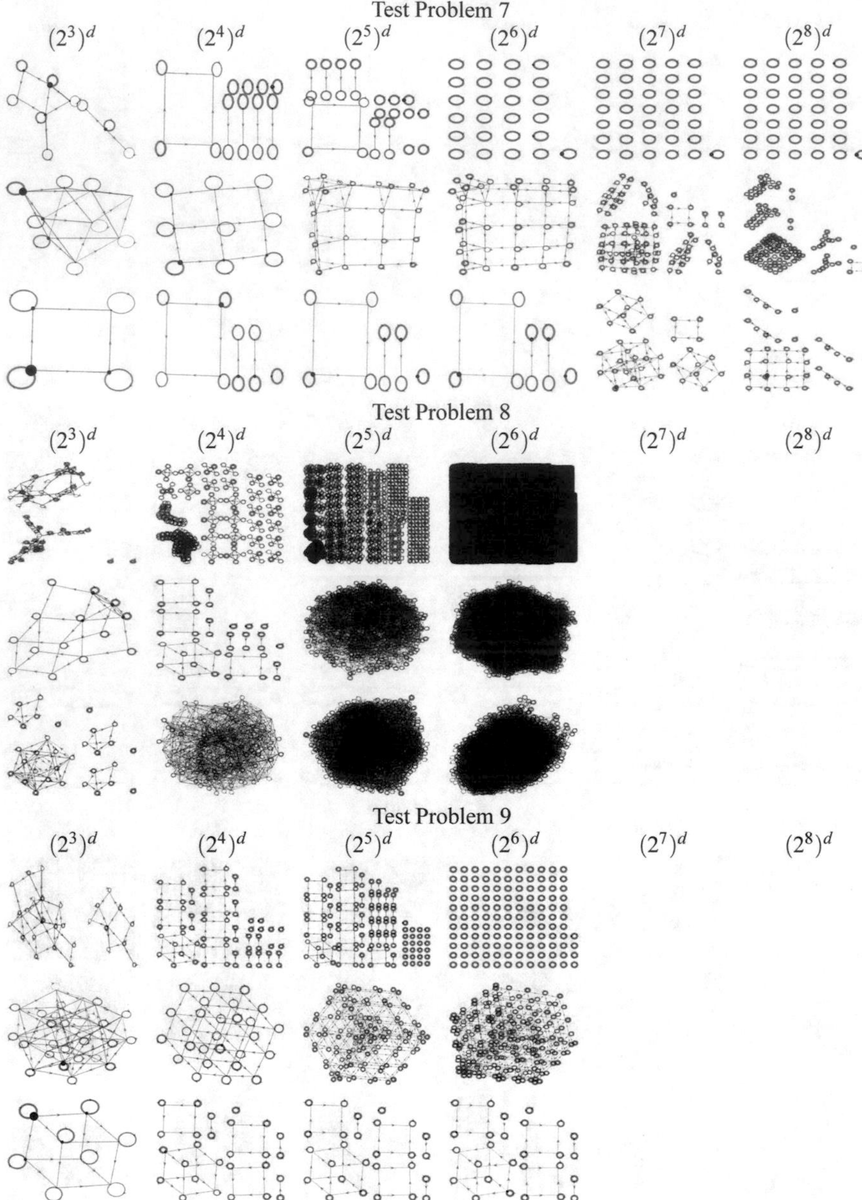

Fig. 17 Exact LON of test problems 7–9 using integer encoding (top row), binary encoding (middle row), and Gray encoding (bottom row). Escape edges for hamming distance ≤ 1 for edges weights

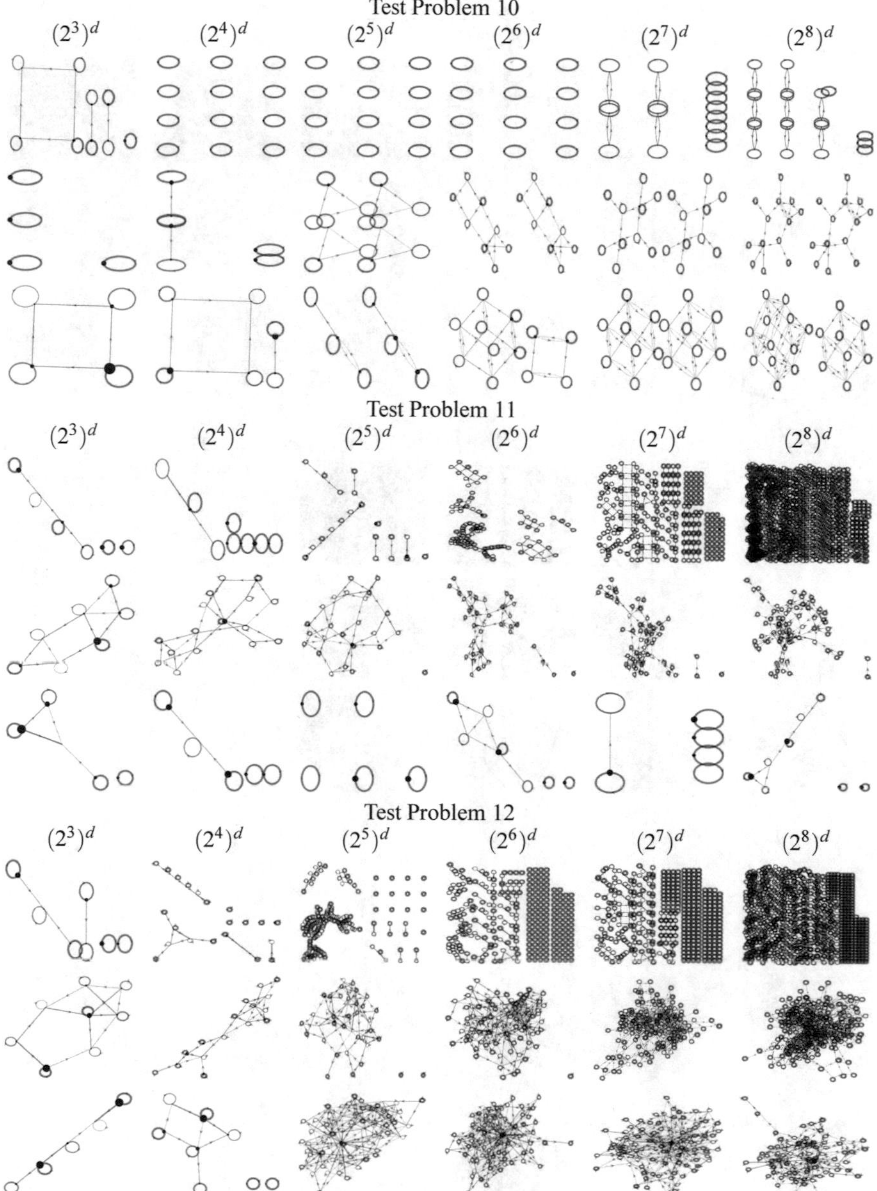

Fig. 18 Exact LON of test problems 10–12 using integer encoding (top row), binary encoding (middle row), and Gray encoding (bottom row). Escape edges for hamming distance ≤ 1 for edges weights

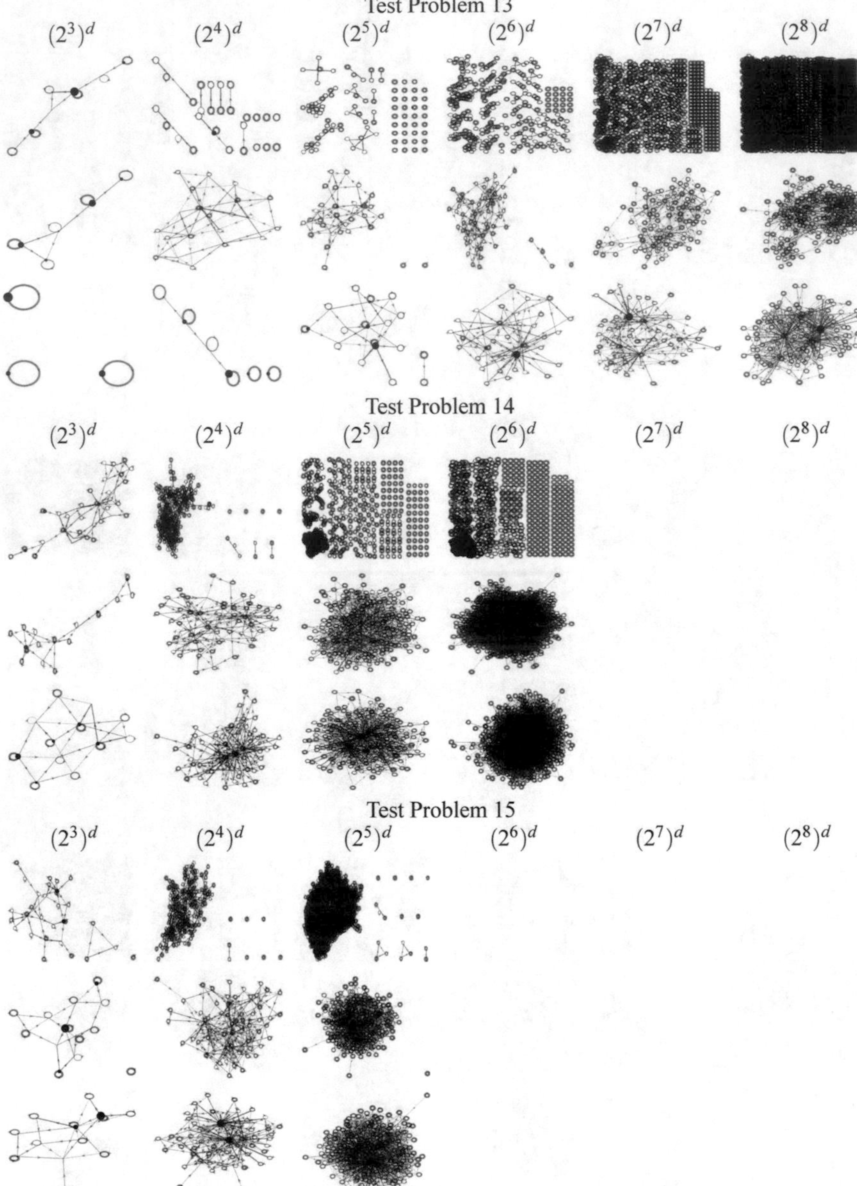

Fig. 19 Exact LON of test problem 13–15 using integer encoding (top row), binary encoding (middle row), and Gray encoding (bottom row). Escape edges for hamming distance ≤ 1 for edges weights

climber on the continuous space gets trapped. *However*, note how not only the Gray encoded landscape (with a larger but not excessive neighborhood for a design) has the same number or fewer modes than the integer neighborhood, but also these are largely connected—enabling a search to transition between these basins. This is even with the quite restrictive connectivity definition of the escape edge, i.e. once a local optima is located, hillclimb from the optima's neighbors to search for other optima.

Test problems 7 and 9 are curious in that although the Gray encoding has fewer modes, the escape edge LON contains a number of disconnected subgraphs, whereas the binary encoding neighborhood has paths possible between all modes. This would suggest that, for these problems at least, a binary neighborhood *may* be more beneficial than a Gray encoded one—depending on the search moves used.

4 Conclusion

In this chapter, we have explored the effect of changing the resolution and representation used when optimizing a problem, along with various visualizations of multimodal problems that the practitioner may wish to use to explore the landscape of a problem. As part of this, we presented the first extensive examination of the IEEE CEC 2013 benchmark problems using local optima networks. Open-source MATLAB code is freely available for these visualizations at https://github.com/fieldsend. It should be noted that for many problems, the generation of an exact LON is intractable (if you can fully enumerate the space to create such a network, then you don't need to search it using an optimizer!). However, in these cases you may approximate the landscape using sampling.

One approach the practitioner may take is to use a LON generated for their problem at a low resolution to get an understanding of the landscape. As shown here, depending on the problem, the landscapes can change considerably with the resolution employed, so in some situations this approach may be misleading. However, where the LON is relatively static under different resolutions, identifying the high-quality regions in a very coarse resolution before searching in this restricted space at higher fidelities is likely to be beneficial, as this is an effective route to search space reduction.

As mentioned earlier in this chapter—on current computing architectures—*all* continuous-variable problems are tackled using a limited bit-string encoding. The number of local optima experienced may therefore be seen to change not just depending on the coarser resolution used in a genetic algorithm, but also in a real-valued optimization depending on whether (for instance) 32-bit or 64-bit representations are used. The development of the LON framework for continuous spaces is an open area of research, as the step size required for greedy search on a 64-bit resolution is prohibitive for the hill-climbing approach commonly used for basin and edge definition; however, there is recent work on gradient approximation-based optimizers for this [1].

The ‘holy grail’ of optimizer design is automatically determining the best neighborhood function to simplify the landscape of the problem at hand (to, for instance, reduce modes or increase the basin sizes of global optima) for a given neighborhood size. Unfortunately, such a general algorithm is yet to be fully realized, but research in this area is gaining increasing traction as *landscape-aware heuristic search*.[2]

Acknowledgements This work was financially supported by the Engineering and Physical Sciences Research Council [grant number EP/N017846/1]. The author would like to thank Sébastien Vérel and Gabriela Ochoa for providing inspirational invited talks on LONs at his institution during this grant, and also Ozgur Akman, Khulood Alyahya, and Kevin Doherty.

References

1. Adair, J., Ochoa, G., Malan, K.M.: Local optima networks for continuous fitness landscapes. In: Proceedings of the Genetic and Evolutionary Computation Conference Companion (GECCO ’19). Association for Computing Machinery, New York, NY, USA, pp. 1407–1414 (2019)
2. Almutairi, A.T., Fieldsend, J.E.: Automated and surrogate multi-resolution approaches in genetic algorithms. In: 2019 IEEE Symposium Series on Computational Intelligence, Xiamen, China, 6th - 9th Dec pp. 2066–2073 (2019)
3. Alshawawreh, J.A.: Multi-scale optimization using a genetic algorithm. Western Michigan University, Kalamazoo (2011)
4. Babbar, M.: Multiscale parallel genetic algorithms for optimal groundwater remediation design. Ph.D. thesis, University of Illinois at Urbana-Champaign (2002)
5. Babbar, M., Minsker, B.S.: Groundwater remediation design using multiscale genetic algorithms. J. Water Resour. Plan. Manag. **132**(5), 341–350 (2006)
6. Chakraborty, U.K., Janikow, C.Z.: An analysis of Gray versus binary encoding in genetic search. Infor. Sci. **156**(3–4), 253–269 (2003)
7. Chan, T.F., Cong, J., Shinnerl, J.R., Sze, K., Xie, M., Zhang, Y.: Multiscale optimization in VLSI physical design automation. In: Multiscale Optimization Methods and Applications, pp. 1–67. Springer, Berlin (2006)
8. Daolio, F., Vérel, S., Ochoa, G., Tomassin, M.: Local optima networks of the quadratic assignment problem. In: 2010 IEEE Congress on Evolutionary Computation (CEC), pp. 1–8. IEEE (2010)
9. Fieldsend, J.E.: Computationally efficient local optima network construction. In: GECCO 2018 Companion - Proceedings of the 2018 Genetic and Evolutionary Computation Conference Companion, pp. 1481–1488 (2018)
10. Lopez Jaimes, A., Coello Coello, C., MRMOGA, : Parallel evolutionary multiobjective optimization using multiple resolutions. In: Proceedings of the IEEE Congress on Evolutionary Computation, vol. 3, pp. 2294–2301. IEEE (2005)
11. Jin, Y.: Surrogate-assisted evolutionary computation: recent advances and future challenges. Swarm Evolut. Comput. **1**(2), 61–70 (2011)
12. Jones, D.R., Perttunen, C.D., Stuckman, B.E.: J. Optim. Theory Appl. **79**, 157–181 (1993)
13. Kim, Y., De Weck, O.: Variable chromosome length genetic algorithm for structural topology design optimization. In: Proceedings of 45th AIAA/ASME/ASCE/AHS/ASC Structural Dynamics and Materials Conference, Palm Springs, California (2004)
14. Kim, D.S., Jung, D.H., Kim, Y.Y.: Multiscale multiresolution genetic algorithm with a golden sectioned population composition. Int. J. Num. Methods Eng. **74**(3), 349–367 (2008)

---

[2] With international workshops running since 2016, see, e.g. http://www.cs.stir.ac.uk/events/gecco-lahs2018/.

15. Li, H., Deb, K.: Challenges for evolutionary multiobjective optimization algorithms for solving variable-length problems. In: Proceedings of the IEEE Congress on Evolutionary Computation, pp. 2217–2224. IEEE (2017)
16. Li, X., Engelbrecht, A., Epitropakis, M.G.: Benchmark, : Functions for CEC'2013 Special Session and Competition on Niching Methods for Multimodal Function Optimization, Technical Report, Evolutionary Computation and Machine Learning Group. RMIT University, Australia (2013)
17. Mitchell, M.: An introduction to genetic algorithms. MIT Press, Cambridge (1998)
18. Nguyen, T.H., Paulino, G.H., Song, J., Le, C.H.: A computational paradigm for multiresolution topology optimization (MTOP). Struct. Multidiscip. Optim. **41**(4), 525–539 (2010)
19. Nguyen, T.H., Paulino, G.H., Song, J., Le, C.H.: Improving multiresolution topology optimization via multiple discretizations. Int. J. Numer. Methods Eng. **92**(6), 507–530 (2012)
20. Ochoa, G., Tomassini. M., Vérel, S., Darabos, C.: A study of NK landscapes' basins and local optima networks. In: Proceedings of the 10th Annual Conference on Genetic and Evolutionary Computation, pp. 555–562. ACM (2008)
21. Ochoa, G., Verel, S., Daolio, F., Tomassini, M.: Local optima networks: a new model of combinatorial fitness landscapes. In: Richter, H., Engelbrecht, A. (eds.) Recent Advances in the Theory and Application of Fitness Landscapes, pp. 233–262 (2014)
22. Park, J., Sutradhar, A.: A multi-resolution method for 3D multimaterial topology optimization. Comput. Methods Appl. Mechan. Eng. **285**, 571–586 (2015)
23. Preen, R.J., Bull, L.: Toward the coevolution of novel vertical-axis wind turbines. IEEE Trans. Evolut. Comput. **19**(2), 284–294 (2015)
24. Sinha, E., Minsker, B.S.: Multiscale island injection genetic algorithms for groundwater remediation. Adv. Water Resour. **30**(9), 1933–1942 (2007)
25. Sun, W., Dong, Y.: Study of multiscale global optimization based on parameter space partition. J. Global Optim. **49**(1), 149–172 (2011)
26. Veerapen, N., Daolio, F., Ochoa., G.: Modelling genetic improvement land- scapes with local optima networks. In: Proceedings of the Genetic and Evolutionary Computation Conference Companion (GECCO '17), pp. 1543–1548. ACM (2017)
27. Vérel, S., Daolio, F., Ochoa, G., Tomassini, M.: Local Optima Networks with Escape Edges. In: Hao, J.-K., Legrand, P., Collet, P., Monmarché, N., Lutton, E., Schoenauer, M. (eds.) Artificial Evolution, pp. 49–60. Springer, Berlin (2012)
28. Vérel, S., Ochoa, G., Tomassini, M.: Local optima networks of NK land- scapes with neutrality. IEEE Trans. Evolut. Comput. **15**(6), 783–797 (2011)

Finding Representative Solutions in Multimodal Optimization for Enhanced Decision-Making

Andreas Miessen, Jaromił Najman, and Xiaodong Li

Abstract Many real-world optimization problems are multimodal by nature, and there may exist a large number of optimal solutions. Despite having the same or similar objective values, solutions can still differ in terms of technical feasibility or the preferred range of their decision variable values. Therefore, it is more desirable to employ optimization methods capable of offering several optimal solutions to the Decision Maker (DM). Existing niching methods aim to find all possible solutions in a single optimization run, resulting in possibly too many options to choose from. Due to limited resources available for evaluating solutions in practice, the DM, however, might only be interested in finding a few sufficiently different solutions quickly. This work aims to facilitate this decision-making process by providing only a number of *representative* solutions to the DM. This way, the DM is not overloaded with superfluous information, resulting in faster and better decision-making. This paper proposes a novel niching method, Suppression Radius-based Niching (SRN), based on the principle of *suppression radius* to determine representative niching areas. The proposed method is especially appealing for real-world scenarios where reducing the number of function evaluations is crucial due to the high computational costs of evaluations.

A. Miessen (✉) · J. Najman
Process Systems Engineering (AVT.SVT), RWTH Aachen University, Forckenbeckstraße 51, 52074 Aachen, Germany
e-mail: andreas.miessen@rwth-aachen.de

J. Najman
e-mail: jaromil.najman@avt.rwth-aachen.de

X. Li
School of Science (Computer Science and Software Engineering), RMIT University, Melbourne, VIC 3001, Australia
e-mail: xiaodong.li@rmit.edu.au

M. Preuss et al. (eds.), *Metaheuristics for Finding Multiple Solutions*, Natural Computing Series,
https://doi.org/10.1007/978-3-030-79553-5_3

1 Introduction

Optimization as the process of finding the best possible solutions from a set of alternatives is widely used in various industries, ranging from engineering to finance sectors [11, 41]. In real-world scenarios, optimization problems are often very complex and multiple optimal solutions may exist. In order to have alternatives to choose from, a Decision Maker (DM) is often interested in finding a number of optimal or close to optimal solutions. Furthermore, through the information about the spread of solutions in the problem space, a better understanding of the problem under study can be gained. As a result, finding a solution close to the expected global optimum becomes more likely. The possibility of choosing from several alternative solutions is also more desirable, since some factors to assess the quality of a solution may not be captured in the mathematical formulation of the problem. As described in the well-publicized *Second Toyota Paradox*, having a set of design alternatives can result in making better cars faster and more cheaply [1]. However, having too many solutions to choose from overloads the DM with unnecessary information [38], which may lead to deterioration of the decision-making process [22]. The DM might be interested in only a few sufficiently different solutions [25]. In some real-world applications, such as structural optimization [27] and scheduling [5], the number of optimal solutions could be numerous. Evaluating all these solutions can be computationally expensive. Instead of finding all possible optimal solutions, search effort should be concentrated on a small number of desired solutions, as specified by the DM. This way, significant computational costs can be saved. This work aims to support the decision-making process by asking the DM to specify a single parameter, i.e., the number of representative optimal solutions (usually a small number), and use this information to obtain results faster by focusing the search only in representative regions of the search space. Without assuming any prior knowledge, a simple way to define "solution representativeness" could be the volume of the search space covered by the solutions as well as the homogeneity of their distribution.

The first methodologies to find multiple optima in a single optimization run, so-called niching methods, were developed in the 1970s, such as fitness sharing [17] and crowding [10]. Since then, many other niching methods have been proposed, such as deterministic crowding [28], derating [3], restricted tournament selection [18], parallelization [4], clustering [39], stretching and deflation [31], clearing and speciation [24, 32].

The goal of a typical niching method is to find as many solutions as possible. As explained above, from the DM's point of view, in most real-world applications this is not necessary. This paper proposes a Suppression-Radius-based Niching (SRN) algorithm, which is able to obtain a user-defined number of representative solutions. The *representativeness* guarantees that the obtained solution points are sufficiently different and representative with respect to the entire search space. By incorporating the DM's knowledge, search effort can be concentrated early on to the representative areas, which in turn leads to a more effective and efficient search. Consequently, the proposed SRN needs substantially less computational effort to obtain similar quality solutions. Reducing function evaluations is critical since solution evaluations in some computational models can be extremely time consuming [8].

To authors' knowledge, SRN is the first niching algorithm aiming to support the decision-making process by producing only representative solutions, thereby not overloading the DM with too many alternatives to choose from. The novelty of SRN lies in the fact that the DM only needs to supply one input parameter value—a *desired number* of representative solutions. As a framework for any population-based metaheuristic, its major advantage, apart from the enhanced decision making, lies in the efficiency due to its focused search around a number of representative areas of the search space. Furthermore, SRN can also be used to automatically discover the number of potential global optima, if the DM is unable to provide a value for such input parameter.

The paper is organized as follows. Section 2 gives an overview about related work and concepts. The proposed SRN algorithm is explained in Sect. 3, after which an extensive experimental analysis is elaborated in Sect. 4. The paper concludes with Sect. 5.

2 Related Work

Multimodal Optimization (MMO) can be achieved by using niching methods, which were originally designed to enhance diversity for an Evolutionary Algorithm's (EA) population. More specifically, niching methods can be effectively used to locate multiple optimal solutions in a single optimization run. In the following, an overview of the most important niching methods is given, as well as concepts used later in the proposed SRN algorithm. These concepts, Hopkins-statistic and adaptable non-maximal suppression, are, respectively, used to identify a cluster tendency in the population, and to identify best-fit and spatially well-spread individuals.

2.1 Classic Niching Methods

Like Evolutionary Computation (EC), niching is also a nature inspired principle. The notions of *niches* and *species* can be found in literature on natural ecosystems where *niche* is defined as a subspace of the environment in which a *species* is a class or type of individuals that is able to evolve. Therefore, "niches are divisions of an environment, while species are divisions of the population" [21]. A species is a group of individuals of similar biological features who are capable of interbreeding among themselves, but not able to interbreed with individuals from a different group. Since each niche is defined through its limited resources, which are shared among species members occupying that niche, over time different subpopulations evolve filling different niches. In EC, niching methods can be divided into two categories, *sequential (temporal) niching* and *parallel (spatial) niching*. Sequential niching locates multiple niches temporally by solving a single problem in each run and combining the results eventually. To avoid converging to the same area in a future run, the fitness landscape around the located optimum is depressed. Therefore, the fitness value is derated according to the distance to the found optimum [3]. Parallel niching aims to form and maintain multiple niches simultaneously within a single optimization run, with a single population. Parallel niching methodologies are considered as superior in terms of computational cost and effectiveness on problems of varying difficulty level [29]. Some of the best-known parallel techniques are briefly outlined here. *Fitness sharing* is one of the most-widely used niching methods. Originally introduced by Holland [19], the principle of sharing limited resources among individuals was adopted by Goldberg and Richardson [17] to divide a population into several subpopulations based on the similarity of individuals. To encourage diversity within the population, an individual's fitness value is degraded according to the presence of other individuals in its vicinity. It has been proven to be a useful niching method if the characteristics of the problem are known a priori. However, without prior knowledge of the underlying problem, it can be difficult to set appropriate niching parameter values, e.g., niche radius [16]. In real-world scenarios, this knowledge may not be available prior to optimization. Further improvements of the original sharing technique led to the development of *dynamic fitness sharing* [9], *dynamic niching sharing* [30], and *clearing* [32]. While in the sharing technique the fitness value of overcrowded individuals is degraded, the *crowding* method [10] is based on the idea of competition among similar individuals. The new offspring is compared to a random sample of the current population, and the most similar individual of the sample is replaced. A parameter, *Crowding Factor (CF)*, is used to specify the sample size. Choosing the right value for CF is crucial and has a large impact on the quality of the results [28]. Mahfoud made several modifications to the original *crowding* and introduced *deterministic crowding*, which is able to find and maintain multiple peaks. Motivated from the drawbacks of the original *fitness sharing*, the *clearing* method was introduced in [32]. *Clearing* is based on the *winner takes it all* principle. The fittest member of a niche remains the only surviving individual with full access to limited resources while the less fit individuals are removed within a *clearing radius* σ_{clear}. However, the results are still highly dependent on the *clearing radius* σ_{clear} which needs to be specified based on prior knowledge of the problem. In nature, a *species* is the smallest independent evolutionary unit whose survival solely depends on the existence of the *niche* it is filling. Inspired by this strong interdependence, the concept of *speciation* was introduced in [24]. Multiple species are adaptively formed and separately evolved. As a result, multiple optima can be found.

2.2 Recent Development

In recent years, metaheuristics (other than EAs) such as Particle Swarm Optimization (PSO) and Differential Evolution (DE) have also been used to design niching algorithms [25]. Promising niching DEs have recently been developed and are briefly described here. The results of these niching DEs are compared with that of the proposed SRN in the experimental study in Sect. 4.

A simple and yet effective niching DE was introduced in [14]. This so-called *DE/nrand/1* changes the mutation scheme by taking the nearest neighbor of the target vector as base vector which induces niching behavior. However, as with most niching DEs, finding and maintaining a large number of optima highly depends on the population size. The *dADE/nrand/1* algorithm proposed by [12] overcomes this dependency by incorporating a dynamic archive into the DE/nrand family. Instead of changing the base vector to induce niching behavior, the neighborhood-based niching DE [33] modifies the difference vector to achieve a similar effect. The difference vector is not calculated from two randomly selected individuals but rather from similar individuals which are located within the neighborhood of each base vector. The neighborhood is defined for each base vector by its m closest individuals in terms of Euclidean distance. This parameter m needs to be selected by the user and depends on the individual problem characteristics. Another approach based on fitness and proximity information a niching DE with so-called local information sharing was proposed in [6], where individuals that are fitter and closer to the target vector are preferred. Due to this probabilistic parent selection scheme, a niching effect is achieved by minimizing the tendency of changing basin of attractions while still encouraging exploration. The abovementioned approaches to inducing niching behavior in DE result in an increase of the algorithm's complexity. To address this issue, Zhang et al. introduced a locality sensitive hashing-based approach *Fast-NCDE* which leads to a fast niching algorithm [42]. Nevertheless, these niching methods assume that all global optimal solutions must be found. Furthermore, they do not make use of the knowledge (or preference information) of a DM. We believe that in real-world situations these are important factors that need to be considered. Herein the proposed SRN niching method primarily focuses on the DM's perspective. SRN incorporates the DM's knowledge into the optimization process leading to a more effective and efficient search that facilitates decision making.

2.3 Differential Evolution

Differential Evolution (DE) was first introduced by Storn and Price [35]. DE belongs to the category of metaheuristics and is getting more attention due to its simple implementation and reliable performance. DE is a population-based global optimization method, which generates new points through perturbations of existing points. Generally, a *base vector* is mutated, recombined with its parent, so-called *target vector*, and then compared to its target vector for entering the new generation. DE comprises following four phases:

2.3.1 Initialization

DE starts its global search with initializing a random population of Np members. Each population member is a real-valued vector whose components represent D variables in the decision space $\mathcal{X} \subseteq \mathbb{R}^D$. A generalized notation of the ith population member in generation g can be written as $\boldsymbol{x}_g^i = \left[x_{g,1}^i, x_{g,2}^i, \ldots, x_{g,D}^i\right]$.

2.3.2 Mutation

In the mutation phase, each population member is mutated according to specific rules. The standard DE mutation scheme is defined as follows:

$$\boldsymbol{v}_g^i = \boldsymbol{x}_g^{r_0} + F\left(\boldsymbol{x}_g^{r_1} - \boldsymbol{x}_g^{r_2}\right), \tag{1}$$

where $\boldsymbol{x}_g^{r_0}$ is the base vector, $\boldsymbol{x}_g^{r_1} - \boldsymbol{x}_g^{r_2}$ the *difference vector* from two randomly selected population members, $F \in (0, 1]$ the so-called *scaling factor*, and $\boldsymbol{v}_g^i$ is the resulting *mutant*. The indices r_0, r_1, and r_2 are mutually exclusive and different from the running index i ($r_0, r_1, r_2 \in \{1, 2, \ldots, Np\} \setminus \{i\}$) of all members of the generation g. A common value for the scaling factor is $F = 0.5$ [12]. There exists extensive research about the right choice of the scaling factor. A small F emphasizes exploitation and can lead to premature convergence, a large F emphasizes exploration leading to a decrease in the performance.

The choice of the base vector has great impact on the search behavior. As described by Storn, the search can be focussed on exploitation by choosing the current best point as the base vector ($r_0 = best$) to produce the following *DE/best/1* variant [34]:

$$\boldsymbol{v}_g^i = \boldsymbol{x}_g^{best} + F\left(\boldsymbol{x}_g^{r_1} - \boldsymbol{x}_g^{r_2}\right) \tag{2}$$

Furthermore, without introducing any additional niching parameters, an automatic niching behavior can be induced as suggested in [14]. By selecting the nearest neighbor of the target vector $\boldsymbol{x}^i$ as the base vector ($r_0 = NN_i$), we can produce a simple yet effective niching DE algorithm, *DE/nrand/1*[1]:

$$\boldsymbol{v}_g^i = \boldsymbol{x}_g^{NN_i} + F\left(\boldsymbol{x}_g^{r_1} - \boldsymbol{x}_g^{r_2}\right) \tag{3}$$

2.3.3 Crossover

The crossover phase aims to combine the parameters, i.e., vector elements, of target vector and mutant to a single offspring, a so-called *trial vector*. A binomial crossover has become the preferred strategy [40]. Here, the *crossover probability* Cr is set to

[1] DE/nrand/1 is a simple modification of Price & Storn's original Matlab code DeMat. Available at https://github.com/mikeagn/DeMatDEnrand.

a constant value between 0 and 1. Then D uniformly distributed random numbers are generated (one for each vector component), and are compared to Cr to decide which component is to be included in the new *trial vector* $\boldsymbol{u}^i_g$:

$$u^i_{g,j} = \begin{cases} v^i_{g,j}, & \text{if } rand_{i,j}(0,1) \leq Cr \text{ or } j = j_r \\ x^i_{g,j}, & \text{otherwise,} \end{cases} \tag{4}$$

where j_r is a randomly selected index from $\{1, 2, \ldots, D\}$ which ensures that at least one component of the mutant is assigned to the new trial vector. The closer the crossover probability is to 1, the more likely a component of the mutant $\boldsymbol{v}^i_g$ is chosen. A common value for the crossover probability is $Cr = 0.9$ [12].

2.3.4 Selection

In the selection phase, the new trial vector is compared to its target vector in order to decide which one will enter the new generation. A simple comparison between their objective values constitutes the decision:

$$\boldsymbol{x}^i_{g+1} = \begin{cases} \boldsymbol{u}^i_g, & f(\boldsymbol{u}^i_g) > f(\boldsymbol{x}^i_g) \\ \boldsymbol{x}^i_g, & \text{otherwise,} \end{cases} \tag{5}$$

where f is a single-objective function to be maximized.

2.4 *Hopkins-Statistic*

The Hopkins-statistic [20] measures *cluster tendency* of a set of data points. Cluster tendency is a well-known concept in the field of cluster analysis for determining the presence or absence of a clustering structure in a given data set [37]. The idea is to compare the distances between two sets of sample points, a randomly chosen subset $X_1 \in X$, and a set of pseudo points Y_1 uniformly distributed, to their nearest neighbors in the full data set X. If the sets of distances are very similar, the data points are not clustered. More precisely, the H-value is defined as

$$H = \frac{\sum_{j=1}^{M} d_j}{\sum_{j=1}^{M} d_j + \sum_{j=1}^{M} \delta_j}, \tag{6}$$

where δ_j denotes the Euclidean distance between $x_j \in X_1$ to its nearest neighbor in $X \setminus \{x_j\}$, d_j the distance between $y_j \in Y_1$ to its nearest neighbor in X. M is the sample size, typically chosen between 5 and 10% of the data [2, 23]. In this paper, a sample size of 10% is used.

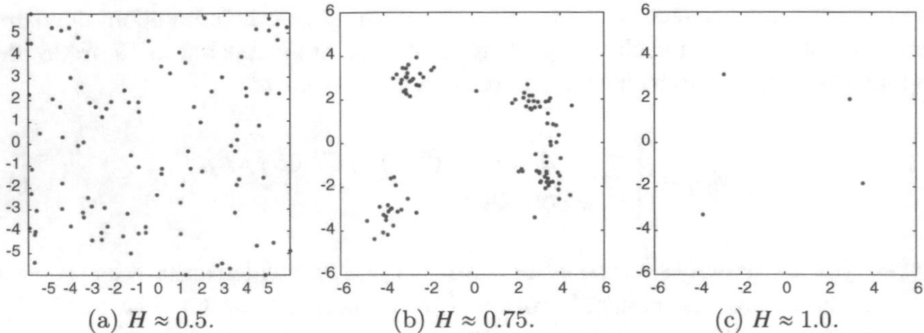

(a) $H \approx 0.5$. (b) $H \approx 0.75$. (c) $H \approx 1.0$.

Fig. 1 Different point distributions of 100 points with low, medium, and high cluster tendency

When data set X contains clearly identifiable clusters, it is assumed that the distances δ_j are very small due to an aggregation of points. Hence, the H-value becomes larger. The upper bound is $H = 1$ for total clustering. On the other hand, assuming no tendency of clustering and thereby randomly distributed points in X, δ_j is assumed to be on average as large as d_j, which results in $H = 0.5$ (see Fig. 1a). Consequently, larger values of H represent a stronger manifestation of clustering structure in the given data set. Figure 1 shows three examples of point distributions of 100 data points. Figure 1a shows randomly distributed points, thus $H \approx 0.5$. In Fig. 1b, a clear cluster tendency is visible, yielding $H \approx 0.75$. When all 100 points converge to four distinct clusters, with all points within each cluster being in the same location (so-called distributed convergence [25]), as shown in Fig. 1c, $H \approx 1.0$ results. This is an important feature allowing us to use the H-value to measure clustering tendency in niching.

2.5 *Adaptable Non-maximal Suppression*

Adaptable Non-Maximal Suppression (ANMS) is a concept which originated from the field of image processing [7]. ANMS can be used to find the n best points which are well distributed across a set of data points. The quality of a point is represented by a scalar value. Originally used in multi-image matching, it was designed to reduce the number of interest points for comparing multiple images with each other. The ANMS method can be adapted to the needs of population-based metaheuristics: instead of using the corner strength, the fitness value is taken to set a weight for each point.

As the first step in ANMS, a so-called *suppression radius (sr)* is computed for each point. Once computed, all points are sorted according to their sr values in a descending order. To determine the n best-fit and spatially well-spread points, we simply take the first n points of the sorted list. sr is calculated as below:

Step 1: Starting with an initial radius $sr = 0$, the fitness (i.e., weight) of the considered point p is compared to the value of its closest neighbor p_{NN_1}. If

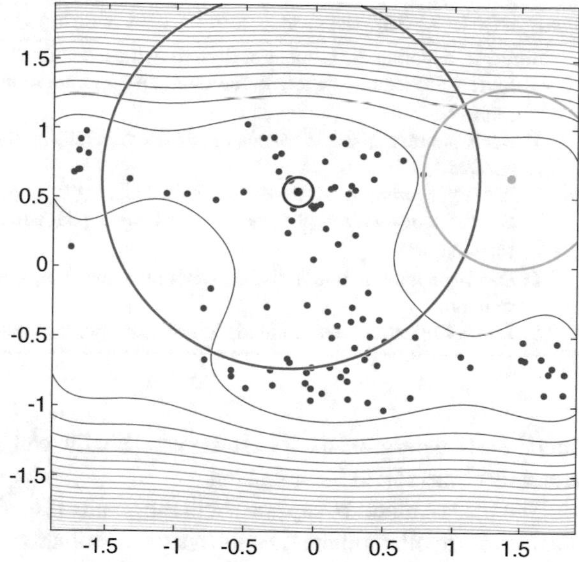

Fig. 2 Each point has a different sr value. Three colored circles around three points are shown here. Each point has a sr value equal to the radius of the same colored circle. Note that the circle represents an expanded circle from the point until a dominant point is encountered

Table 1 Ranked sr and objective values (ov) of an exemplary maximization problem

Ranks of sr	Ranks of ov	ov
...	...	...
2	5	0.95
...	...	...
5	100	−2.20
...	...	...
62	12	0.84
...	...	...

the fitness value of p is better (i.e., larger in case of maximization) than the fitness value of p_{NN_1}, we compare p to the next nearest neighbor p_{NN_2}. This process is repeated until a neighbor p_{NN_k} with a better or equal fitness value is encountered.

Step 2: The radius sr is set as the distance between the considered point p and the dominant neighbor point p_{NN_k}. sr is different for each point.

Figure 2 shows an example of three points for a maximization problem (in purple, yellow, and red color) and their corresponding dominant areas represented by circles, each with a sr value set to its radius. These sorted sr values are displayed in Table 1. After having computed the sr value for each point of the data set, in our context meaning the entire population, the points are ranked according to their sr values in a descending order. Hence, the first entry of the list has the largest sr, in fact being always infinity because the fittest individual is not dominated by any other population

Algorithm 1: Suppression-Radius-based Niching (SRN)

Input: n, denoting the number of desired optimal solutions, supplied by the DM
Output: n sufficiently different, representative optimal solutions finally found

Phase I

1: Run a population-based metaheuristic (such as DE) until the cluster tendency threshold is reached.
2: Identify n *representative solutions* by calculating suppression radius sr.
3: Determine n *representative areas* by calculating degradation radius dr.

Phase II

4: Divide population into n distinct subpopulations (i.e., *species*), one for each representative solution.
5: Run optimization independently to find one optimal solution in each representative area.

member. To determine the n best-fit and spatially well-spread points, simply the first n points of the sorted list are taken.

The scalar-valued sr captures both the quality (i.e., fitness) of a solution as well as its distance to other points (i.e., the spatial distribution). In order to have a large sr, it is not enough to only have a relatively good fitness value, but the solution also needs to be located far away from other better solutions. As shown in Table 1, assuming the 5th best solution ranked by its objective value (ov) might have the 2nd largest sr. The 12th best solution with an objective value almost as good, however, is located in the vicinity of a superior solution and has therefore a much smaller sr, as seen in Fig. 2. A much inferior point in terms of ov, in this example the worst solution of all 100, which has no superior point in its vicinity, has a larger (the 5th largest) sr value.

3 Suppression-Radius-Based Niching (SRN)

Herein, a new DE niching algorithm SRN is proposed which is able to obtain a given number of *representative* optimal solutions, based on a number supplied by the DM. The idea is to confine the search for optimal solutions to the required number n, instead of finding too many solutions. This way, a more focused search can be carried out with much better search efficiency. The approach consists of two phases. In Phase I, the emphasis lies on exploring the search space and developing a tendency toward possible peaks. Once a tendency to cluster building is observed, n representative areas are determined using the suppression radius sr. In Phase II, we then focus the search on the n representative areas to find one optimal solution within each identified area, resulting in n representative optimal solutions. Algorithm 1 shows the pseudocode of the proposed SRN.

3.1 Phase I—Identifying Representative Areas

In order to identify *n representative areas*, the only information known is the location of each individual in the decision space as well as its objective value. Based on this, *n representative areas* are to be determined by detecting *n* representative solutions, so-called *representatives*, first, and defining their immediate vicinities as the targeted areas, subsequently.

3.1.1 Finding Representative Solutions

Since the initial population is randomly distributed, little information about the landscape can be deduced. Observations on the dynamic of the DE algorithm suggest, however, that after a number of iterations, individuals tend to cluster around optimal and suboptimal solutions [13, 15, 36]. This behavior can be used to gain knowledge about the landscape to identify representative areas. The idea is to let the DE run until enough information can be deduced from the current population to identify representative areas. Using the mutation strategy of *DE/nrand/1*, at each generation the cluster tendency is examined by measuring the H-value (Sect. 2.4). Once a threshold value is reached, it can be deduced that the random distribution of individuals from the initial population has moved closer toward possible optimal solution areas due to the induced niching behavior. Although it is still far from convergence, the probability of obtaining quality solution points in the vicinity of a real optimum has increased drastically after the first few iterations. In fact, after reaching a certain threshold of cluster tendency, the solutions reached so far provide valuable information about real optima, which can be used to focus the search. In this research, this threshold is set to $H = 0.67$. It is empirically determined and has proven to be a reasonable trade-off between an adequate level of *distributed convergence* and an early enough detectable clustering tendency. Figure 3 shows an initial population and its evolved population after just 4 generations, reaching the H-threshold ($H = 0.67$). Due to the stochastic nature of the H-measure, induced by choosing a random sample of solutions each time, the average over 100 H-value results is taken [14]. Then, the ANMS scheme is applied subsequently to determine the n best-fit and spatially well-distributed points.

3.1.2 Defining Representative Areas

In general, it is assumed that a global optimum is located in the vicinity of the *representative solution* r_i. Consequently, a *representative area* can be defined within which a potential global solution is sought while not impacting the overall representativeness of the obtained (representative) solutions. Here, we use a *degradation radius* (dr) to define the representative area which guarantees that computing representative solutions via their suppression radii in a future generation would not lead

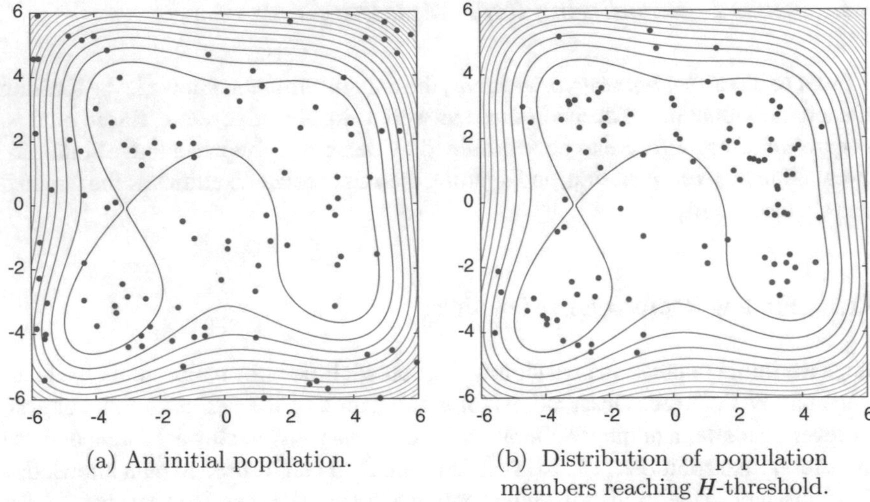

(a) An initial population.

(b) Distribution of population members reaching H-threshold.

Fig. 3 An evolving population for the Himmelblau function with tendency to form clusters, reaching H-threshold ($H = 0.67$) after 4 generations

to different *representative areas*, as long as the updated representative in the new generation lies within this initial radius. This means that the possible new *representative* r_{i^*}, defined as the new best solution found so far, should not decrease the other representatives' suppression radii in such way that different solutions would be determined as representatives. The maximum distance between a new best solution and the old representative needs to be chosen such that none of the suppression radii of the current n selected solutions will become smaller than the next best $(n + 1)$th suppression radius sr_{n+1} (see Fig. 4). As a consequence, the degradation radius (dr) is computed as

$$dr_i = \frac{d_i^{nn} + sr_{n+1}}{2} - sr_{n+1} = \frac{d_i^{nn} - sr_{n+1}}{2}, \tag{7}$$

where d_i^{nn} denotes the distance from the current representative i to its nearest neighboring representative. Equation (7) considers the worst case scenario of how far the current representative r_i can be moved without influencing the representativeness of others nor its own. Figure 5a shows the degradation radii when applying the above explained principle to the example shown in Fig. 3 with $n = 3$. The degradation radius dr_i is set individually for each representative r_i. Those representatives with a larger distance to their nearest neighbor have a larger radius and therefore a larger area without degradation. A larger radius increases the probability of finding the real optimum while decreasing the convergence speed slightly.

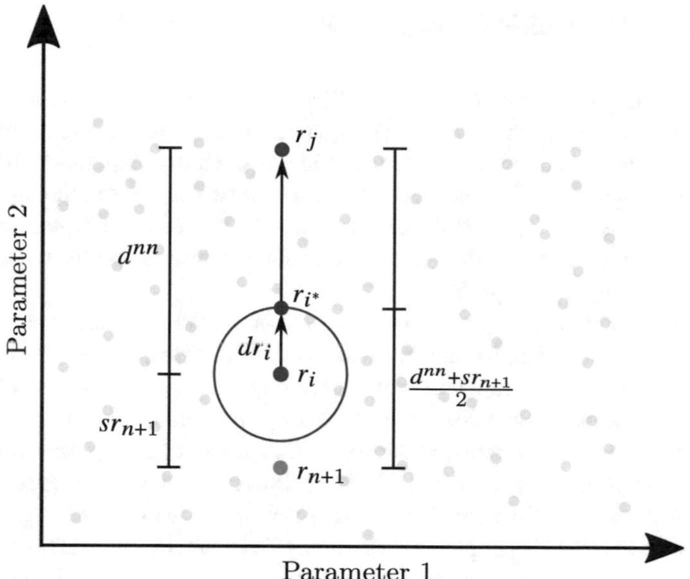

Fig. 4 Derivation of degradation radius dr_i guaranteeing representativeness

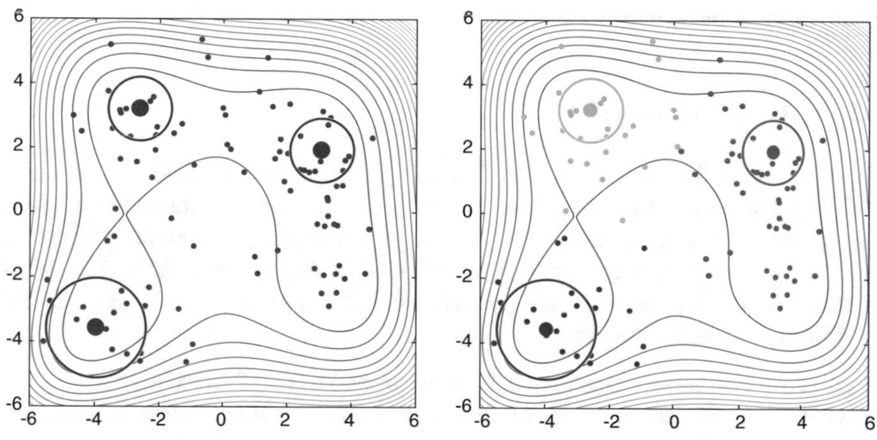

(a) Resulting representative areas (purple circles of radius dr) around obtained initial representatives (bigger red dots).

(b) Resulting species after reaching the H-threshold. All points with the same color belong to the same species.

Fig. 5 Representative areas defined by circles of radius dr around representatives. Species affiliation of each population member for the Himmelblau function with $n = 3$ after 4 generations

3.2 Phase II—Guided Search Toward Representative Areas

After identifying representative areas in Phase I, this information is now used to guide the search toward these areas. The goal is to obtain n representative optimal solutions, one in each representative area, without wasting computational resources on superfluous solutions. Consequently, Phase II focuses on exploitation of the determined areas. The population is divided into n subpopulations, i.e., *species*, which independently evolve toward their own optima. Due to these n detached optimization processes, this step can be perfectly parallelized.

The requirement on an optimal solution is to be sufficiently different in order to be representative for the entire solution space. For this reason, the search aims to find each species' optimal solution within the determined *representative area* around the initial representative, even when other global optima within the same species but outside the defined region exist. If the final obtained solution lies within this area, the representativeness is guaranteed. Hence, solutions located in this area should be preferred, whereas solutions outside this area should be disfavored.

There are two strategies implemented in Phase II enhancing the feature of focusing the search on the representative areas. One strategy manipulates the mutation scheme to produce solutions in the vicinity of the representative. The other introduces a new fitness degradation scheme which impacts the selection process by favoring solutions closer to the representative.

3.2.1 Division into Species

When dividing the population into distinct species, every population member is uniquely assigned to the species of one representative. Thereby, n mutually exclusive and collectively exhaustive species, with respect to the population, are created. In order to guarantee that an initial representative is also the best individual in its species, which is important for the guided exploitation later, the species are assigned as follows. For each individual:

- Step 1: Determine its closest representative as well as the distance to it.
- Step 2: If the distance to the determined representative is smaller than the representative's suppression radius, the individual is assigned to this species.
- Step 3: Else, continue with second closest representative and check if the individual lies within the representative's suppression radius. If not, repeat the same process with the next closest representative until the individual lies within the representative's suppression radius.

It is guaranteed that every individual is assigned to one species because every individual lies within the suppression radius of the best representative, which is always infinity. The case of overlapping suppression radii is unambiguously covered by the above mentioned steps affiliating individuals to their closest representative.

3.2.2 Enhanced-Exploitation Mutation Scheme

Due to the determination of representative areas, the search can now be concentrated in these regions of interest rather than continuing to explore the entire search space. To exploit the representative areas, *DE/best/1* is used to take the current representative as the base vector for every mutation operation:

$$\boldsymbol{v}_g^i = \boldsymbol{x}_g^{rep} + F\left(\boldsymbol{x}_g^{r_1} - \boldsymbol{x}_g^{r_2}\right), \tag{8}$$

where $\boldsymbol{x}_g^{rep}$ is the species' representative in the current generation g taken as the base vector, and everything else remains the same as in Eq. (1). This mutation scheme (Eq. (8)) is applied for each species separately. For simplicity, the species index is not included. Additionally, the scaling factor F is randomized and defined as $F = rand_i\,(0, 1)$, a random number between 0 and 1 for the ith target vector, in contrast to a typically static chosen value (e.g., $F = 0.5$ as in [12]). The reason for the randomization is that now, a very small F may occur, leading to newly created mutants closer to the current representative than with a static F. Consequently, the probability to find a better solution in the vicinity of the current representative is improved leading to a higher accuracy of the eventually obtained optima.

3.2.3 Fitness Degradation Scheme

In addition to the enhanced-exploitation mutation scheme, the second strategy modifies the selection scheme to favor the selection of nearby solutions. By introducing an adaptable fitness degradation scheme, individuals inside the representative areas are favored by degrading individuals' fitness values which lie outside the representative area, i.e., outside the degradation radius. The initial degradation radius dr is set as explained in Sect. 3.1.2. Within this radius, the original fitness value is kept intact so that all globally optimal solutions within this radius have an equal chance to become the new representative for this species. Globally optimal solutions outside this radius are disfavored and therefore their original fitness f^{og} is degraded smoothly according to the following formula:

$$f^{new}\left(\boldsymbol{x}\right) = \begin{cases} f^{og}\left(\boldsymbol{x}\right) - \left|f^{og}\left(\boldsymbol{x}\right) \cdot \left(1 - g\left(\boldsymbol{x}\right)\right)\right|, & d\left(\boldsymbol{x}\right) > dr \\ f^{og}\left(\boldsymbol{x}\right), & \text{otherwise.} \end{cases} \tag{9}$$

The degradation function $g\left(\boldsymbol{x}\right)$ is defined as

$$g\left(\boldsymbol{x}\right) = e^{-\frac{d(\boldsymbol{x})^2}{2\sigma^2}}, \tag{10}$$

where $d\left(\boldsymbol{x}\right)$ denotes the Euclidean distance between $\boldsymbol{x}$ and $\boldsymbol{x}^{rep}$ and σ the standard deviation.

The idea of the fitness degradation scheme is based on the idea of the Gaussian normal distribution $f\left(z \mid \mu, \sigma^2\right) = \frac{1}{\sqrt{2\pi\sigma^2}} e^{-\frac{(z-\mu)^2}{2\sigma^2}}$ representing a smooth bell-shaped curve. For our purpose, only the symmetric half-bell curve for $z \geq 0$ is relevant (zero mean $\mu = 0$), and modified in a way that $f\,(z = 0) = 1$ and keeping its natural characteristic of being strictly monotonically decreasing while converging asymptotically to 0 for $z \to \infty$. z is defined as the distance to the representative, and σ is defined as the distance to the representative within which 68.28% of the individuals are located.

As defined in Eq. (9), this modified half-bell curve is used as a factor degrading the original fitness smoothly according to the distance of the considered individual to the species' representative. The ratio of *new fitness* to *original fitness* remains 1 as long as the individual lies within the degradation radius (dr), and decreases in the shape of the half-bell curve for individuals outside dr. The fitness degradation automatically adapts to the convergence of the (sub)population by computing σ for every generation anew. Using the same example as above, Fig. 6a shows the subpopulation with its dr (purple circle), Fig. 6b the fitness degradation of the individuals according to the distance to their subpopulation's representative, both at the start of Phase II.

With each generation, the subpopulation will converge closer to its representative. If dr remains constant, at some point during the optimization, all individuals lie inside the static dr, and no fitness degradation takes place. As a consequence, the convergence slows down drastically and finding the global optimum with high accuracy becomes more unlikely and more costly. Hence, dr should ideally decrease according to the converging subpopulation in order to preserve the fitness degradation and thereby an effective and efficient search. By using information about the individuals' distribution, dr can be automatically adapted in each iteration according to

$$dr = \frac{2}{3}\sigma \quad \text{if } \sigma \leq dr. \tag{11}$$

Figure 6 shows the subpopulation (left column) at different generations during Phase II and the resulting fitness degradation (right column). In each generation, the fitness degradation curve is adapted according to the newly computed σ. It can be seen that after 6 generations, $\sigma \leq dr$, leading to the adaption of the degradation radius to $dr = \frac{2}{3}\sigma$ to preserve fitness degradation for the converging subpopulation.

The total number of iterations depends on the optimization run and does not need to be specified by the user. The algorithm is terminated when there is no improvement on any of the n representatives for a certain number of iterations. It can be assumed that each species has converged into a single point, i.e., a local or global optimum. In SRN, the user is only required to specify one parameter n, the desired number of optima. Since we assume no prior knowledge about the optimization problem under study, the population size Np can be computed adaptively as a function of the user-specified number n, according to the following:

$$Np = \max\left(20 \cdot n, 100\right) \tag{12}$$

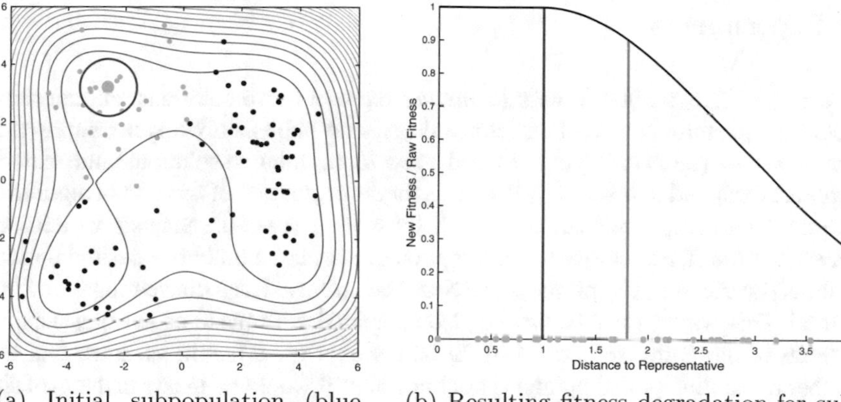

(a) Initial subpopulation (blue dots) with initial *dr* (purple circle) after 4 generations.

(b) Resulting fitness degradation for subpopulation after 4 generations.

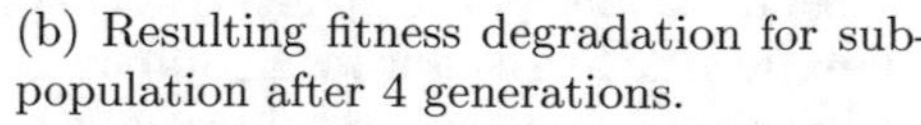

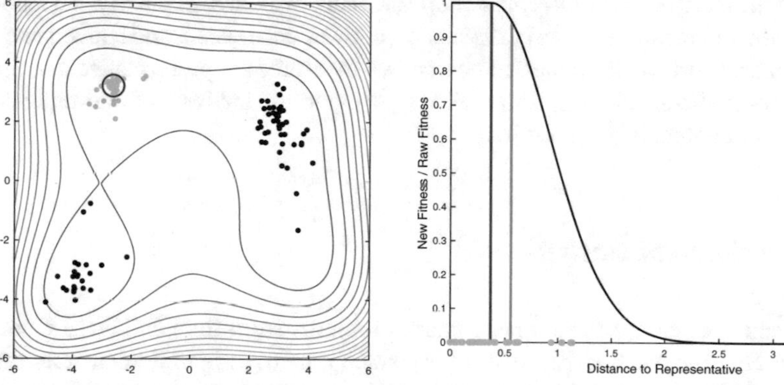

(c) Subpopulation (blue dots) with initial *dr* (purple circle) after 6 generations.

(d) Resulting fitness degradation for subpopulation after 6 generations.

Fig. 6 Converging subpopulation at different stages during the optimization process with its resulting fitness degradation. The purple line indicates dr, and the orange line denotes σ

Thus, it is to ensure that each species comprises in average of 20 individuals, and the population size in total is at least 100.

In SRN, the search is effectively focused on finding one optimal solution in the vicinity of the representative area, resulting in one representative solution for each species. Hence, SRN is able to meaningfully incorporate the information about the number of desired optima supplied by the DM. As a result, computational resources can be efficiently used to produce a desired number of solutions with optimal, sufficiently different solutions to the DM.

4 Experiments

To evaluate SRN's effectiveness in finding representative solutions, an extensive experimental study is carried out. Since there is no published work examining this niching aspect (i.e., aiming only to find a few representative solutions), the experiments are designed to first establish a basis for the evaluation of this sort of algorithm with similar assumptions (see Sect. 4.2). In the second part of the analysis, we demonstrate that even if we remove the assumption of having a number of desired optima specified by the user, the proposed SRN can be still used as a conventional niching method. This way, its results can be also compared with those of existing niching methods in literature (see Sect. 4.4). In both scenarios, it is still assumed that the number of existing global optima is not known, as it would be in any real-world situation. To establish a basis for algorithms aiming to find representative solutions, in the first part of the experiments, a modified performance measure as well as a newly introduced measure on *solution representativeness* are provided. Secondly, the algorithm is evaluated regarding to its performance, convergence speed, and convergence development toward cluster centers. The first two measures are performance metrics used in state-of-the-art niching competitions [26], and the analysis of convergence toward cluster centers is based on [14].

4.1 Experimental Design

SRN is evaluated using selected benchmark problems from the CEC'2013 Competition on Niching Methods in MMO [26], since our focus is to demonstrate the usefulness of SRN and reveal insights on solution representativeness for enhancing decision making. We selected 12 different test cases including functions up to 5 dimensions of the decision space as well as landscapes with various characteristics. Table 2 provides the details of these selected test functions. In all test cases here, an average over 100 runs is taken to minimize the influence of result fluctuations.

The number of representative solutions is determined in two parts of the experimental study as the following:

- In the first part of the experimental study for measuring representativeness and performance, every test case is evaluated with four different desired numbers of optimal solutions $n \in \{3, 5, 7, 9\}$ to cover a reasonable range of choices made by the decision maker.
- In order to make proper comparisons with existing niching algorithms, on convergence speed and convergence development (toward cluster centers), in the second part of the experimental study, the number of existing global optima is automatically estimated by the SRN proposed in Sect. 4.3. Thereby, the need to specify the number n of desired optima a priori is removed. This can be beneficial when the DM does not wish to specify n.

Table 2 Test functions with different characteristics (#GO and #LO denote the number of global and local optima, respectively)

Functions	1D	2D	3D	5D	#GO	#LO
F_1 Five-Uneven-Peak Trap	x				2	3
F_2 Equal Maxima	x				5	0
F_4 Himmelblau		x			4	0
F_5 Six-Hump Camel Back		x			2	2
F_6 Shubert		x	x		$D \cdot 3^D$	742
F_7 Vincent		x	x		6^D	0
F_8 Modified Rastrigin		x			12	0
F_{11} Composition Func. 3		x	x	x	6	Many

The population size in both scenarios is automatically adapted rather than manually chosen to fit the underlying problem. We assume that the DM has no prior knowledge of the problem as it would be in a real-world situation. For the first scenario, taking into consideration the DM's desired number of optima n, the population size Np is calculated as described above in Eq. (12). In the second scenario, where no such information about the number of desired optima is provided, Np is chosen according to the following:

$$Np = 2^D \cdot 50 \tag{13}$$

with D as the dimension of the decision space. This is a simple yet reasonable rule for determining population sizes for low-dimensional (2D and 3D) benchmark functions [14, 15, 42].

As for comparison, the published results of several recently proposed niching DEs, as described in Sect. 2.2, are used. Namely, DE/nrand/1 [14], dADE/ nrand/1 [12], NCDE [33], and Fast-NCDE [42] are compared with the proposed SRN.

SRN is evaluated in four different categories: representativeness, performance, convergence speed, and convergence development toward cluster centers. For *representativeness*, a new metric is defined since there exists no consensus on the exact definition of representativeness. Generally, without prior knowledge of the location of optima, it is desirable to cover as much volume as possible in the decision space. However, only measuring the volume, e.g., a multi-dimensional convex hull around the solutions, does not cover points located inside and how well these solutions are distributed. Ideally, a scalar metric could be introduced to capture the coverage of the volume by the found solutions as well as the points' distribution. Considering this, a *representativeness metric* comprising of the average distance to each nearest neighboring representative $\overline{Dist}$ as well as their standard deviation $St.D.$ is proposed. These two scalar values function as an indication for representativeness and are defined as follows:

$$\overline{Dist} = \frac{1}{n}\sum_{i=1}^{n} d_i^{nn}, \tag{14}$$

$$St.D. = \sqrt{\frac{\sum_{i=1}^{n}\left(d_i^{nn} - \overline{Dist}\right)^2}{n-1}}, \tag{15}$$

with d_i^{nn} the Euclidean distance between representative r_i and its nearest neighboring representative. By measuring the average distance, the covered area is factored in, where larger values are considered as better. The desire to have similar distances between found solutions to ensure that these solutions are as different as possible is captured by the standard deviation. Values close to zero stand for very similar distances from each representative to its nearest neighbor.

4.2 Incorporating a User Specified Number of Optima

The first part of the experiments, investigating solution representativeness and performance, requires the DM to supply a desired number of optima, as an input to SRN.

4.2.1 Quantitative Representativeness

As explained in the above, the average distance to the closest representative ($\overline{Dist}$) as well as its standard deviation ($St.D.$) can be used to evaluate the solution representativeness. Tests were executed for the number of desired optima $n \in \{3, 5, 7, 9\}$ for all test functions in Table 2. These scenarios cover both cases where n is higher and lower than the number of known global optima/peaks (NKP). Figure 7 shows three illustrative examples, F_8 with twelve global optima and no local optima covering the most realistic case of $n < NKP$ (top left figure), F_4 with four global optima covering the case where $n < NKP$ for $n = 3$ (top right figure), and F_2 with five global optima covering the scenarios of $n = NKP$ for $n = 5$ (bottom figure). Each small dot represents the result of one run, the large dot of the same color represents the average over all 100 runs. The different colors represent different scenarios for n.

In Fig. 7a, where $n < NKP$ for all n, one can see that the higher the number of desired optima, the more consistent the results between different runs are. When n becomes closer to the number of known existing global optima, there are fewer possibilities to obtain n out of NKP representatives resulting in more similar results. Figure 7b illustrates the scenario for the smallest chosen $n = 3$ ($n < NKP$), and for all larger n ($n > NKP$). As seen above, if $n < NKP$ but is close to the known number of global optima ($n = 3$, $NKP = 4$), the results are very consistent. The reason again is that there are only very few different choices to pick $n = 3$ out of $NKP = 4$

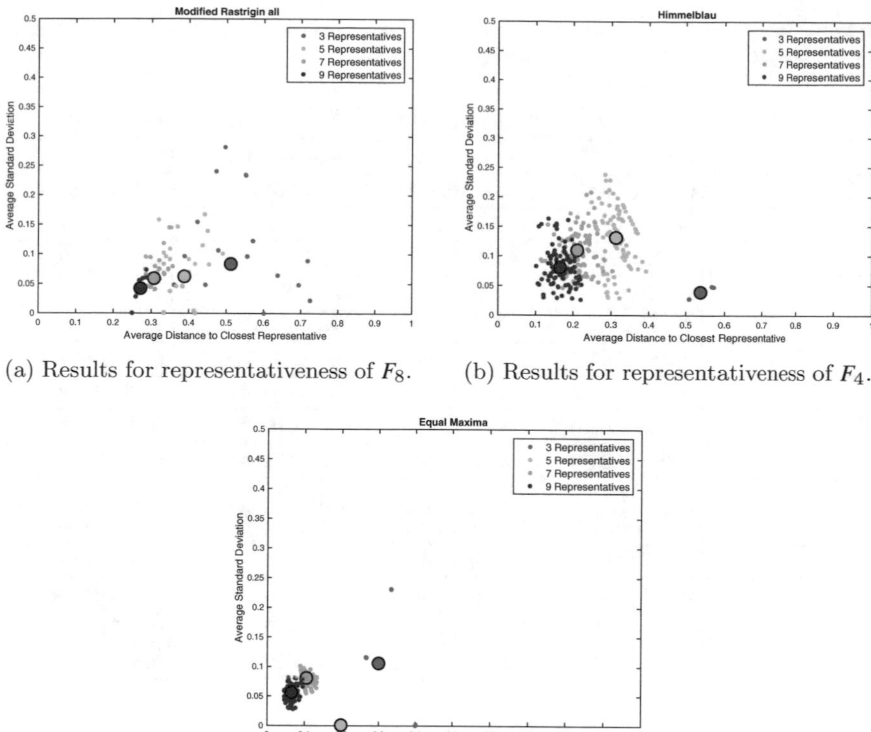

(a) Results for representativeness of F_8.

(b) Results for representativeness of F_4.

(c) Results for representativeness of F_2.

Fig. 7 Results for representativeness of different scenarios, including F_2 (1D), F_4 (2D), F_8 (2D), for $n = 3, 5, 7, 9$. For each n, 100 runs are computed and illustrated as the same colored (small) dots. The average over all 100 runs is represented through the bigger dot of the same color

possible representatives, assuming that the algorithm does locate the existing $n = 4$ peaks consistently, which is shown in the performance analysis in Sect. 4.2.2. If $n > NKP$ but is still close ($n = 5$, $NKP = 4$), the consistency becomes worse. 4 out of 5 points are still found consistently, but the 5th point in this case may differ for a different run. Figure 7c shows a similar scenario, with a different case of matching the number of desired optima n with the number of known true optima NKP for $n = 5$. Here, the results of all 100 successful runs are identical since always all 5 global optima are found. The jump to the inconsistent results is seen for the first $n > NKP$, which matches the scenario of Fig. 7b.

Full results on solution representativeness for all functions can be found in Table 3. The average distance to the nearest neighbor $\overline{Dist}$ as well as the standard deviation $St.D.$ are normalized to the unit space for better comparability. Larger values for $\overline{Dist}$ stand for wider spread representatives. Smaller values for $St.D.$ represent a more equidistant distribution of representative solutions.

Table 3 Representativeness measured by average distance to nearest neighbor $\overline{Dist}$ and standard deviation $St.D.$ (normalized to unit space) of the proposed SRN

n	F_1 (1D)	F_2 (1D)	F_4 (2D)	F_5 (2D)
	$\overline{Dist}$ $(St.D.)$	$\overline{Dist}$ $(St.D.)$	$\overline{Dist}$ $(St.D.)$	$\overline{Dist}$ $(St.D.)$
3	0.468 (0.110)	0.301 (0.105)	0.540 (0.039)	0.394 (0.028)
5	0.217 (0.047)	0.200 (0.000)	0.313 (0.130)	0.320 (0.102)
7	0.126 (0.072)	0.106 (0.080)	0.211 (0.109)	0.222 (0.114)
9	0.092 (0.056)	0.066 (0.056)	0.166 (0.080)	0.173 (0.095)
n	F_6 (2D)	F_7 (2D)	F_8 (2D)	F_{11} (2D)
	$\overline{Dist}$ $(St.D.)$	$\overline{Dist}$ $(St.D.)$	$\overline{Dist}$ $(St.D.)$	$\overline{Dist}$ $(St.D.)$
3	0.504 (0.061)	0.543 (0.080)	0.513 (0.083)	0.584 (0.141)
5	0.364 (0.066)	0.402 (0.061)	0.389 (0.062)	0.389 (0.120)
7	0.310 (0.023)	0.321 (0.058)	0.308 (0.058)	0.302 (0.086)
9	0.300 (0.017)	0.273 (0.070)	0.272 (0.041)	0.252 (0.071)
n	F_6 (3D)	F_7 (3D)	F_{11} (3D)	F_{11} (5D)
	$\overline{Dist}$ $(St.D.)$	$\overline{Dist}$ $(St.D.)$	$\overline{Dist}$ $(St.D.)$	$\overline{Dist}$ $(St.D.)$
3	0.627 (0.075)	0.678 (0.076)	0.836 (0.052)	1.065 (0.153)
5	0.489 (0.070)	0.545 (0.088)	0.606 (0.109)	0.853 (0.158)
7	0.428 (0.071)	0.467 (0.079)	0.471 (0.129)	0.695 (0.166)
9	0.394 (0.065)	0.418 (0.061)	0.393 (0.126)	0.613 (0.168)

4.2.2 Performance

This performance analysis pursues two goals. First, SRN's performance behavior in relation to different n is analyzed. Second, detailed results are shown to build a basis to compare future algorithms with similar characteristics. All test functions according to Sect. 4.1 are tested for five accuracy levels $\epsilon \in \{1.0\text{e-}01, 1.0\text{e-}02, 1.0\text{e-}03, 1.0\text{e-}04, 1.0\text{e-}05\}$ and four different $n \in \{3, 5, 7, 9\}$. For the sake of brevity only selected results are shown in Fig. 8.

In general, functions with distinct global optima and only a few or no local optima are solved with a better performance. Local optima might mislead the algorithm and causes premature convergence, resulting in fewer global optima found (PR) and a lower Success Rate (SR). As described in [38], the higher the so-called *ruggedness* of a function, the harder it is to optimize. Naturally, it can be seen that the higher the desired level of accuracy is, the lower the peak ratio and success rate. However, this decrease is subtle even for functions with sharp peaks and only more visible in higher dimensions.

The impact of the provided number of desired optima n on the performance results is negligible for the majority of considered cases. Independently of a chosen n being smaller or larger than the actual number of existing global optima, the proposed SRN delivers consistent results. This means that it is robust to the user's input and even in scenarios where $n > NKP$, the results are not disturbed from a less-than-ideal input,

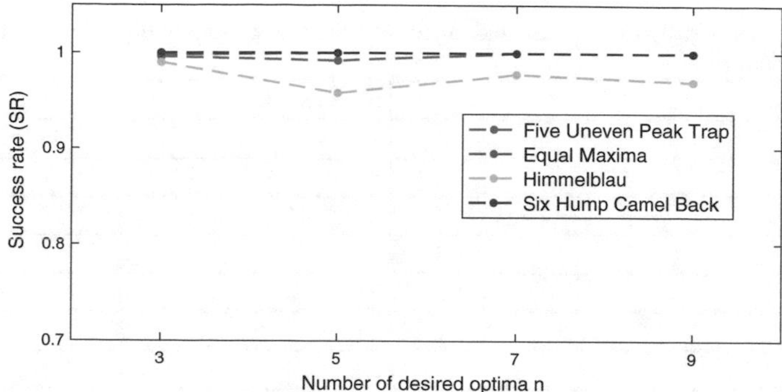

Fig. 8 Consistent Success Rate (SR) for 1D and simple 2D benchmark test functions even for $n > NKP$

as shown in Fig. 8. Robustness to the input parameter is crucial since for a plausible approach in real-world scenarios, we assume that the DM has no prior knowledge of the problem and its optimal solutions.

4.3 Automatic Estimation of the Number of Optima

It is possible that the DM might not wish to specify a number of desired optima, hence the assumption of a provided n needs to be removed. In this section, we propose a method to estimate the number of global optima that are to be found. Based on the information provided by the suppression radius as well as the individuals' fitness values after Phase I, the number of global optima, which will eventually be found, can be detected. The idea is to recognize a pattern in the ranks of the individuals' suppression radii sr and the ranks of their (original, non-degraded) fitness values. Therefore, two sorted lists are created, one with the suppression radii of all individuals in a descending order, denoted by SL, the other with all individuals sorted according to their objective values in a descending order (in case of maximization), denoted by OL. The rank of an individual in SL or OL refers to the position of the individual in the respective sorted list. An example is shown in Table 4, presenting one individual's ranks in both SL and OL in each row.

As described in Sect. 3.1.1, representative solutions, i.e., solutions expected to lie close to a global optimum, are determined by choosing the individuals with the highest suppression radii, corresponding to the first entries in SL. The idea is that solutions close to a global optimum should also have a reasonably good objective value, thus must be one of the top entries in OL. One would expect a correlation between a large suppression radius and a good objective value. However, individuals with a rather poor objective value can still have a large suppression radius due to their distant location away from any other (fitter) individual. As shown in the example in

Table 4 Ranks of suppression radii and objective values. Big jump in objective value rank after first 4 individuals

Rank in SL	Rank in OL
1	1
2	8
3	4
4	2
5	74
6	94
...	...

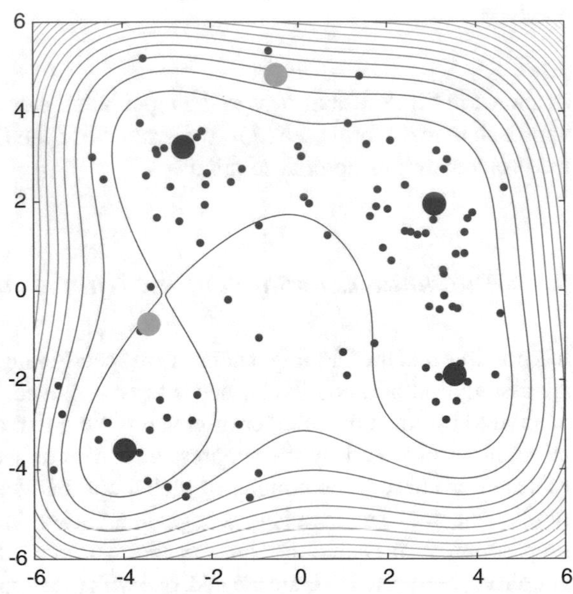

Fig. 9 Four global optima detected (red points). Individuals with 5th and 6th best sr values (green points) have much lower objective value ranks

Table 4, this is recognizable by a jump in the rank of the objective values in OL. Here, the objective value's rank of the four individuals with the largest suppression radii sr is high. All four individuals belong to the best ten individuals of the population of 100. The individual with the 5th largest sr, however, associates with the 74th best objective value. Deriving from the above observation, this individual is located rather remotely and therefore far away from a global optimum. In Fig. 9, the two green points represent the individuals with the 5th and 6th largest sr, respectively. Both are located distant from any of the four global optima.

This observation can be used to estimate the number of global optima the population comprises. By ascertaining a significant difference between the ranks in OL (right column in Table 4) of two consecutive individuals according to their suppression radii, the number of global optima is determined. Starting with the first two

Algorithm 2: Determining estimated number of global optima NKP^{est}

Input: Vector $\boldsymbol{\delta}$ with differences of ranks in OL between two consecutive individuals
Output: Estimated number of global optima NKP^{est}

```
for i from 1 to Np − 1 do
    if δ_i > 0.3 · Np then
        NKP^est = i
        Stop.
    end if
end for
```

entries, the difference δ of their ranks in OL is computed by

$$\delta_i = rank(OL, i+1) - rank(OL, i) \tag{16}$$

with $rank(OL, i)$ being the objective value rank of the individual with the ith largest suppression radius sr (right column of Table 4), calculated for all $i \in \{1, 2, \ldots, Np-1\}$ (left column of Table 4). The estimated number of global optima NKP^{est} is determined by the first entry in δ greater than 30% of the population size Np according to Algorithm 2. This way, the drop of the rank in OL (as seen from row four to row five in Table 4) can be identified. The threshold of 30% is empirically determined.

Thus, the number of global optima the algorithm is able to find is estimated before the algorithm has actually converged toward any optima. This methodology to estimate the number of global optima is used for the next section, evaluating experiments without the specification of the number of desired optimal solutions (or representatives) by the DM.

4.4 No Specification of Number of Representatives

For the second part of the experimental study, the assumption of a provided n is removed, using the methodology proposed in the previous section to automatically estimate the number of global optima. This way, results considering performance, convergence speed, and convergence development can be compared to existing niching algorithms, to see if all global optima can be found. However, one key advantage of SRN is that it does not assume that the total number of optima is known a priori as part of the performance measure PR and SR. This information is discovered using Eq. (16).

Table 5 Performance results—Peak Ratio (PR) and Success Rate (SR)—for SRN and four state-of-the-art niching DEs for $\epsilon = 1.0\text{e-}04$

Algorithm	F_1 (1D)		F_2 (1D)		F_4 (2D)	
	PR	SR	PR	SR	PR	SR
DE/nrand/1	1.000	1.000	1.000	1.000	1.000	1.000
dADE/nrand/1	1.000	1.000	1.000	1.000	1.000	1.000
NCDE	1.000	1.000	1.000	1.000	1.000	1.000
Fast-NCDE	1.000	1.000	1.000	1.000	1.000	1.000
SRN	0.985	0.970	0.996	0.980	1.000	1.000
Algorithm	F_5 (2D)		F_6 (2D)		F_7 (2D)	
	PR	SR	PR	SR	PR	SR
DE/nrand/1	1.000	1.000	0.717	0.010	0.400	0.000
dADE/nrand/1	1.000	1.000	0.984	0.780	0.823	0.000
NCDE	1.000	1.000	0.110	0.000	0.667	0.000
Fast-NCDE	1.000	1.000	0.999	0.980	0.673	0.000
SRN	1.000	1.000	0.487	0.000	0.263	0.000
Algorithm	F_8 (2D)		F_{11} (2D)		F_6 (3D)	
	PR	SR	PR	SR	PR	SR
DE/nrand/1	0.815	0.160	0.667	0.000	0.095	0.000
dADE/nrand/1	0.967	0.140	0.667	0.000	0.431	0.000
NCDE	1.000	1.000	0.590	0.000	0.553	0.000
Fast-NCDE	1.000	1.000	0.673	0.000	0.915	0.000
SRN	0.981	0.800	0.495	0.000	0.179	0.000
Algorithm	F_7 (3D)		F_{11} (3D)		F_{11} (5D)	
	PR	SR	PR	SR	PR	SR
DE/nrand/1	1.000	1.000	0.667	0.000	0.663	0.000
dADE/nrand/1	1.000	1.000	0.667	0.000	0.667	0.000
NCDE	0.265	0.000	0.663	0.000	0.660	0.000
Fast-NCDE	0.274	0.000	0.667	0.000	0.667	0.000
SRN	0.064	0.000	0.662	0.000	0.647	0.000

4.4.1 Performance

The benchmark functions described in Sect. 4.1 are now used to measure the performance, i.e., Peak Ratio (PR) and Success Rate (SR), according to [26]. The proposed SRN is compared to four leading niching DE algorithms. The results of the compared algorithms are taken from their original publications, DE/nrand/1 [14], dADE/nrand/1 [12], and results for NCDE and Fast-NCDE are both from [42]. Table 5 shows all algorithms' results, including the results for SRN.

Generally, SRN's performance is similar to the state-of-the-art niching DEs, reaching a success rate of 1.0 for simple 2D functions F_4 and F_5. Only in the more complex

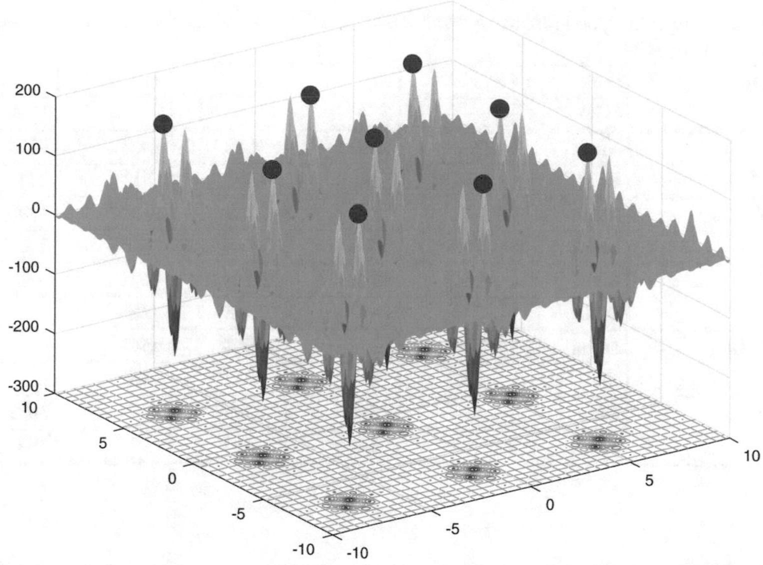

Fig. 10 F_6 Shubert 2D function where one optimum per each pair of optima is found, resulting in $PR = 0.5$

scenarios, where most algorithms do not succeed in finding all optima, is SRN's peak ratio inferior. However, the motivation of the proposed SRN was not to find all possible optimal solutions, but rather well-distributed representative ones. As seen for the Shubert function F_6 (2D), although $SR = 0.000$, meaning that in none of the 100 runs all 18 peaks are found, the peak ratio $PR = 0.487$, which means in average about 50% of the existing peaks are found. With the knowledge of the landscape (nine pairs of two close peaks), the results indicate that SRN is able to find one peak within each pair (see Fig. 10). Thus, only sufficiently different global optima are found, which is consistent with the purpose of SRN. This explains why the success rate might not tell the full story, and actually reasonable results are obtained as far as finding representative solutions is concerned. The main reason for not reaching the highest success rate and peak ratio in general is the information obtained in Phase I. Only the optima which are deduced from the information gained in Phase I can be found with the focused search in Phase II. If the metaheurstic used in Phase I, namely DE/nrand/1, is not able to identify all existing optima, they will not be found in the exploitation-focused Phase II. Therefore, we do not generally expect that the peak ratio and success rate to exceed those of DE/nrand/1.

4.4.2 Convergence Speed

To evaluate the convergence speed of SRN, it is compared to the same four niching DE algorithms, namely DE/nrand/1 [14], dADE/ nrand/1 [12], NCDE [33], and

Table 6 Convergence speed results to reach accuracy level ϵ = 1.0e-04 for SRN and four state-of-the-art niching DEs

Algorithm	Func.	F_1	F_2	F_4	F_5	F_8
	#GO	2	5	4	2	12
DE/nrand/1	Mean	22886	1552	13610	3806	9858
	St. D.	2689	386	1399	619	833
dADE/nrand/1	Mean	20202	1800	12703	3567	12904
	St. D.	2788	586	1668	652	2169
NCDE	Mean	356	2902	12416	4204	–
	St. D.	234	1413	3588	751	–
Fast-NCDE	Mean	**342**	2548	12064	6548	–
	St. D.	**137**	814	957	1135	–
SRN	Mean	1088	**550**	**2489**	**1658**	**3039**
	St. D.	2199	**349**	**134**	**177**	**785**

Fast-NCDE [42]. Since non-successful runs strongly impact the convergence speed (the maximum number of iterations is taken for those runs), we focus on those test functions where the majority of algorithms have a reasonably SR. This way, the convergence speed analysis gives insight about how fast the algorithm can solve the problem. Table 6 shows the results from all five algorithms for five selected benchmark test functions. The number of function evaluations needed for each algorithm to reach the given accuracy level of ϵ = 1.0e-04 is measured. The first number denotes the average over 100 runs (Mean), the second number denotes the standard deviation (St.D.), also measured in runs.

SRN clearly outperforms existing niching DEs in all functions with exception of F_1, where SRN still outperforms DE/nrand/1 and dADE/nrand/1 by a factor of 18. For the functions F_2, F_4, F_5, and F_8, SRN uses two to five times fewer function evaluations than the second best algorithm. For some functions, the standard deviation of SRN is significantly higher relatively to the average function evaluations, i.e., for F_1 and F_2, which is caused by unsuccessful runs. In short, SRN has a great convergence speed advantage over the compared niching DEs. This can be explained through the focused search in Phase II, exploiting only the representative areas and therefore finding the global optima efficiently.

4.4.3 Convergence Development Toward Cluster Centers

The convergence development, i.e., the speed in which individuals converge toward the cluster centers, is evaluated by measuring the H-value in each iteration. As opposed to the previous section, not only the end result (when convergence is reached) is considered, but rather the development toward cluster centers over iterations. The results for four benchmark test functions are illustrated in Fig. 11.

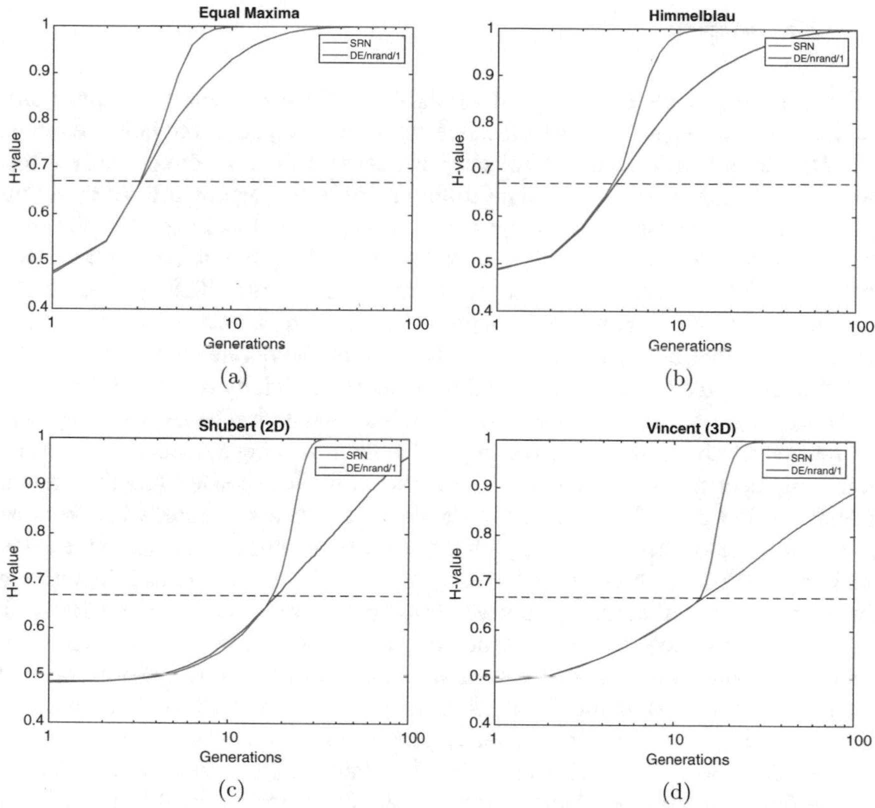

Fig. 11 H-values over generations as measures of convergence development toward cluster centers, averaged over 100 runs plotted over 100 generations. SRN outperforms DE/nrand/1 in all benchmark test functions. The results of four selected functions are shown here

SRN is compared to DE/nrand/1, which forms the basis of SRN in Phase I. Following this, the convergence of the two algorithms until reaching the H-threshold of 0.67 (dotted line) is almost identical, and only marginal differences are observed due to stochastic fluctuations. Once the H-threshold is reached, SRN switches to Phase II, and starts to focus on exploiting the identified representative areas. Due to the enhanced-exploitation mutation and the fitness degradation schemes, the convergence is speeded up, visible by a much steeper slope of the blue line representing SRN. In all benchmark test functions (also the ones not shown in Fig. 11), SRN outperforms DE/nrand/1 clearly by reaching full convergence ($H = 1$) many iterations earlier. The more complex the function is, the more significant the advantage of SRN. Reaching convergence at an earlier stage leads to a substantive saving in function evaluations. As the cost of function evaluation in real-world applications can be very expensive, it is crucial to reduce the number of function evaluations in order to obtain results in an acceptable time.

5 Conclusions

Existing niching methods aim to find all possible optimal solutions of a given multimodal optimization problem, which can result in too many solutions being returned. The DM faces the challenge of selecting a suitable solution from possibly a vast number of alternatives, which might differ significantly regarding to their technical feasibility or variable value ranges. The proposed SRN facilitates the decision-making process by providing solely a small variety of representative solutions well distributed in the solution space. Representative areas are identified by computing a so-called suppression radius for each point. Thus, the n best-fit and spatially well-spread points can be identified at an early stage of the optimization process. The search is focused on finding an optimal solution in the vicinity of each of the identified areas by introducing a self-adaptive fitness degradation scheme. This way, only the desired number of solutions is computed without wasting computational cost on obtaining superfluous information. Results show that SRN consistently finds representative solutions well distributed in the decision space in all scenarios. Without any knowledge of the underlying problem, SRN automatically adapts to the given landscape finding the best-fit and spatially well-distributed solutions. It is robust to the user input, i.e., the desired number of optimal solutions (supplied by the DM), without relying on knowledge of the number of true global optima. While maintaining a similar quality of solutions, optima are found with substantively fewer function evaluations than existing niching algorithms. Due to the independent optimization of the representative areas, it is perfectly parallelizable and the computational time can be reduced even further. Nevertheless, the quality of the representative solutions can be further improved through enhancing the metaheuristic used in Phase I. As the SRN concept is a general framework, any population-based metaheuristic can potentially be employed. Due to its self-adaptability, the proposed SRN is suitable for real-world scenarios where no prior knowledge about the problem exists. Through SRN's convergence speed advantage, it is especially appealing for scenarios where each function evaluation is computationally costly.

Acknowledgements The first author would like to thank Aachen University and RMIT University for supporting this research project in 2018 during his visit at the ECML group of RMIT University.

References

1. Ward, A., Liker, J.K., J.J.C., Sobek, D.K.: The second Toyota paradox: how delaying decisions can make better cars faster. Sloan Manag. Rev. **36**(3), 43 (1995)
2. Banerjee, A., Dave, R.N.: Validating clusters using the hopkins statistic. In: 2004 IEEE International Conference on Fuzzy Systems (IEEE Cat. No.04CH37542), vol. 1, pp. 149–153 (2004)
3. Beasley, D., Bull, D.R., Martin, R.R.: A sequential niche technique for multimodal function optimization. Evolut. Comput. **1**, 101–125 (1993)
4. Bessaou, M., Pétrowski, A., Siarry, P.: Island model cooperating with speciation for multimodal optimization. In: Schoenauer, M., Deb, K., Rudolph, G., Yao, X., Lutton, E., Merelo, J.J.,

Schwefel, H.P. (eds.) Parallel Problem Solving from Nature PPSN VI, pp. 437–446. Springer, Berlin (2000)
5. Billaut, J.C., Hebrard, E., Lopez, P.: Complete characterization of near-optimal sequences for the two-machine flow shop scheduling problem. In: Beldiceanu, N., Jussien, N., Pinson, É. (eds.) Integration of AI and OR Techniques in Contraint Programming for Combinatorial Optimzation Problems, pp. 66–80. Springer, Berlin (2012)
6. Biswas, S., Kundu, S., Das, S.: Inducing niching behavior in differential evolution through local information sharing. IEEE Tran. Evolut. Comput. **19**(2), 246–263 (2015)
7. Brown, M., Szeliski, R., Winder, S.: Multi-image matching using multi-scale oriented patches. In: 2005 IEEE Computer Society Conference on Computer Vision and Pattern Recognition (CVPR'05), vol. 1, pp. 510–517 (2005)
8. Carrese, R., Li, X.: Preference-Based Multiobjective Particle Swarm Optimization for Airfoil Design, pp. 1311–1331. Springer, Berlin (2015)
9. Cioppa, A.D., Stefano, C.D., Marcelli, A.: Where are the niches? Dynamic fitness sharing. IEEE Trans. Evolut. Comput. **11**(4), 453–465 (2007)
10. De Jong, K.A.: An analysis of the behavior of a class of genetic adaptive systems. Ph.D. thesis, Ann Arbor, MI, USA (1975)
11. Dimopoulos, C., Zalzala, A.M.S.: Recent developments in evolutionary computation for manufacturing optimization: problems, solutions, and comparisons. IEEE Trans. Evolut. Comput. **4**(2), 93–113 (2000)
12. Epitropakis, M.G., Li, X., Burke, E.K.: A dynamic archive niching differential evolution algorithm for multimodal optimization. In: 2013 IEEE Congress on Evolutionary Computation. pp. 79–86 (2013)
13. Epitropakis, M.G., Plagianakos, V.P., Vrahatis, M.N.: Balancing the exploration and exploitation capabilities of the differential evolution algorithm. In: 2008 IEEE Congress on Evolutionary Computation (IEEE World Congress on Computational Intelligence). pp. 2686–2693 (2008)
14. Epitropakis, M.G., Plagianakos, V.P., Vrahatis, M.N.: Finding multiple global optima exploiting differential evolution's niching capability. In: 2011 IEEE Symposium on Differential Evolution (SDE), pp. 1–8 (2011)
15. Epitropakis, M.G., Tasoulis, D.K., Pavlidis, N.G., Plagianakos, V.P., Vrahatis, M.N.: Enhancing differential evolution utilizing proximity-based mutation operators. IEEE Trans. Evolut. Comput. **15**(1), 99–119 (2011)
16. Goldberg, D.E., Goldberg, D.E., Deb, K., Horn, J.: Massive multimodality, deception, and genetic algorithms (1992)
17. Goldberg, D.E., Richardson, J.: Genetic algorithms with sharing for multimodal function optimization. In: Proceedings of the Second International Conference on Genetic Algorithms on Genetic Algorithms and Their Application, pp. 41–49. L. Erlbaum Associates Inc., Hillsdale, NJ, USA (1987)
18. Harik, G.: Finding multimodal solutions using restricted tournament selection. In: Proceedings of the Sixth International Conference on Genetic Algorithms, pp. 24–31. Morgan Kaufmann (1995)
19. Holland, J.H.: Adaptation in Natural and Artificial Systems. University of Michigan Press, Ann Arbor (1995)
20. Hopkins, B., Skellam, J.G.: A new method for determining the type of distribution of plant individuals. Ann. Botany **18**(70), 213–227 (1954)
21. Horn, J.: The nature of niching: genetic algorithms and the evolution of optimal, cooperative populations. Technical Report (1997)
22. Kuksov, D., Villas-Boas, J.M.: When more alternatives lead to less choice. Market. Sci. **29**(3), 507–524 (2010)
23. Lawson, R.G., Jurs, P.C.: New index for clustering tendency and its application to chemical problems. J. Chem. Inf. Comput. Sci. **30**(1), 36–41 (1990)
24. Li, J.P., Balazs, M.E., Parks, G.T., Clarkson, P.J.: A species conserving genetic algorithm for multimodal function optimization. Evolut. Comput. **10**(3), 207–234 (2002)

25. Li, X., Epitropakis, M.G., Deb, K., Engelbrecht, A.: Seeking multiple solutions: an updated survey on niching methods and their applications. IEEE Trans. Evolut. Comput. **21**(4), 518–538 (2017)
26. Li, X., Engelbrecht, A., Epitropakis, M.G.: Benchmark functions for cec'2013 special session and competition on niching methods for multimodal function optimization. Technical Report, Evolutionary Computation and Machine Learning Group, RMIT University, Melbourne, Australia (2013), http://titan.csit.rmit.edu.au/46507/cec13-niching/competition/
27. Luh, G.C., Lin, C.Y.: Optimal design of truss-structures using particle swarm optimization. Comput. Struct. **89**(23–24), 2221–2232 (2011)
28. Mahfoud, S.W.: Crowding and preselection revisited. In: Manner, R., Manderick, B. (eds.) Parallel Problem Solving From Nature, pp. 27–36. North-Holland (1992)
29. Mahfoud, S.W.: A comparison of parallel and sequential niching methods. In: Proceedings of the Sixth International Conference on Genetic Algorithms, pp. 136–143. Morgan Kaufmann (1995)
30. Miller, B.L., Shaw, M.J.: Genetic algorithms with dynamic niche sharing for multimodal function optimization. In: Proceedings of IEEE International Conference on Evolutionary Computation, pp. 786–791 (1996)
31. Parsopoulos, K.E., Vrahatis, M.N.: On the computation of all global minimizers through particle swarm optimization. IEEE Trans. Evolut. Comput. **8**(3), 211–224 (2004)
32. Petrowski, A.: A clearing procedure as a niching method for genetic algorithms. In: Proceedings of 1996 IEEE International Conference on Evolutionary Computation, Nayoya University, Japan, May 20–22. pp. 798–803 (1996)
33. Qu, B.Y., Suganthan, P.N., Liang, J.J.: Differential evolution with neighborhood mutation for multimodal optimization. IEEE Trans. Evolut. Comput. **16**(5), 601–614 (2012)
34. Storn, R.: On the usage of differential evolution for function optimization. In: NAFIPS'96, pp. 519–523. IEEE (1996)
35. Storn, R., Price, K.: Differential evolution - a simple and efficient heuristic for global optimization over continuous spaces. J. Global Optim. **11**(4), 341–359 (1997)
36. Tasoulis, D.K., Plagianakos, V.P., Vrahatis, M.N.: Clustering in evolutionary algorithms to efficiently compute simultaneously local and global minima. In: 2005 IEEE Congress on Evolutionary Computation, vol. 2, pp. 1847–1854 (2005)
37. Theodoridis, S., Koutroumbas, K.: Pattern Recognition, 4th edn. Academic, New York (2008)
38. Weise, T., Chiong, R., Tang, K.: Evolutionary optimization: Pitfalls and booby traps. J. Comput. Sci. Technol. **27**(5), 907–936 (2012)
39. Yin, X., Germay, N.: A fast genetic algorithm with sharing scheme using cluster analysis methods in multimodal function optimization. In: Albrecht, R.F., Reeves, C.R., Steele, N.C. (eds.) Artificial Neural Nets and Genetic Algorithms, pp. 450–457. Springer Vienna, Vienna (1993)
40. Zaharie, D.: Influence of crossover on the behavior of differential evolution algorithms. Appl. Soft Comput. **9**(3), 1126–1138 (2009)
41. Zenios, S.A.: Financial Optimization. Cambridge University Press, Cambridge (2002)
42. Zhang, Y.H., Gong, Y.J., Zhang, H.X., Gu, T.L., Zhang, J.: Toward fast niching evolutionary algorithms: a locality sensitive hashing-based approach. IEEE Trans. Evolut. Comput. **21**(3), 347–362 (2017)

Lifting the Multimodality-Fog in Continuous Multi-objective Optimization

Pascal Kerschke and Christian Grimme

Abstract Multimodality plays a key role as one of the most challenging problem characteristics in the common understanding of solving optimization tasks. Based on insights from the single-objective optimization domain, local optima are considered to be (deceptive) traps for optimization approaches such as gradient descent or different kinds of neighborhood search. Consequently, as continuous multi-objective (MO) problems usually combine multiple (often multimodal) single-objective problems, multimodality is considered an important challenge for MO problems as well. In fact, even very simple MO problems possess a multimodal landscape due to the interactions among its objectives. Thus, modern benchmarks name multimodality an important problem characteristic, while at the same time, heuristic algorithms (like evolutionary algorithms) are expected to be almost mandatory for handling multimodality in an effective way. Here, we continue our previous work by (1) formally defining multimodality in the continuous MO setting, (2) provide techniques for visualizing landscapes of continuous MO problems—not only in the objective but also in the decision space—to improve the intuition regarding continuous MO multimodality, and (3) analyze MO problems based on examples from an extensive test-bed. Thereby, we provide the tools for displaying and detecting basins of attraction, as well as superpositions of local optima, in the decision space of the landscapes. Most important, and maybe unexpected, we are able to show that multimodality in continuous MO optimization differs largely from our understanding of multimodality in the single-objective domain: for simple MO optimization approaches, local efficient sets are often no traps. Instead, locality can even be exploited to slide toward the global efficient set.

P. Kerschke (✉)
Big Data Analytics in Transportation, TU Dresden, 01062 Dresden, Germany
e-mail: pascal.kerschke@tu-dresden.de

C. Grimme
Data Science: Statistics and Optimization, University of Münster, Leonardo-Campus 3, 48149 Münster, Germany
e-mail: christian.grimme@uni-muenster.de

M. Preuss et al. (eds.), *Metaheuristics for Finding Multiple Solutions*, Natural Computing Series,
https://doi.org/10.1007/978-3-030-79553-5_4

Keywords Multi-objective optimization · Multimodal optimization · Continuous optimization · Fitness landscapes · Local search · Exploratory landscape analysis

1 Introduction

Every text book on optimization sooner or later presents a standard figure to the reader, which depicts the functional landscape of a (made up) continuous one-dimensional optimization problem that comprises some local and one or more global optima, sometimes even saddle points. All specificities of this problem are labeled and explained in order to show the reader what difficulties might occur in the area of optimization. With this figure in mind, many theoretical derivations and practical solution approaches are discussed.

These images also show that local optima pose a challenge to optimization approaches, as many search methods can get trapped therein. Multimodality—the presence of multiple local or global optima—is rightly considered a challenge in single-objective optimization, as is ultimately demonstrated by this book. Thanks to this challenge, evolutionary algorithms became popular solution heuristics [34] for single-objective optimization problems as they are able to escape local optima and converge to a global optimal solution, in principle.

Evolutionary algorithms explore and exploit a given decision space by perturbing and recombining previously determined solutions within an evolutionary loop. It is certainly not surprising that the practical improvement of mutation (for exploration) and recombination (for exploitation) operators [34], as well as their theoretical foundation [4], has been driven by understanding the algorithms' behavior in decision and objective space together—basically with the concept of the described figure from above in mind. For problems with unknown structure, research strives for the identification of features in decision space [11, 23, 24, 32] to enable the creation of rule sets and the educated selection of algorithm parameters or algorithms themselves [22, 26].

Both the development of evolutionary operators and *Exploratory Landscape Analysis* (ELA, [27, 31]) are based on a notion of the so-called *function* or *fitness landscape* [40]. Plotting scalar objective values for a two-dimensional problem is the most common approach for visualizing multimodality, identifying basins of attraction of local optima, or finding difficulties for algorithms (e.g., ridges or valleys).

For continuous multi-objective (MO) problems, however, a standard notion of "function landscapes" is not available because the image of decision variables is not scalar anymore but instead comprises several objective values. This severely hinders an informative glance into the functional relationship of decision and objective space. Interestingly, two effects on research in MO optimization arise from this:

1. The single-objective concepts of locality and multimodality are often directly transferred to the MO case. This leads to the inherent assumption, that "local

optima" in MO problems are traps as well and that multimodality is *at least* as challenging as in single-objective optimization.
2. Due to the lack of visualization and an intuitive concept of the relation of decision and objective space, the latter is almost never employed in solution strategies. For MO evolutionary algorithms, research focuses mainly on selection mechanisms, not on exploration or exploitation in the decision space.

In this paper, we specifically address the first effect of transferring (continuous) single-objective understanding and interpretation of multimodality to the (continuous) MO case and show that the property of multimodality has to be considered differently for continuous MO problems. We even show empirically that local optima can be helpful in finding the global solution faster. Here the second effect comes into play: we propose a simple mechanism in which we apply our gained insights.

This paper is based on previous research, which started in 2016 with a definition for MO problems [28] and a technique for visualizing multimodal MO landscapes [21, 29]. Both aspects are briefly introduced in Sect. 3, where Sect. 3.1 details the theoretical background, Sect. 3.2 summarizes the visualization method. This method is then applied to current continuous MO benchmark problems in Sect. 4. More precisely, we provide a feeling on "how these problems look like" in Sect. 4.1 and categorize the observations in Sect. 4.2. In Sect. 5, we discuss implications and show, how the insights from these visual analyses can be used to solve continuous multimodal MO problems by exploiting locality. Finally, Sect. 6 concludes this work.

We believe that this work can lift some of the fog that covers continuous multimodal MO landscapes and may contribute to a new perspective on problems from this class.

2 Related Work

The generation of function landscapes for continuous single-objective problems enables—besides simple visualization—a figurative description of properties directly exposed by the continuous depiction of function values. As such, mountains, valleys, plateaus, or ridges are identified. On the one hand, these geographical notions allow an immediate common understanding of the described structures. On the other hand, they comprise an implicit mathematical interpretation as local optima, modality, areas of similar quality, or discontinuities, respectively.

For a concise discussion of the literature background for this work, we should clearly highlight the distinction of *continuous* and *combinatorial* MO optimization. The first type is in the focus of this work. For this area considerations on landscapes and their visualization are rather new. For the latter, however, a larger corpus of literature exists.

The area of continuous MO optimization ignores the topic of landscape almost completely. To the best of the authors' knowledge, only their own works [17, 18, 21,

25, 28] (which are described in detail later on) and very few visualization approaches by Fonseca [14], as well as Tušar and Filipič [35, 36] exist.

Tušar [35], as well as Tušar and Filipič [36] review multiple visualization techniques for MO results and extend these techniques. Their works provide a good tool set for analyzing algorithm behavior but provide almost no means for acquiring insights into a complete MO problem landscape.

The only approach for actually visualizing a MO landscape is given by Fonseca [14]. In his work, a so-called "cost-landscape" based on population rankings is computed. This gives—at least for the global efficient set's perimeter—some impression of the MO functional landscape. Note that Fonseca designed his approach to show difficulties, e.g., the influence of constraints of decision maker preferences, for search algorithms around the global solution set. Consequently, it is not designed and used in the more general context of visualizing an entire landscape including locally efficient sets.

Interestingly, the area of combinatorial MO optimization is very strong in using the analogies from visualized landscapes [12]. Garrett and Dasgupta [16] describe how families of landscapes can be generated for a given MO problem by successively scalarizing the objective functions using different weighting vectors. Measures for single-objective landscape analysis are then generalized for those families of landscapes. In the works of Knowles and Corne [30], as well as Garrett and Dasgupta [16], problem specific knowledge—e.g., on the *quadratic assignment problem* (QAP) or the *traveling salesperson problem* (TSP)—is used to identify features for combinatorial landscapes. In a later work, Garrett [15] considers so-called plateau structures imposed by the domination relation, while Rosenthal and Borschbach [33] estimate modality, ruggedness, correlation and plateaus specifically for MO molecular landscapes for biochemical optimization problems. For a more general perspective in combinatorial MO optimization, Aguirre et al. [1, 2] transfer the principle of NK-landscapes to the MO case and several following articles build on this definition of interaction of genes, which is understood as ruggedness of the problem [8, 38].

All approaches ignore the classical visualization of landscapes, but directly present abstract geographically labeled features. This is not surprising for combinatorial optimization, as those problems are normally of high dimensionality while landscape visualizations are restricted to low dimensions. Additionally, the notion of neighborhood of solutions poses a problem for the visualization of the decision space in the combinatorial case. The lack of visualization in continuous space is the more astonishing as both problems do not exist there. At least most artificial test problems (e.g., ZDT [42], DTLZ [10], bi-objective BBOB [7, 37], or the CEC 2019 test suite [41]) can be reduced to a low number of dimension and few objectives (two or three for adequate visualization). Certainly, the reduction of dimensionality and objectives reduces the problem complexity, still, we expect, that visualization is helpful to understand structures and their following results in many cases.

3 Multimodality in MO Optimization

As one of the goals of this work is an improved understanding of multimodality in continuous MO optimization, we first briefly define our employed terminology. Afterward, a (colorful) visualization approach is introduced, which does not only reveal the local and global optima in the decision and objective space but also captures the interaction effects between the different objectives. As a result, this method provides valuable insights and thus lays the basis for further analyses.

3.1 Theoretical Foundations

A MO function $\mathbf{f} : \mathscr{X} \rightarrow \mathbb{R}^p$ with $\mathbf{f}(\mathbf{x}) = (f_1(\mathbf{x}), \ldots, f_p(\mathbf{x}))^T \in \mathbb{R}^p$ is a collection of p single-objective functions $f_i : \mathscr{X} \rightarrow \mathbb{R}$, $i = 1, \ldots, p$, which jointly map observations $\mathbf{x} \in \mathscr{X}$ from a d-dimensional continuous decision space $\mathscr{X} \subseteq \mathbb{R}^d$ into a p-dimensional continuous objective space. In the following, we further consider (without loss of generality) all single objectives f_i as minimization problems and additionally employ the notion of *Pareto dominance* for our definitions.

Definition 1 For two vectors $\mathbf{a} = (a_1, \ldots, a_n) \in \mathbb{R}^n$ and $\mathbf{b} = (b_1, \ldots, b_n) \in \mathbb{R}^n$, which are both to be minimized, vector **a** dominates vector **b**, if and only if $a_i \leq b_i$ for all $i \in \{1, \ldots, n\}$, including at least one element $j \in \{1, \ldots, n\}$ in which $a_j < b_j$. Alternatively, this can be written as: $\mathbf{a} \prec \mathbf{b}$.

Below, we define *connected components*, which are an important aspect for our interpretation of multimodality from a MO point of view.

Definition 2 A set $A \subseteq \mathbb{R}^n$ is called connected if and only if there do not exist two open and *disjoint* subsets $U_1, U_2 \subseteq \mathbb{R}^n$ such that $A \subseteq (U_1 \cup U_2)$, $(U_1 \cap A) \neq \emptyset$, and $(U_2 \cap A) \neq \emptyset$. Further let $B \subseteq \mathbb{R}^n$. A subset $C \subseteq B$ is a connected component of B if and only if $C \neq \emptyset$ is connected, and $\nexists\, D$ with $D \subseteq B$ such that $C \subset D$.

In other words, one set is a connected component of another one, if the former is the largest non-empty and connected subset of the latter. Note that a set can consist of multiple (mutually disconnected) connected components.

Using these connected components, we can finally define *local efficient sets*, i.e., the MO pendant of local optima from single-objective problems.

Definition 3 An observation $\mathbf{x} \in \mathscr{X}$ is called locally efficient if there is an open set $U \subseteq \mathbb{R}^d$ with $\mathbf{x} \in \mathbf{U}$ such that there is no (further) point $\tilde{\mathbf{x}} \in (U \cap \mathscr{X})$ that fulfills $\mathbf{f}(\tilde{\mathbf{x}}) \prec \mathbf{f}(\mathbf{x})$. The set of all locally efficient points of $\mathscr{X}$ is denoted $\mathscr{X}_{LE}$, and each connected component of $\mathscr{X}_{LE}$ forms a local efficient set (of **f**).

A special case of local efficiency is the so-called *Pareto efficiency* or *global efficiency*, i.e., the MO version of a global optimum.

Definition 4 An observation $\mathbf{x} \in \mathscr{X}$ is said to be *Pareto efficient* or *globally efficient*, if and only if there exists no further observation $\tilde{\mathbf{x}} \in \mathscr{X}$ that dominates $\mathbf{x}$. The set of all global efficient points is denoted $\mathscr{X}_E$, and each connected component of $\mathscr{X}_E$ is a *(global) efficient set*.

Given that local and global optimality are not only phenomena of the decision space, we also define their respective pendants in the objective space.

Definition 5 The image (under **f**) of a problem's local efficient set is called *local (Pareto) front*, whereas the respective image of the union of global efficient sets is called *(global) Pareto front*.

Note that if the function **f** is continuous on a (local or global) efficient set, its respective front—i.e., the local or Pareto front—is also connected.

3.2 Visualizing Landscapes of Multi-objective Gradients

In analogy to single-objective problems, where the problem's gradient is zero in the position of a local or global optimum, the normalized MO gradient—and hence its length—is zero in each of its (locally or globally) efficient points.[1] Initiated by this property, this work's authors proposed a gradient-based view onto MO landscapes (cf. [21]), which is described in the following.

Although the majority of optimization algorithms perform their actions in the decision space—such as mutations and/or recombinations performed by evolutionary algorithms—their behavior is usually (if at all) visualized in the objective space. However, in case of single-objective problems, we do not simply focus on its (one-dimensional) objective space, but instead rather look at the entire problem landscape, e.g., by means of three-dimensional surface/perspective plots or at least two-dimensional contour plots and/or heatmaps. Of course, it is much more difficult to mutually visualize the decision space with two (or even more) objectives rather than with just a single objective; nonetheless it is not impossible. As we will show in Sect. 4, such a visualization provides valuable insights into—and thereby clearly improves our understanding of—such problems, which in turn can be used for developing simple, though very promising, MO optimization algorithms.

Basically all we need for such a joint visualization of decision and objective space is a utility function, which aggregates valuable information of the objective space into a single dimension. The method that was presented in Ref. [21], the so-called *gradient field heatmaps*, uses the cumulated lengths of the normalized gradients toward the respective attracting local efficient set.

Intuitively, points that are local optima in (at least) one of the single objectives should also be locally efficient from a MO point of view due to the dominance

[1] Note that this zero length property only applies to MO gradients whose components are normalized and equally weighted as this leads to tied impacts in local efficient points.

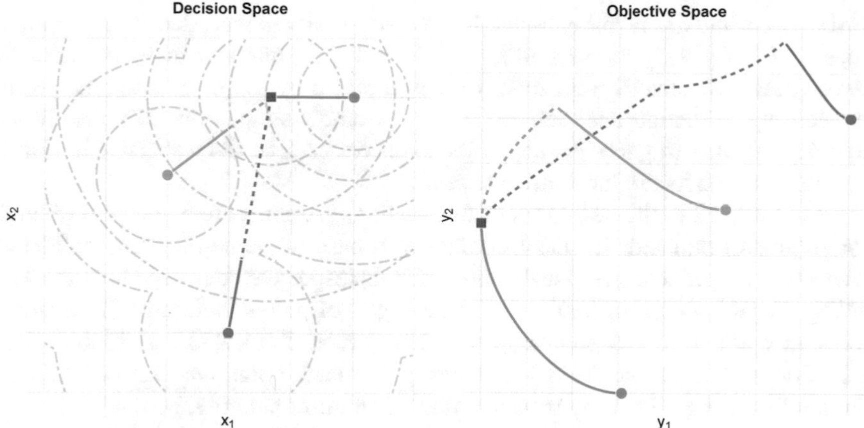

Fig. 1 Example of a simple bi-objective mixed-sphere problem with one (square) and three (circles) peaks, respectively. The left image displays both objectives separately by means of contour lines (yellow and light pink dot-dashed lines), as well as the corresponding local and global efficient sets (solid parts along the linear connections of peaks between both objectives). The plot on the right hand side displays the corresponding images of the efficient sets, i.e., the local fronts, in the objective space

(none of its neighbors can be superior w.r.t. the respective objective). Then, due to the interaction effects between the objectives, one would expect the efficient sets to be located somewhere along the (not necessarily linear) connection between local optima from the different objectives. In the most simple scenario one could think of a bi-objective problem in which each objective consists of a single, sphere-shaped peak. Its efficient set would be identical to the straight connection of the two peaks. The next step would be a problem, in which one objective contains a few sphere-shaped peaks as for instance depicted in Fig. 1. The landscape of the first objective[2] is indicated by yellow circle-like contour lines around its single peak (indicated by a pink square), and the landscape of the second objective[3] contains three peaks (marked by blue, green and red circles), which already lead to interaction effects between them as the light-pink contour lines of this landscape indicate. For each of the peaks from the latter objective, there exists a direct connection to the peak of the first objective. Noticeably, only the blue line in the top right of the decision space is completely solid—indicating that all points along that line are locally (actually globally) efficient—whereas the green and red lines from the remaining two peaks start as solid lines in their corresponding peaks, but are continued as dashed lines for their remainders. These dashed parts of the lines indicate that the points along that part are no longer locally efficient. For instance, the points along the dashed part

[2] $f_1(x_1, x_2) = 1 - \left(1 + 4 \cdot \left((x_1 - 2/3)^2 + (x_2 - 1)^2\right)\right)^{-1}$.

[3] $f_2(x_1, x_2) = 1 - \max\{g_1, g_2, g_3\}$ with $g_i(x_1, x_2, h, c_1, c_2) = h / \left(1 + 4 \cdot \left((x_1 - c_1)^2 + (x_2 - c_2)^2\right)\right)$ and $h = 1.5$, $c_1 = 0.5$ and $c_2 = 0$ (for g_1), $h = 2$, $c_1 = 0.25$ and $c_2 = 2/3$ (for g_2), and $h = 3$, $c_1 = 1 = c_2$ (for g_3).

of the green line are rather attracted by the blue peak (as indicated by the contour lines). Similarly, only the very first part along the red line is actually attracted by its red peak, whereas the remainder of it is attracted by the green and blue peaks, respectively. Although this scenario is still extremely simple, one can already see how difficult it is to guess the actual locations for each of the respective basins of attraction (of the local/global efficient sets).

As mentioned in the beginning of this section, gradients are a potential (though not sufficient) indicator for local optimality in case of single-objective problems. Therefore, we propose the visualization of their respective counterparts in the MO setting. For this purpose, we first construct a grid of points in the two-dimensional decision space. In our experiments we use between 250×250 and 1000×1000 points depending on the desired granularity. For each point, we then estimate the single-objective gradients by means of numerical differentiation, i.e.,

$$\nabla f_i(\mathbf{x}) \approx \lim_{\varepsilon \to 0} \frac{f_i(\mathbf{x} + \varepsilon) - f_i(\mathbf{x} - \varepsilon)}{||\varepsilon||}, \qquad i \in \{1, \ldots, p\}.$$

The sum of the two normalized single-objective gradients results in the *bi-objective gradient*. Due to the normalization of its two gradients, the effects of the two objectives on the joint bi-objective gradient are balanced. It further allows to interpret the length of the latter w.r.t. the closeness of the underlying point to a local efficient set: if both gradients point in the same direction, the bi-objective gradient will be of length two, whereas in cases in which both gradients point in opposite directions, the resulting bi-objective gradient will be of length zero. In the latter case, we have found an efficient point. The visualization of these bi-objective gradients (or at least a subset of them)—denoted *gradient field*—looks similar to a vector field as known from differential equations. By looking at the alignment of the arrows, one can already imagine the location and size of each efficient set's basin of attraction. Starting in a random point of the decision space, one simply follows the path of gradients until one reaches an efficient point—similar to an object in an ocean, which follows the water flow until it reaches a quiet area. Intuitively, the set of points that led to the same (local or global) efficient set forms the respective basin of attraction.

Exploiting the aforementioned principle, we now construct the *gradient field heatmap*, whose "height" simply represents the cumulated length of the gradients along the path through the grid toward its attracting locally or globally efficient point. The procedure for computing the cumulated path lengths is as follows. For each point in the grid, we check whether its gradient is (approximately) zero[4] as this would indicate local efficiency and thus a height of zero. Otherwise, we move from the current cell in the grid to its adjacent cell that is located in the direction (right, top right, top, top left, ...) of the bi-objective gradient. These moves from cell to cell (each time toward the respective most promising neighbor) are repeated until an

[4] Due to numerical imprecision, we defined a point to be locally or globally efficient, if the length of its bi-objective gradient is at most 10^{-3}.

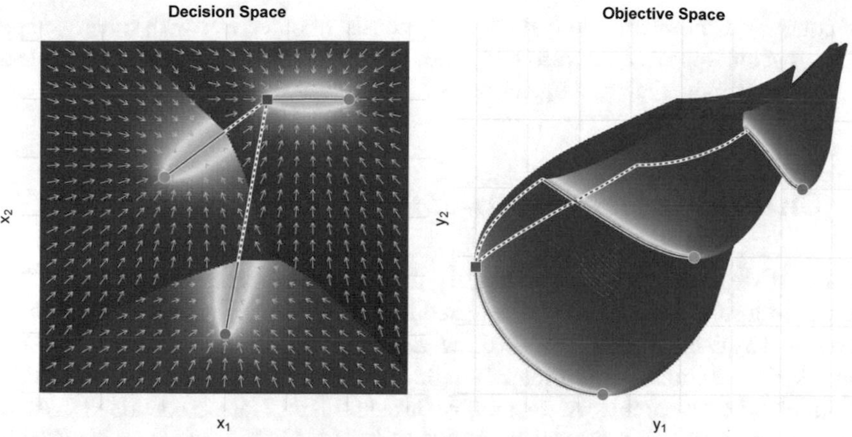

Fig. 2 *Gradient field heatmap* for the bi-objective mixed-sphere problem of Fig. 1 in the decision space (left), along with its corresponding image in the objective space (right). Blue areas (covered by a solid pink line) mark the problem's efficient sets. The attraction basins (of the efficient sets) become clearly visible (**a**) due to the gradient field of the bi-objective gradients (displayed by arrows), and (**b**) by means of their surrounding ridges

efficient point is reached. For all points/cells along that path we then cumulate the lengths of the considered gradients.

The resulting heatmap for the previously discussed scenario is shown in the left image of Fig. 2. Each efficient set comes with its own basin of attraction in which points that are far away from the efficient sets are colored in (dark) red and points in the proximity of the efficient sets are colored in yellow and blue. The latter points even form a valley around the respective efficient set. Furthermore, in contrast to most single-objective problems, transitions between adjacent basins are not smooth but rather discontinuous—as revealed by hard ridges along the borders of each basin. Interestingly, the global efficient set, i.e., the connection of the blue circle and the pink square, is the only efficient set which is not cut by a ridge. In fact, if an efficient set is cut by a ridge, it has to be *locally* efficient, as it is superimposed by a more attractive basin, whereas an efficient set without a ridge indicates (parts of) a potentially *global* efficient set. In case it is the only ridge-free efficient set of the problem, it also guarantees to be a global efficient set.

In addition to our proposed visualization of the decision space, we also use (the coloring of) the cumulated path lengths of the gradients to display the local and global fronts within the objective space. As the right image of Fig. 2 shows, each of the three fronts is indeed locally (or even globally) efficient—the blue and yellow points from the valleys that surround the respective efficient sets in the decision space, are all located in the dominated upper right part of the respective fronts.

Here, we introduced our approach—for reasons of simplicity and traceability—by means of a very simple and lowly multimodal MO problem, and thereby showed its ability to efficiently visualize interaction effects between the different objectives.

Of course, MO problems likely do not only consist of objectives with sphere-shaped peaks, but our approach can also be applied to any other shape and even to problems consisting of more than two objectives.

4 On the Properties of State-of-the-Art Benchmarks

In a next step, we apply the previously described methodology of gradient field heatmaps for the visualization of selected MO problems. A first set of problems is composed by using the *Multiple Peaks Model 2* (MPM2) generator devised by Wessing [39] for constructing multimodal single-objective problems and combining them to multimodal MO problems. Second, we take problems that are contained in established benchmark suites like ZDT [42] and DTLZ [10]. Those contain specifically constructed problems with different degrees of "locality" and deceptiveness. They are also the probably most frequently applied benchmarks in competitive literature on the development of evolutionary MO algorithms. Finally, we extend our observations by using candidates of a very recent benchmark – the *bi-objective black-box optimization benchmark* (BBOB) [7, 37]. This benchmark is constructed by combining multimodal single-objective functions from the BBOB set [19] in the same fashion as we combine MPM2 problems to generate a bi-objective problem. Note that we restrict all problems to the bi-objective case as well as to two-dimensional decision spaces to ensure visualization.

4.1 A Visual Overview

The images in Fig. 3 show the joint visualization of a unimodal and a multimodal MPM2 function. The resultant bi-objective problem comprises two local and a single global optimum. The local and global optima of the objectives are indicated by symbols of different shape (i.e., the shape of the symbol indicates the objective that it belongs to). The left hand side of the figure shows the decision space. Therein, we find two local efficient sets and a global efficient set. Their domination behavior in the objective space can be observed in the depiction on the right hand side.

The shading from dark red via yellow toward blue gives an impression of the MO (gradient-field-based) basin of attraction for each (local or global) efficient set, which expose similar characteristics as the sphere-based problem in Fig. 2. The most interesting observation is again the occurrence of ridges, which mark a sharp distinction of the basins of attraction for (locally) efficient sets. We also observe that only local efficient sets (and their basins of attraction) are cut by ridges, whereas global efficient sets are surrounded by a "complete" basin of attraction.

For a slightly more complex MPM2 problem, consisting of two multimodal functions, we find a similar visualization in Fig. 4. Again, we see a representation of local and global optima as well as ridges that divide the basins of attraction. Interestingly,

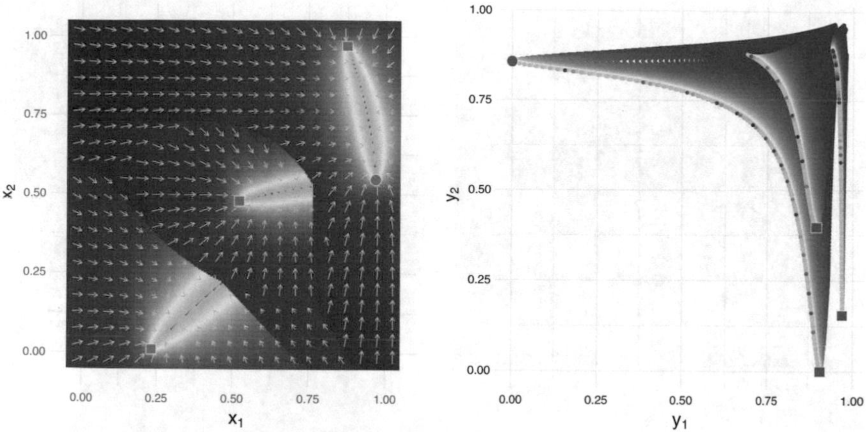

Fig. 3 Combination of two rather simple MPM2 problems (parameters for the MPM2 generator: $n_1 = 1, s_1 = 4, n_2 = 3, s_2 = 8$ and `shape = "ellipse"`)

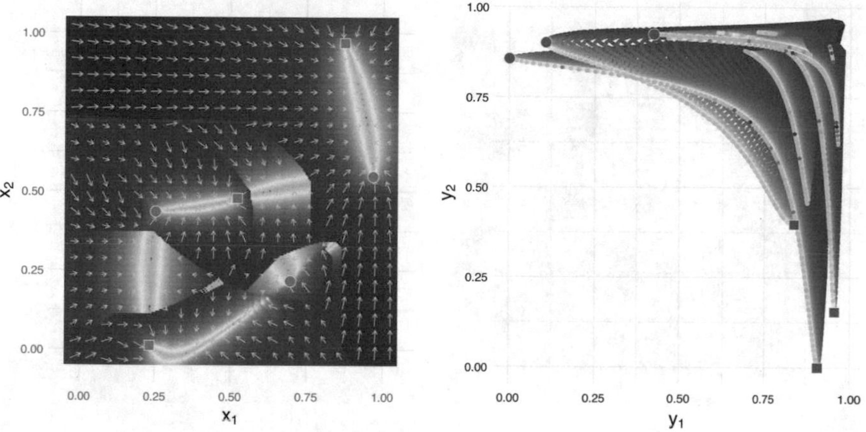

Fig. 4 Complex problem consisting of two multimodal MPM2 problems (parameters for the MPM2 generator: $n_1 = 3, s_1 = 4, n_2 = 3, s_2 = 8$ and `shape = "ellipse"`)

we observe a crossing of first-order fronts in the objective space such that a part of the respective efficient set becomes Pareto-optimal while another part is dominated. At the same time, we identify two efficient sets, which are surrounded by complete basins (without ridges). This can be interpreted as a kind of independency of partially global (and partially local) efficient sets in decision space.

For problems ZDT1 and DENT (see Fig. 5), as well as the problem DTLZ2 (see Fig. 6, left hand side), the decision space looks rather simple. We only observe a single efficient set that is located at the boundary of the decision space for ZDT1. For DENT we observe a diagonal set of which the middle part maps to the concave part of its well-known Pareto-front.

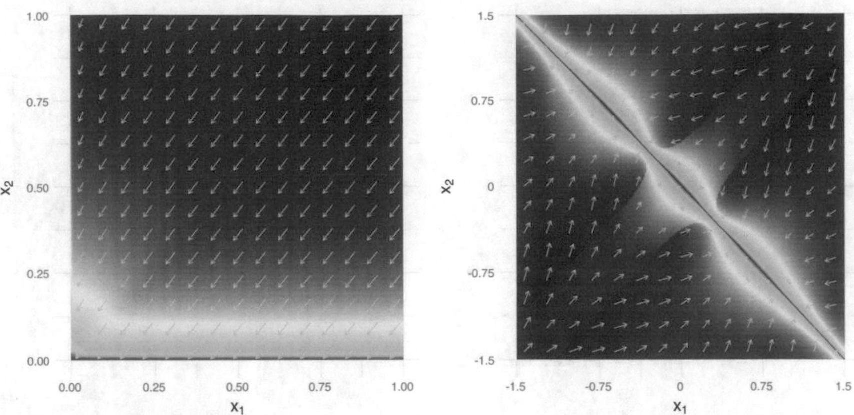

Fig. 5 Gradient heat map depiction of the decision spaces for problems ZDT1 (left) and DENT (right). As both problems do not comprise local efficient sets, we omit the visualizations of their well-known objective spaces

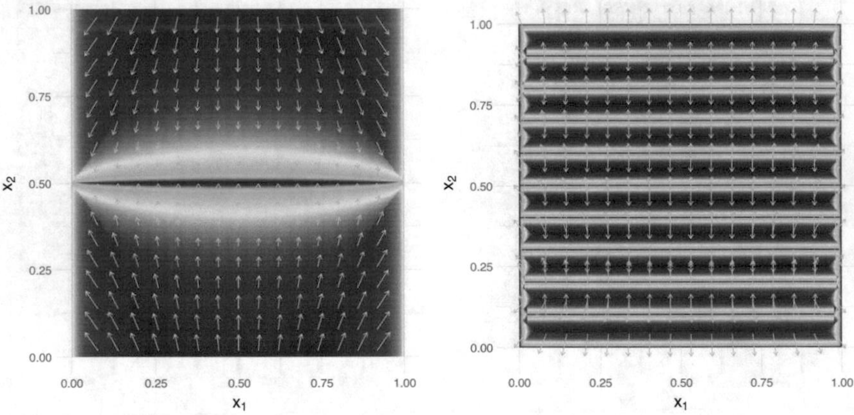

Fig. 6 Gradient heat map depiction of the decision spaces for problems DTLZ2 (left) and DTLZ3 (right). Only DTLZ3 comprises a regular pattern of local fronts that are represented by dominated spheres analog to the Pareto-optimal front. Thus, we omit a depiction of the objective space

For DTLZ2, we observe a single global efficient set, however, the surrounding gradient field shows some deceptiveness to the extremal solutions of the efficient set. This is rooted in the concave design of this problem. In contrast, DTLZ3 exposes multiple but rather regular local efficient sets. Each set maps to a dominated front that is similar to the optimal one—but far off in the dominated objective space.

All of the above findings are, of course, consequences from the general construction principles of ZDT and DTLZ [10, 42].

The investigation of exemplary instances from the bi-objective BBOB in Figs. 7 and 8 provides insights into what can be called "highly multimodal" MO prob-

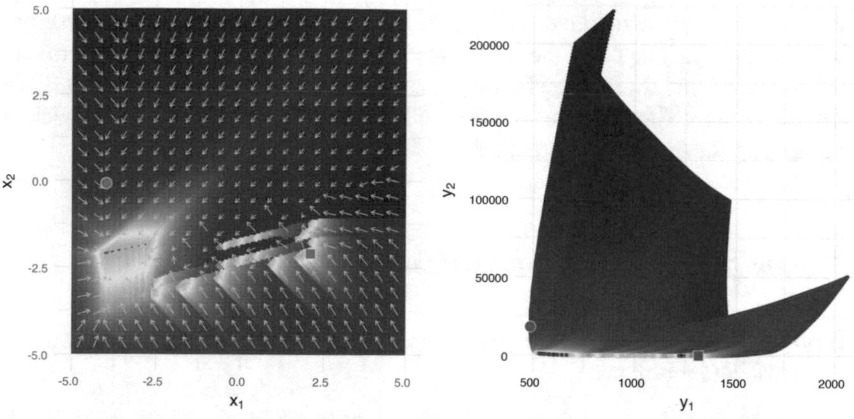

Fig. 7 Gradient heat map depiction of the decision (left) and objective space (right) for the bi-objective BBOB problem FID = 39 (IID = 5), composed from the *Sharp Ridge* and the *Schwefel* function (FID1 = 13, IID1 = 11 and FID2 = 20, IID2 = 12)

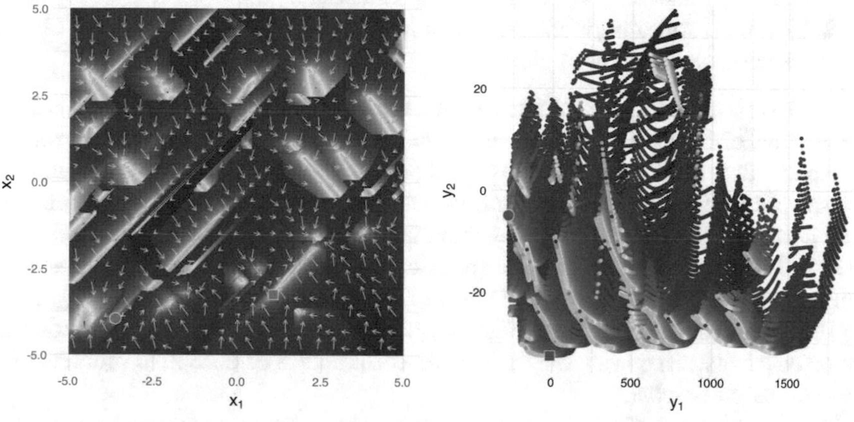

Fig. 8 Gradient heat map depiction of the decision (left) and objective space (right) for the bi-objective BBOB problem FID = 40 (IID = 1), composed from the *Sharp Ridge* and *Gallagher's Gaussian 101-me Peaks* function (FID1 = 13, IID1 = 3 and FID2 = 21, IID2 = 4)

lems. Constructed from multimodal single-objective functions, the multiple local and global optima emerge to many local and (some) global efficient sets. Again, we observe multiple local efficient sets with basins of attraction that superimpose each other. This produces ridges at which the attracting basins rapidly change. Global efficient sets are (as before) surrounded by complete basins of attraction. For information purposes, we additionally depicted the known global optima of the underlying single-objective BBOB problems into the search and objective space.

In Fig. 7 an isolated global efficient solution is denoted by the global optimum of the "sharp ridge function" located in the red-shaded area without a visible basin of attraction. However, a detailed analysis of the area and a closer look at the global efficient points (displayed in the objective space) have shown that this effect is due to the nature of the "sharp ridge function" used in generating the bi-objective BBOB problem instance. The ridge poses a difficulty to the inherent numerical imprecision in computing the gradient heatmaps for visualization.

4.2 Interpretation and Categorization

The observations made above reveal two important aspects for the visualization of multimodal MO problems:

1. Using the combined gradients of the objective functions for computing the distance of each point from the next attracting (local) efficient set, contributes to an intuitive understanding of the MO problem landscape. From observations we find that the basins of attraction are superimposed. This leads to so-called ridges, i.e., boundaries that divide adjacent basins of attraction.
2. Basins that surround globally efficient solutions are not superimposed. Global efficient sets are not cut by ridges.

From the evaluation of multiple benchmarks (comprising the entire ZDT, DTLZ, MPM2, and bi-objective BBOB benchmarks), these observations appear to be robust.

Apart from these general findings, we are able to discriminate two preliminary classes of benchmark problems from visualization and observation of the gradient field heatmaps. The first class of benchmark problems contains, e.g., instances from ZDT or DTLZ. These problems are specifically constructed by intertwining several configurable single-objective problems with known properties [10, 42]. They allow for the layered creation of local efficient sets and enable explicit setups for convexity or concavity. That way, they allow the construction of deceptive problems for gradient-based descent.

The second class of problems is resulting from a simple construction principle: the combination of single-objective multimodal problems. Therein, multimodality is achieved by combining at least two single-objective functions of which at least one is multimodal [7, 37]. We may argue, that this approach is rather congruent with the interpretation of MO real-world problems and we can follow Huband et al. [20] in their reasoning that problems like ZDT and DTLZ over-represent specifically adjusted and defined properties.

Overall, the visual investigation of multimodal MO problems and the superposition of local efficient sets (which is present in both classes) reveals a surprising insight: multimodality may be a less challenging property than expected. In contrast to single-objective optimization, we may even be able to exploit local efficient solutions to converge to the global efficient set. As a consequence, the two identified classes may only represent different classes of difficulty to such exploitation mechanisms.

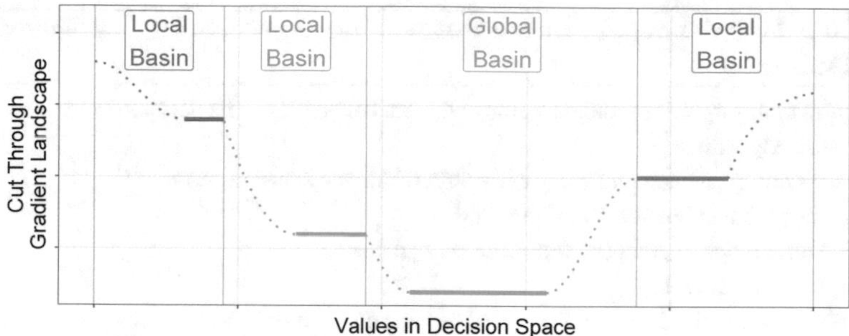

Fig. 9 Schematic view on superimposed basins of attraction. The vertical orange lines represent the ridges between adjacent basins, solid horizontal lines (within the basins) indicate the respective efficient sets and the dotted lines represent the path toward the attracting efficient set

5 How Multi-objective Optimization Algorithms Can Capitalize from Basins of Attraction

As discussed in the previous sections, our gradient-based view onto MO problems revealed some interesting findings. Two of its most intriguing ones are (1) the basins of attraction are superimposed, which right away leads to (2) local (i.e., non global) efficient sets are interrupted by ridges, whereas their global counterparts are not interrupted. Instead, the latter remain complete and thus connect peaks from two (interacting) objectives. Figure 9 provides a schematic view on this interaction among the basins. This principle in turn leads to a very important observation:

> *In contrast to single-objective optimization, in which local optima often are deceptive traps for an optimization algorithm, they can be very handy in MO optimization as* ***they often enable to escape basins*** *that belong to non-global efficient sets.*

This "escape mechanism" is very simple: given a continuous multimodal MO problem, we could start in a random point of its landscape and follow the MO gradients toward a local efficient set—similar to a child, which is riding its sled downhill until it stops in the valley. Once the child (i.e., the algorithm) reaches this valley (i.e., the local efficient set), it finds itself on a plateau with two potential ways out (of the valley). It then moves (along the plateau) through the valley in the direction of one of the two exits. Eventually, it reaches a dead end, i.e., a position where all directions (except for the one that led to this location) lead uphill. In this case, the child returns to its starting point within the valley and from there moves into the opposite direction. There, it will likely reach a ridge and—by passing it—enter the adjacent basin, where it again can slide downhill toward the next valley. This procedure can be repeated until both ends of the detected valley are dead ends. In that case, the child—more precisely the optimization algorithm—has either reached (at least a part of) the

Algorithm 1 (High-Level) Pseudocode of the *Multi-Objective Gradient Sliding Algorithm* (MOGSA)

```
Require multi-objective problem f along with the boundaries of its decision space
Initialize Algorithm
Choose starting position uniformly at random within the decision space
While no global efficient set was found do
    Find local efficient point (by following MO gradients)
    Explore efficient set
    Check whether the found set is a global or local efficient set
End While
Return archive of evaluated points
```

global efficient set or a local trap.[5] In case the global efficient set of the problem is connected, the algorithm will find the entire set. Otherwise, it needs to be restarted from another initial starting position on the landscape.

As figuratively described above, our proposed algorithm—denoted *MO gradient sliding algorithm* (MOGSA, [17, 18])—in its very essence consists of three steps as summarized in Algorithm 1. At first, the algorithm starts in a random location of the decision space and follows the MO gradients toward a (locally or globally) efficient point. From this point on, MOGSA explores the efficient set and afterward checks, whether the efficient set is a local or global optimum (or local trap). The latter is indicated by "dead ends" in both ends of the valley, whereas the former is revealed by a ridge that is abruptly cutting the set (and thereby "opens the door" toward a stronger attracting and hence more promising basin). If a (potentially) global efficient set was found, the algorithm terminates, otherwise it continues with a (multi-objective) gradient descent from the point that indicated the movement across the ridge.

Despite its simplicity, some first—exclusively exploratory—experimental studies, in which we compared our gradient-based approach to two state-of-the-art evolutionary algorithms, namely NSGA- II [9] and SMS- EMOA [3], have confirmed the effectiveness of MOGSA. We want to emphasize that this is not at all meant to be a comparative benchmark study; instead we simply want to show that the gained insights from the previous sections can be very helpful for algorithm design.

As for instance shown in Fig. 10, our approach exploits the considered mixed-peaks problem[6] in a straightforward manner and hence strives much faster toward the problem's efficient sets (and fronts, respectively) than its two contenders. Of course, this goal-oriented behavior—almost like within a tunnel vision—implies that MOGSA explores the decision space a lot less compared to NSGA- II or SMS-EMOA. Still, due to the low amount of performed function evaluations, MOGSA could simply be restarted from a different location in the decision space until the

[5] Note that this case of local traps is rather uncommon based on our extensive experimental studies and observations across popular benchmark problems.

[6] Note that all test functions within this paper were created using the `R`-package `smoof` [6].

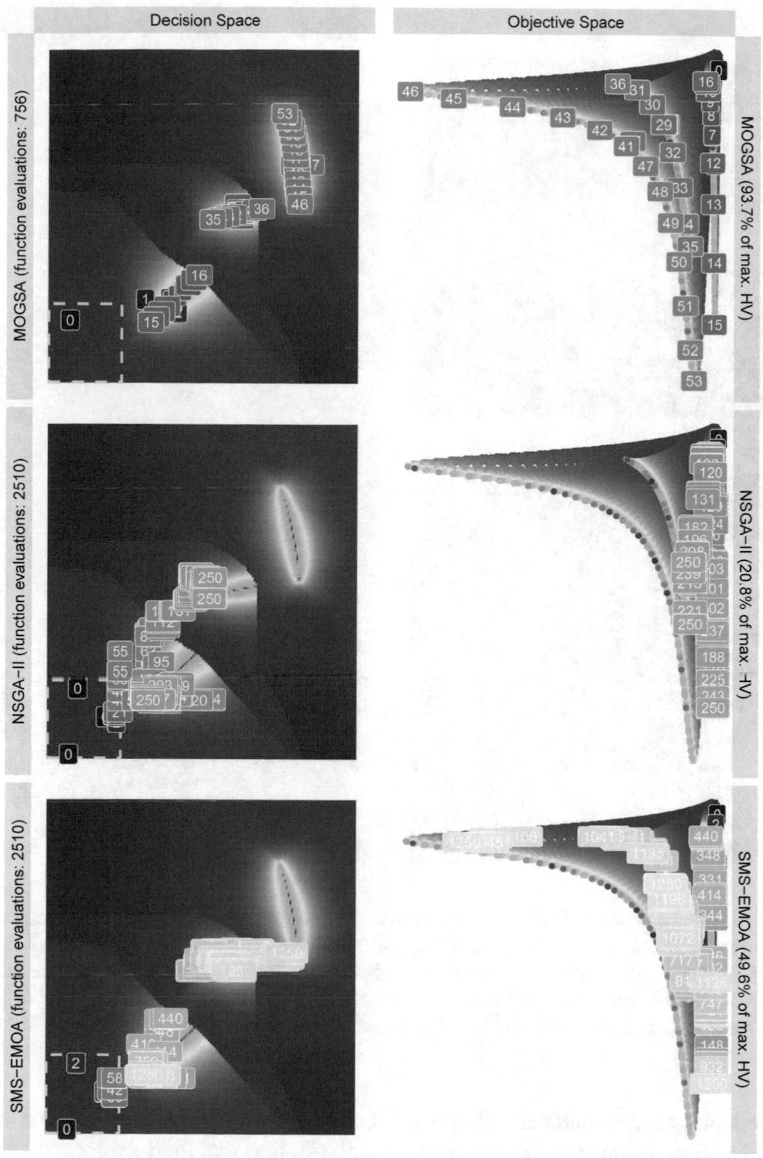

Fig. 10 Exploratory comparison of the search behavior of MOGSA (top row), NSGA- II (middle) and SMS- EMOA (bottom) in the decision (left column) and objective space (right). The underlying problem is a mixed-peak problem with one and three elliptically shaped peaks per objective, respectively. NSGA-II was executed for 250 generations with a population of $\mu = 5$ individuals and $\lambda = 5$ offspring, whereas SMS- EMOA ran for 1 250 generations with population size $\mu = 5$ and $\lambda = 1$ offspring per generation. Both algorithms thus evaluated each objective 1255 times. MOGSA on the other hand only required 756 evaluations in total (already including the ones for approximating the gradient). All solvers were initialized in the bottom left (grey dashed box) of the decision space and the hypervolumes were computed using the problem's nadir as reference point

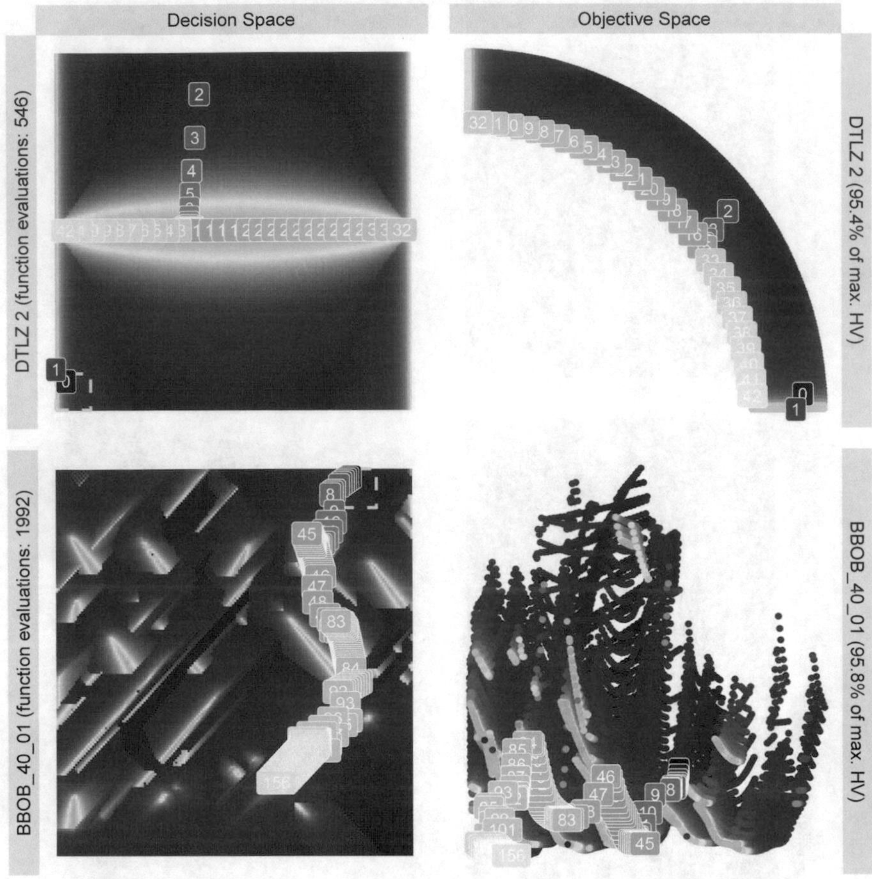

Fig. 11 Search behavior of MOGSA visualized in the decision and objective space (left and right, respectively) based on two challenging bi-objective problems. Top: the deceptive DTLZ 2, which partially misleads optimizers to its boundaries. Bottom: a highly multimodal instance from the bi-objective BBOB benchmark (FID: 40, IID: 1). Noticeably, MOGSA again was able to quickly strive toward the respective global efficient sets and distribute uniformly along the Pareto fronts

available budget is exhausted. Thereby, it could explore the problem from a variety of angles while ensuring to capture the global efficient sets (and fronts).

Comparing the performances of the algorithms—under this particular setup—by means of function evaluations, as well as covered hypervolume, shows that MOGSA only required approximately 30% of the budget that was given to its two contenders, while at the same time covering more than 90% of the theoretically possible hypervolume. In contrast to that, after consuming the entire budget of 2510 function evaluations per algorithm, SMS- EMOA and NSGA- II only managed to cover 50% and 20%, respectively, of the (theoretically possible) hypervolume.

Once again, we want to emphasize that the shown problem, as well as the chosen algorithms, were just used for exemplary and comparative purposes. Certainly, this does not imply that MOGSA is superior to any of them—let alone to all state-of-the-art solvers across all problems. Instead, our intention rather is to confirm the applicability and expected behavior of our method. Therefore, we also did not tune any algorithm's hyperparameters (including MOGSA) to the problem at hand, but simply used the R-implementations of SMS- EMOA and NSGA-II provided by `ecr` [5] and reduced the population size (from its default value 100) to $\mu = 5$. The motivation for the latter parameter "configuration" was to avoid an extreme computational overhead and thus disadvantage for the two evolutionary algorithms.

In order to confirm that the search behavior of MOGSA does not only work well on the admittedly rather simple mixed-peaks problem from Fig. 10, we also tested it on further benchmark problems. Figure 11 displays MOGSA's optimization paths in the decision and objective space for DTLZ2 (top row) and a highly multimodal instance from bi-objective BBOB (bottom row). The latter comprises of the *Sharp Ridge* and *Gallagher's Gaussian 101-me Peaks* functions, i.e., problems 13 and 21 from the classical single-objective BBOB [13, 19], representing the classes "unimodal functions with high conditioning" and "highly multimodal functions with a weak global structure."

The deceptive structure of DTLZ2 (see images in the upper row of Fig. 11) initially lures MOGSA to the problem's boundaries. However, after detecting this "trap," a simple restart from a different location within the decision space allowed the algorithm to quickly move toward the efficient set. Having reached this efficient point, it is able to spend the majority of its budget for the exploration of the set.

Even more interesting is the second problem from Fig. 11, i.e., the highly multimodal bi-objective BBOB problem. Although the problem itself looks extremely rugged, MOGSA is able to cope with that structure and simply "slides" through the landscape from (local) basin to (local) basin until it reaches the problem's global efficient set. This search behavior again confirms one of the key findings from the previous sections:

In MO optimization, local optima are often no obstacles or even traps for (local) search algorithms. Instead, they can provide the means to efficiently maneuver through the landscape toward the problem's global efficient set(s).

6 Conclusion

Continuous optimization is in general considered to be a well-studied research area. However, although there exist numerous works dealing with multimodal *or* MO optimization, only a handful of them explicitly targets both domains simultaneously. As a result, many assumptions made on the complexity or difficulty of multimodal MO (continuous) optimization problems rely purely on findings from their single-objective counterparts. Within this work, we want to decrease the nescience on this particular topic and, by looking underneath this "multimodality-fog," show that (in case of MO optimization) multimodality should no longer be regarded as an insurmountable obstacle. In fact, we even show that an improved understanding of such rugged landscapes might facilitate the search for the global optimum of MO problems.

In order to raise the awareness for these insights, we first defined multimodality in the MO sense and introduced a methodology, which enables the visualization of MO problems in the decision *and* objective space. By means of this visualization technique, which we denoted *gradient field heatmaps*, the interactions of the problem's objectives are clearly revealed in both spaces. In case of the decision space, the interactions lead to multiple basins of attractions, which are separated from each other by ridges. Based on the structure of these basins, (potentially) global and certainly local efficient sets, i.e., the MO pendant to global and local optima of single-objective problems, can be distinguished from each other.

The proposed visualization approach does not only allow to distinguish local and global optima within a problem but also helps to classify common benchmark problems (by visual means) into separate classes. On the one hand, there are benchmark problems that were explicitly constructed for the purpose of MO optimization (e.g., DTLZ and ZDT), whereas recently proposed benchmarks such as bi-objective BBOB, or the MO MPM2 problems, are simply concatenations of separate single-objective functions.

Interestingly, the visualization of the problems also revealed that each optimum possesses its own basin of attraction and that the basins are superimposed. More precisely, they enable movement into the basin's efficient set and from there toward the next basin until (at least a part of) the global optimum—or local trap—was found. Based on these findings, we proposed a simple, yet often effective, MO gradient sliding (optimization) algorithm (MOGSA), which maneuvers through the local optima toward the problem's global optimum.

For future works, we intend to further analyze preconditions for the existence of local traps and global optima as ridge-free efficient sets. These insights shall be used to improve the proposed algorithm and afterward compare its performance against several state-of-the-art optimization algorithms in an extensive benchmark study.

Acknowledgements The authors acknowledge support from the *European Research Center for Information Systems (ERCIS)*.

References

1. Aguirre, H.E., Tanaka, K.: Insights on properties of multiobjective MNK-landscapes. In: Proceedings of the IEEE Congress on Evolutionary Computation (CEC), vol. 1, pp. 196–203. IEEE, New York (2004). https://doi.org/10.1109/CEC.2004.1330857
2. Aguirre, H.E., Tanaka, K.: Working principles, behavior, and performance of MOEAs on MNK-landscapes. Eur. J. Oper. Res. **181**(3), 1670–1690 (2007). https://doi.org/10.1016/j.ejor.2006.08.004

3. Beume, N., Naujoks, B., Emmerich, M.T.M.: SMS-EMOA: multiobjective selection based on dominated hypervolume. Eur. J. Oper. Res. **181**(3), 1653–1669 (2007). https://doi.org/10.1016/j.ejor.2006.08.008
4. Beyer, H.G.: The theory of evolution strategies. In: Natural Computing Series (NCS). Springer, Berlin (2001). https://doi.org/10.1007/978-3-662-04378-3
5. Bossek, J.: ecr 2.0: a modular framework for evolutionary computation in R. In: Proceedings of the Genetic and Evolutionary Computation Conference (GECCO) Companion, pp. 1187–1193 (2017). https://doi.org/10.1145/3067695.3082470
6. Bossek, J.: smoof: single- and multi-objective optimization test functions. R J. **9**(1), 103–113 (2017)
7. Brockhoff, D., Tran, T.D., Hansen, N.: Benchmarking numerical multiobjective optimizers revisited. In: Proceedings of the 17th Annual Conference on Genetic and Evolutionary Computation (GECCO), pp. 639–646. ACM, New York (2015). https://doi.org/10.1145/2739480.2754777
8. Daolio, F., Liefooghe, A., Verel, S., Aguirre, H.E., Tanaka, K.: Global vs local search on multi-objective NK-landscapes: contrasting the impact of problem features. In: Proceedings of the 17th Annual Conference on Genetic and Evolutionary Computation (GECCO), pp. 369–376. ACM, New York (2015). https://doi.org/10.1145/2739480.2754745
9. Deb, K., Pratap, A., Agarwal, S., Meyarivan, T.: A fast and elitist multiobjective genetic algorithm: NSGA-II. IEEE Trans. Evol. Comput. **6**(2), 182–197 (2002). https://doi.org/10.1109/4235.996017
10. Deb, K., Thiele, L., Laumanns, M., Zitzler, E.: Scalable test problems for evolutionary multi-objective optimization. In: Advanced Information and Knowledge Processing (AI & KP), pp. 105–145. Springer, Berlin (2005). https://doi.org/10.1007/1-84628-137-7_6
11. Derbel, B., Liefooghe, A., Verel, S., Aguirre, H., Tanaka, K.: New features for continuouse exploratory landscape analysis based on the SOO tree. In: Proceedings of the 15th ACM/SIGEVO Conference on Foundations of Genetic Algorithms, pp. 72–86. ACM, New York (2019)
12. Ehrgott, M., Klamroth, K.: Connectedness of efficient solutions in multiple criteria combinatorial optimization. Eur. J. Oper. Res. **97**(1), 159–166 (1997). https://doi.org/10.1016/S0377-2217(96)00116-6
13. Finck, S., Hansen, N., Ros, R., Auger, A.: Real-parameter black-box optimization benchmarking 2010: presentation of the noiseless functions. Tech. Rep., INRIA (2010)
14. da Fonseca, C.M.M.: Multiobjective genetic algorithms with application to control engineering problems. Ph.D. thesis, Department of Automatic Control and Systems Engineering, University of Sheffield (1995)
15. Garrett, D.: Plateau connection structure and multiobjective metaheuristic performance. In: Proceedings of the IEEE Congress on Evolutionary Computation (CEC), pp. 1281–1288. IEEE, Piscataway, NJ (2009). https://doi.org/10.1109/CEC.2009.4983092
16. Garrett, D., Dasgupta, D.: Multiobjective landscape analysis and the generalized assignment problem. In: Maniezzo, V., Battiti, R., Watson, J.P. (eds.) Proceedings of the Second International Conference on Learning and Intelligent Optimization (LION), Lecture Notes in Computer Science (LNCS), vol. 5313, pp. 110–124. Springer, Berlin (2008). https://doi.org/10.1007/978-3-540-92695-5_9
17. Grimme, C., Kerschke, P., Emmerich, M.T.M., Preuss, M., Deutz, A.H., Trautmann, H.: Sliding to the global optimum: how to benefit from non-global optima in multimodal multi-objective optimization. In: AIP Conference Proceedings, pp. 020052-1–020052-4. AIP Publishing (2019). https://doi.org/10.1063/1.5090019
18. Grimme, C., Kerschke, P., Trautmann, H.: Multimodality in multi-objective optimization—more boon than bane? In: Deb, K., Goodman, E., Coello, C.C.A., Klamroth, K., Miettinen, K., Mostaghim, S., Reed, P. (eds.) Proceedings of the 10th International Conference on Evolutionary Multi-Criterion Optimization (EMO), Lecture Notes in Computer Science, vol. 11411, pp. 126–138. Springer, New York (2019). https://doi.org/10.1007/978-3-030-12598-1_11

19. Hansen, N., Finck, S., Ros, R., Auger, A.: Real-parameter black-box optimization benchmarking 2009: noiseless functions definitions. Tech. Rep. RR-6829. INRIA (2009)
20. Huband, S., Hingston, P., Barone, L., While, L.: A review of multiobjective test problems and a scalable test problem toolkit. Trans. Evol. Comp. **10**(5), 477–506 (2006). https://doi.org/10.1109/TEVC.2005.861417
21. Kerschke, P., Grimme, C.: An expedition to multimodal multi-objective optimization landscapes. In: Trautmann, H., Rudolph, G., Kathrin, K., Schütze, O., Wiecek, M., Jin, Y., Grimme, C. (eds.) Proceedings of the 9th International Conference on Evolutionary Multi-Criterion Optimization (EMO), pp. 329–343. Springer, New York (2017). https://doi.org/10.1007/978-3-319-54157-0_23
22. Kerschke, P., Hoos, H.H., Neumann, F., Trautmann, H.: Automated Algorithm Selection: Survey and Perspectives. Evolutionary Computation (ECJ) **27**(1), 3–45 (2019). https://doi.org/10.1162/evco_a_00242
23. Kerschke, P., Preuss, M., Wessing, S., Trautmann, H.: Detecting funnel structures by means of exploratory landscape analysis. In: Proceedings of the 17th Annual Conference on Genetic and Evolutionary Computation (GECCO), pp. 265–272. ACM, New York (2015). https://doi.org/10.1145/2739480.2754642
24. Kerschke, P., Preuss, M., Wessing, S., Trautmann, H.: Low-budget exploratory landscape analysis on multiple peaks models. In: Proceedings of the 18th Annual Conference on Genetic and Evolutionary Computation (GECCO), pp. 229–236. ACM, New York (2016). https://doi.org/10.1145/2908812.2908845
25. Kerschke, P., Trautmann, H.: The R-package FLACCO for exploratory landscape analysis with applications to multi-objective optimization problems. In: Proceedings of the IEEE Congress on Evolutionary Computation (CEC) (2016). IEEE, New York. https://doi.org/10.1109/CEC.2016.7748359
26. Kerschke, P., Trautmann, H.: Automated algorithm selection on continuous black-box problems by combining exploratory landscape analysis and machine learning. Evol. Comput. **27**(1), 99–127 (2019). https://doi.org/10.1162/evco_a_00236
27. Kerschke, P., Trautmann, H.: Comprehensive feature-based landscape analysis of continuous and constrained optimization problems using the R-package Flacco. In: Bauer, N., Ickstadt, K., Lbke, K., Szepannek, G., Trautmann, H., Vichi, M. (eds.) Applications in Statistical Computing, pp. 93–123. Springer, Berlin (2019). https://doi.org/10.1007/978-3-030-25147-5_7
28. Kerschke, P., Wang, H., Preuss, M., Grimme, C., Deutz, A.H., Trautmann, H., Emmerich, M.T.M.: Towards analyzing multimodality of multiobjective landscapes. In: Handl, J., Hart, E., Lewis, P.R., López-Ibáñez, M., Ochoa, G., Paechter, B. (eds.) Proceedings of the 14th International Conference on Parallel Problem Solving from Nature (PPSN XIV), Lecture Notes in Computer Science (LNCS), vol. 9921, pp. 962–972. Springer, Berlin (2016) (Best Paper Award). https://doi.org/10.1007/978-3-319-45823-6_90
29. Kerschke, P., Wang, H., Preuss, M., Grimme, C., Deutz, A.H., Trautmann, H., Emmerich, M.T.M.: Search dynamics on multimodal multiobjective problems. Evol. Comput. **27**(4), 577–609 (2019)
30. Knowles, J.D., Corne, D.W.: Towards landscape analyses to inform the design of hybrid local search for the multiobjective quadratic assignment problem. In: Proceedings of the Second International Conference on Hybrid Intelligent Systems (HIS). Springer, Berlin (2002)
31. Mersmann, O., Bischl, B., Trautmann, H., Preuss, M., Weihs, C., Rudolph, G.: Exploratory landscape analysis. In: Proceedings of the 13th Annual Conference on Genetic and Evolutionary Computation (GECCO), pp. 829–836. ACM, New York (2011). https://doi.org/10.1145/2001576.2001690
32. Muñoz Acosta, M.A., Kirley, M., Halgamuge, S.K.: Exploratory landscape analysis of continuous space optimization problems using information content. IEEE Trans. Evol. Computa. **19**(1), 74–87 (2015). https://doi.org/10.1109/TEVC.2014.2302006
33. Rosenthal, S., Borschbach, M.: A concept for real-valued multi-objective landscape analysis characterizing two biochemical optimization problems. In: Mora, M.A., Squillero, G. (eds.)

Proceedings of the 18th European Conference on Applications of Evolutionary Computation (EvoApplications), Lecture Notes in Computer Science (LNCS), vol. 9028, pp. 897–909. Springer, Berlin (2015). https://doi.org/10.1007/978-3-319-16549-3_72

34. Schwefel, H.P.: Evolution and Optimum Seeking. Wiley, New York (1995)
35. Tušar, T.: Visualizing solution sets in multiobjective optimization. Ph.D. thesis, Jožef Stefan International Postgraduate School (2014)
36. Tušar, T., Filipič, B.: Visualization of Pareto front approximations in evolutionary multiobjective optimization: a critical review and the prosection method. IEEE Trans. Evol. Comput. **19**(2), 225–245 (2015). https://doi.org/10.1109/TEVC.2014.2313407
37. Tušar, T., Brockhoff, D., Hansen, N., Auger, A.: COCO: the bi-objective black box optimization benchmarking (bbob-biobj) test suite. arXiv preprint (2016). https://arxiv.org/abs/1604.00359
38. Verel, S., Liefooghe, A., Jourdan, L., Dhaenens, C.: On the structure of multiobjective combinatorial search space: MNK-landscapes with correlated objectives. Eur. J. Oper. Res. **227**(2), 331–342 (2013). https://doi.org/10.1016/j.ejor.2012.12.019
39. Wessing, S.: Two-stage methods for multimodal optimization. Ph.D. thesis, Technische Universität Dortmund (2015)
40. Wright, S.: The roles of mutation, inbreeding, crossbreeding and selection in evolution. In: Proceedings of the 6th International Congress of Genetics, Ithaca, New York, vol. 1, pp. 356–366. Brooklyn Botanic Garden, Brooklyn, New York, USA (1932)
41. Yue, C., Qu, B., Yu, K., Liang, J., Li, X.: A novel scalable test problem suite for multimodal multiobjective optimization. Swarm Evol. Comput. **48**, 62–71 (2019). https://doi.org/10.1016/j.swevo.2019.03.011
42. Zitzler, E., Deb, K., Thiele, L.: Comparison of multiobjective evolutionary algorithms: empirical results. Evol. Comput. J. **8**(2), 173–195 (2000). https://doi.org/10.1162/106365600568202

Towards Basin Identification Methods with Robustness Against Outliers

Simon Wessing

Abstract An important subtask in multimodal optimization is the identification of the attraction basins of individual optima. The knowledge about these basins can, for example, be used to start an appropriate number of local searches or to classify the problem instance in an exploratory landscape analysis before the optimization is started. In a black-box setting, the identification process necessarily needs a sample of evaluated solutions as input data. As these evaluations are expensive, it would be desirable to reuse previously acquired samples, if existing. In this case, arbitrary mixture distributions of the data must be assumed. Unfortunately, there is no basin identification method currently available that is robust to spatial outliers in the sample and at the same time can provide a ranking and/or a clustering of all the solutions. Topographical selection, which is based on a k-nearest-neighbor graph, is robust against outliers, but does not provide clustering information and determines the number of selected solutions on its own. Nearest-better clustering, on the other hand, can provide a hierarchical clustering but is not very robust to outliers. In this work, we adopt ideas from density-based clustering to develop a new basin identification method. The core idea is to use the number of neighbors within the distance to the nearest-better neighbor as a selection criterion. Experiments show that the new method combines the desirable features of the existing ones.

1 Introduction

In the simplest case, basin identification can be viewed as a kind of subset selection, where from a discrete set of evaluated solutions in the search space of an optimization problem, the subset is sought which resembles most closely the set of local optima of the optimization problem. Extended definitions of the basin identification task may also request an approximation of the shapes of the attraction basins, which requires some kind of modeling, possibly preceded by a clustering procedure. The first basin

---

S. Wessing (✉)
CLK GmbH, Altenberge, Germany
e-mail: mail@simonwessing.de

M. Preuss et al. (eds.), *Metaheuristics for Finding Multiple Solutions*, Natural Computing Series,
https://doi.org/10.1007/978-3-030-79553-5_5

identification method to be developed, topographical selection (TS) by Törn and Viitanen [12], was only aimed at the simpler subset selection. An important aspect of the method is that by using a parametrized heuristic, TS chooses the number of selected solutions on its own. The first clustering procedure specifically for basin identification was nearest-better clustering (NBC) by Preuss et al. [10], which is a hierarchical clustering method via a directed spanning tree. It has the advantage that it can provide a ranking of all solutions, so the number of selected solutions can either be fixed beforehand or determined by a data-dependent heuristic.

In the remainder of the chapter, we introduce NBC and briefly discuss related work. Then we introduce our proposal for a refined basin identification method and develop a parametric model to determine its single parameter. Finally, a validation experiment is conducted and conclusions are drawn.

2 Nearest-Better Clustering

Nearest-better clustering relies on the concept of the nearest-better neighbor

$$\mathrm{nbn}(\mathbf{x}, \mathscr{P}) = \arg\min_{\mathbf{y} \in \mathscr{P}} \{d(\mathbf{x}, \mathbf{y}) \mid f(\mathbf{y}) < f(\mathbf{x})\}$$

of a point $\mathbf{x}$, where $\mathscr{P} \subset \mathbb{R}^n$ is typically a finite set of points with $|\mathscr{P}| = N$, d is a distance measure, and f is the objective function. The distance to this point shall be denoted as the distance to the nearest-better point, $d_{\mathrm{nb}}(\mathbf{x}, \mathscr{P}) := d(\mathbf{x}, \mathrm{nbn}(\mathbf{x}, \mathscr{P}))$. The conventional nearest-neighbor distance will be named $d_{\mathrm{nn}}(\mathbf{x}, \mathscr{P})$ in the following. The actual operation of NBC is described in Algorithm 1. In a first step, it creates a directed graph consisting of edges from points to their nearest-better neighbors. This graph is acyclic and can have at most $N - 1$ edges. The number of $N - 1$ edges is reached only if the best objective value in the sample is unique; in this case, the graph is a spanning tree, otherwise a spanning forest [9, pp. 77–78]. Afterwards, the tree is divided into several connected components by removing "long" edges.

For characterizing edges as long, two heuristics exist, which are called rule 1 and rule 2 in the pseudocode. Rule 1 in its simplest form removes all edges whose length exceeds the mean nearest-neighbor distance d_{ref} by more than a factor ϕ (this is actually a slight variation of the original rule and was proposed in Ref. [15]). Rule 2 is only applied to edges e whose tail v has an indegree $\deg^-(v) \geq 3$. The rule states to cut such an edge e if its length is more than b times longer than the median of the incoming edges of v. The parameter b has been derived by experimentation and is actually dependent on the number of points and the dimension [9, Sect. 4.5]:

$$\begin{aligned} b(N, n) = &(-4.69 \cdot 10^{-4} n^2 + 0.0263 n + 3.66 n^{-1} - 0.457) \cdot \log_{10}(N) \\ &+ 7.51 \cdot 10^{-4} n^2 - 0.0421 n - 2.26 n^{-1} + 1.83\,. \end{aligned}$$

Algorithm 1 Nearest-better clustering

```
Input: points 𝒫 = {x_1, ..., x_N}
Output: clusters in form of connected components of a graph
create an empty, weighted, directed graph G = (V, E)
for all points do
    find nearest better neighbor and create edge to it
end for                                              // G is now a spanning tree or forest
w_max ← φ · d_ref                                    // calculate weight threshold for rule 1
E' ← E                                               // store copy of all edges
for all points do                                    // apply rule 2
    if the point has ≥ 3 incoming edges in E' and 1 outgoing then
        if outgoing edge is > b times longer than median of the incoming then
            delete outgoing edge from E
        end if
    end if
end for
for all edges in E' do                               // apply rule 1
    if if edge length > weight threshold w_max then
        delete edge from E
    end if
end for
return G
```

The aim of rule 2 was to produce a correction yielding more clusters for large random uniform samples on highly multimodal functions, while not detecting more than 1.1 clusters on average in the case of unimodal functions [9, Sect. 4.5]. Apart from $b(N, n)$ being rather involved, there is another potential problem with the number of incoming edges: under the assumption of unique objective values for all solutions in $\mathscr{P}$, $\deg^-(v)$ cannot exceed the kissing number τ_n, the maximal number of non-overlapping unit hyperspheres in the Euclidean space that can be arranged such that they all touch another central unit hypersphere [2, p. 21]. In other words, τ_n is the highest number of points that can have a common nearest neighbor.

Proposition 1 *Let $\mathscr{P} = \{\mathbf{x}_1, \ldots, \mathbf{x}_N\}$, $N < \infty$, and $V = \{v_1, \ldots, v_N\}$ the corresponding nodes in the minimum spanning tree constructed by Algorithm 1. If $\forall i, j \in \{1, \ldots, N\}, i \neq j : f(\mathbf{x}_i) \neq f(\mathbf{x}_j)$, it holds for every node v_i that*

$$\deg^-(v_i) \leq \tau_n .$$

Proof The proof is by contradiction. Assume a point $\mathbf{x}_i$ has a set of neighbors $\mathscr{Q} \subset \mathscr{P}$, with $|\mathscr{Q}| > \tau_n$ and $\forall \mathbf{z} \in \mathscr{Q} : \mathrm{nbn}(\mathbf{z}, \mathscr{P}) = \mathbf{x}_i$, meaning that $\deg^-(v_i) = |\mathscr{Q}|$. From the definition of τ_n, it follows that at least two of these points must be closer to each other than $d_{\mathrm{nn}}(\mathbf{x}_i, \mathscr{Q})$. Let these points be $\mathbf{z}'$ and $\mathbf{z}''$. But according to our premise, we also have $f(\mathbf{z}') \neq f(\mathbf{z}'')$. Thus, one of the two must be the nearest-better neighbor of the other, in contradiction to our assumption that $\mathrm{nbn}(\mathbf{z}', \mathscr{P}) = \mathrm{nbn}(\mathbf{z}'', \mathscr{P}) = \mathbf{x}_i$.

In one dimension, the kissing number is two, so rule 2 in its current definition cannot be used for $n = 1$. Other exactly known values for the kissing number are $\tau_2 = 6$ and

$\tau_3 = 12$, which should be taken into account when changing the required number of incoming edges. Luckily, the number appears to grow exponentially (while the exact values are known only for a few other dimensions) [2, p. 23], so in higher dimensions, we are less restricted.

3 Related Research

Topographical selection (TS) by Törn and Viitanen [12] was developed in the context of global optimization and works by building a directed graph with edges to the k nearest neighbors of each point. Having the edges point towards the better solution, as defined in Ref. [15], the algorithm then selects all points whose nodes v in the graph have an outdegree $\deg^+(v) = 0$. In Ref. [15], TS was experimentally compared to NBC with the result that both showed a quite different behavior. TS performed better in low dimensions and with irregularly sampled points, while NBC seemed preferable in high dimensions. Both methods have a parameter indirectly controlling the number of selected points that needs to be defined by the user. NBC has the advantage that the number of selected points can alternatively be chosen directly because its selection criterion can be used to rank the solutions.

Maree et al. [7] employed NBC as part of a heuristic to determine the parameters of a Gaussian mixture model (GMM). Instead of calculating the nearest-better distances between individual solutions, their proposal is to calculate them between cluster centers. A cluster is regarded as better than another if at least one of its solutions is better than all solutions of the other one. Merging the two clusters with the smallest nearest-better distance in each iteration, an agglomerative clustering is then possible, in contrast to the divisive style of the original NBC. To choose the final clustering from the set of possible solutions, a GMM is fitted to the clusters and evaluated with a criterion called density-fitness rank correlation (DFC). DFC is based on the Spearman rank correlation between probability densities of the GMM and fitness ranks at the sampled solutions. The GMM is used for the variation in an estimation-of-distribution algorithm (EDA) for multimodal optimization. Note that while divisive NBC can be combined just as well with GMM, no systematic comparison between the divisive and agglomerative variant was presented in Ref. [7].

Recently, the idea of basin identification has also been transferred from multimodal optimization to multiobjective optimization by Braun et al. [1]. They show that applying NBC and local search to a scalarizing function of a multiobjective problem is a promising approach to obtain very small, but diverse solution sets. Due to the reduced amount of information, these solution sets are ideal for presenting them to human decision makers.

Basin identification is also loosely related to other research areas where diversity and quality of a subset play a role. One such area is drug discovery, where a diverse subset of promising molecules must be selected from a large database [8]. Another slightly related area is feature selection for machine learning, where irrelevant and redundant features of data sets need to be eliminated to improve the speed and quality of, e. g., classification models [4]. In both mentioned areas, forward selection algorithms with run time $O(mN)$ are favored, where m is the number of selected and N is the total number of solutions. These forward selection algorithms start with an empty set and add solutions sequentially according to a scalar objective aggregated from the two criteria of quality and diversity. As these algorithms are fast, but heavily parameter-dependent due to the scalarization, they should rather be used for cases with large m and N and not for our case of basin identification, where we usually want to carefully select at most a few dozen solutions.

4 Ideas for New Basin Identification Methods

Looking at Algorithm 1, the paradigm is clearly that deleting edges results in clusters, from which representatives may be selected to obtain a subset of solutions. However, this is merely a matter of perception. Alternatively, one could just as well select a number of solutions first and obtain a clustering by cutting their outgoing edges in the nearest-better graph. Figure 1 illustrates this mode of thought by showing two different clusterings of the same graph: the criterion used to select solutions is arbitrary, the clustering follows from the selected set and the nearest-better graph. Thus, we will mostly focus on the subset selection part for the rest of the chapter. Even for this simpler task, only a few alternative algorithms have been tried to date.

There was an impression early on that NBC selects too many solutions if the sample size is large and the dimension low [9, Sect. 4.4]. Therefore, a correction factor depending on these two variables was multiplied with ϕ initially [9, Sect. 4.4]. In Ref. [15], it was shown that the severity of the problem actually mostly depends on the variance of the nearest-neighbor distances in the point set. The reason for this effect is that spatial outliers also have extreme values in the distribution of nearest-neighbor distances. As it holds $d_{\text{nb}}(\mathbf{x}, \mathscr{P}) \geq d_{\text{nn}}(\mathbf{x}, \mathscr{P})$, these points tend to be selected regardless if they are close to a local optimum. An example for this effect

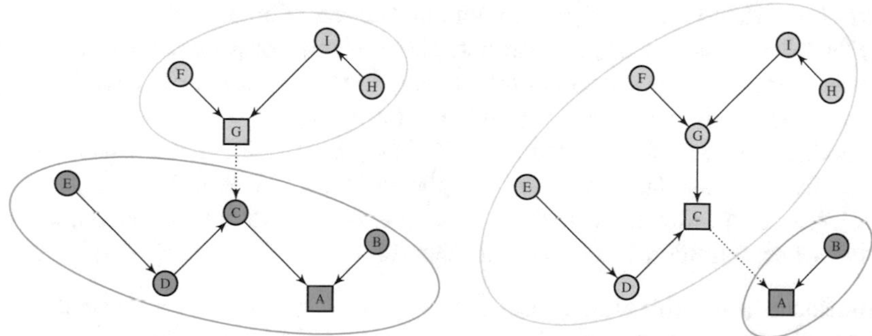

Fig. 1 Examples of how different selected nodes (marked as squares) induce different clusterings in a spanning tree by disregarding outgoing edges of selected nodes

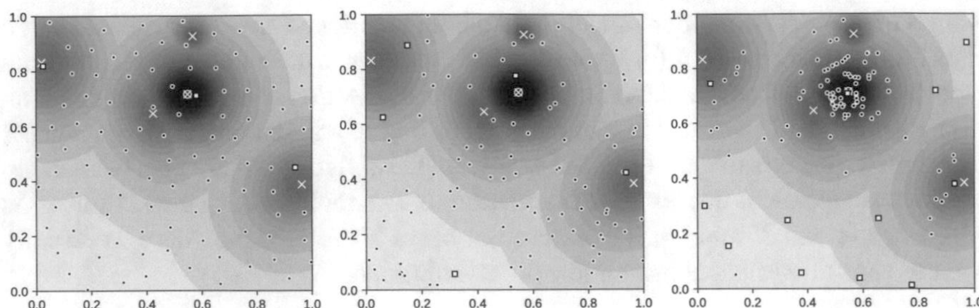

Fig. 2 Examples of nearest-better clustering on a 2D landscape, applied to $N = 100$ highly uniform, random uniform, and clustered points (from left to right). Points selected as potential basin centers are marked with white squares

can be found in Fig. 2, where the uniformity of the points decreases from left to right. What we would ideally like to see is the situation in the leftmost panel, where at least for the three large basins, exactly the best fitting point in the sample is selected. The used point set was generated by maximizing the smallest nearest-neighbor distance in the set with an algorithm called maximin reconstruction [15]. In the middle panel, we see random uniform data, and already two false positive classifications: in the basin on the left, two points are selected, and on the bottom of the panel, one of the worst points to start a local search from is selected. On the most clustered data set on the right, points were generated by a short run of a population-based optimization algorithm. NBC becomes fairly unusable here because the number of selected solutions (twelve) is far greater than the number of basins. This is despite the fact that the figure was generated with a high NBC threshold factor of $\phi \approx 2.41$, the value predicted by the trained NBC regression model from [15]. This model directly controls ϕ depending on n and N and replaced the cumbersome correction factor and rule 2. As Fig. 2 indicates this is not sufficient, we want to explore further ideas to deal with the weakness here.

A first idea is to restrict the selected points to those with at least one incoming edge, in other words, points that are the nearest-better neighbor to at least one other point. Given a fixed value of ϕ, it is obvious the number of points selected this way can only decrease versus the original variant. The restriction only affects spatial outliers because points near optima in densely sampled regions necessarily have incoming edges. We will call this variant NBC+ in the following.

Another idea was first presented in Ref. [14, Sect. 5.2]. It is loosely inspired by density-based spatial clustering of applications with noise (DBSCAN) by Ester et al. [5], more specifically its notion of *core points*. To describe core points, we need the following definition of an ϵ-neighborhood N_ϵ.

Definition 1 (ϵ-neighborhood) Let $B(\mathbf{x}, r) = \{\mathbf{y} \in \mathscr{X} \mid d(\mathbf{x}, \mathbf{y}) < r\}$ be the open ball of radius r around $\mathbf{x} \in \mathscr{X}$. We call the subset of $\mathscr{P}$ lying in $B(\mathbf{x}, \epsilon)$,

$$\begin{aligned} N_\epsilon(\mathbf{x}, \mathscr{P}) &= \mathscr{P} \cap B(\mathbf{x}, \epsilon) \\ &= \{\mathbf{y} \in \mathscr{P} \mid d(\mathbf{x}, \mathbf{y}) < \epsilon\}, \end{aligned}$$

the ϵ-neighborhood of $\mathbf{x}$.

In DBSCAN, a core point is a point $\mathbf{x}$, for which $|N_\epsilon(\mathbf{x}, \mathscr{P})| \geq k$ for some predefined threshold k. To employ this criterion for basin identification, we simply set $\epsilon = d_{\text{nb}}(\mathbf{x}, \mathscr{P})$. Informally, the criterion now favors solutions with many (necessarily worse) solutions that are closer than the nearest-better neighbor. Also for this criterion, spatial outliers are quite unlikely to obtain a good score.

We will assume that the point $\mathbf{x}$ is also counted in its nearest-better ball, so $1 \leq |N_{\text{nb}}(\mathbf{x}, \mathscr{P})| \leq |\mathscr{P}|$. For a given k, we simply have to calculate $|N_{\text{nb}}(\mathbf{x}, \mathscr{P})|$ for each point $\mathbf{x}$. After sorting these values, we can select the best solutions. As indicated in Fig. 1, a clustering of $\mathscr{P}$ can be obtained by using the selected solutions as seeds for the clusters and adding the remaining points according to where their outgoing edge in the nearest-better graph is pointing. So, this concept of reachability in the nearest-better graph is analogous to DBSCAN's density-reachability. We will call the whole approach nearest-better ball clustering (NBBC) to distinguish it from the existing NBC.

Finally, one can also try to combine the approaches from NBC and NBBC in one method. To do this, we normalize the measured nearest-better distances by the reference distance d_{ref} to get a mean value of one and multiply them with $|N_{\text{nb}}(\mathbf{x}, \mathscr{P})|$. The resulting figure can again be compared with some threshold k, which should be in the same order as for NBBC. We will call this variant NBBC+ in the following.

5 Experiments

All of the methods defined so far have one parameter heavily influencing how many solutions are selected. This is either the threshold for the number of neighbors k or the factor ϕ defining the threshold for nearest-better distances. So, to offer usable methods on a wide range of search space dimensions and sample sizes, we have to provide heuristic models suggesting how to set these parameters for a given input combination. We set up controlled experiments to fit appropriate regression models on training data and evaluate their performance on test data.

5.1 Determining Regression Models

Research Question How do good regression models for predicting the right parameters for all the proposed methods look?

Pre-Experimental Planning We loosely replicate the experiment in Ref. [15], which had the analogous task of finding regression models for NBC and TS. The range of possible values for the number of neighbors k and the threshold factor ϕ was refined during some preliminary runs of the experiment and is slightly larger than in Ref. [15].

Task As the identification of basins can be defined as a classification problem, we use the two simple measures precision and recall [6, p. 155] from the domain of information retrieval for evaluation. Recall has also become known as the *peak ratio* in the optimization community [11]. It uses the number of found optima

$$o = |\{\mathbf{x}^* \in \mathscr{O} \mid d_{\text{nn}}(\mathbf{x}^*, \mathscr{P}) \leq r\}|,$$

where $\mathscr{O}$ is the set of desired optima, $\mathscr{P}$ is the approximation set obtained by some algorithm, and r is a user-defined threshold parameter. The peak ratio is then $o/|\mathscr{O}|$, while precision is calculated as $o/|\mathscr{P}|$. By setting k and ϕ to extremely low values, one could get the methods to select almost all solutions and thus obtain arbitrarily high peak ratio values. Thus, we give precision a higher importance than peak ratio in our application and assess the performance in this experiment by a lexicographic ordering with these priorities. If there are several optimal configurations, we average their values of k or ϕ. Linear regression models are fitted to these averaged values of the optimal configurations by using ordinary least squares. The actual regression formula is chosen by manual experimentation considering the coefficient of determination R^2 and the model complexity as criteria.

Setup Table 1 shows the factors of our experiment. We create random artificial test instances with the generator for multimodal objective functions from [13] to be able to control many problem features. Only two factors are observable in the real world (problem dimension n and the number of points sampled N), so the regression models can use them as inputs. The number of local optima $\nu = |\mathscr{O}|$ and the global structure of the objective function are assumed to be non-observable in the real world. To have ground truth, we add all optima positions to the point samples. This way we can set $r = 0$ in the peak ratio. The remaining $N - \nu$ points are chosen by the respective sampling algorithm. We use the same sampling algorithms as in Ref. [15], maximin reconstruction (MmR) [15], the generalized Halton points (GHalton) [3], and random uniform points. These three sampling algorithms differ in the uniformity of the points sampled, which was shown to be an important factor for NBC [15] (cf. Fig 2). Note that the setup allows a few configurations with $N \leq \nu$. These are completely dismissed because the optimal behavior for them would be to select all solutions, which is, however, an unlikely scenario in the real world.

There are 192 different configurations of environmental (observable and non-observable) factors. For each of them, we make fifty stochastic replications using common random numbers. The experimental design is full-factorial on the top-level factors. Additionally, each basin identification method has one nested factor.

Results/Observations Figure 3 shows non-parametric regression models obtained from the combined data sets of all sampling methods. Figures for individual sampling algorithms are not shown here, because they look qualitatively similar, but they were

Table 1 Factors for the experiment in Sect. 5.1

Top-level factor	Type	Symbol	Levels
Problem topology	Non-observable		{random, funnel}
Number of local optima	Non-observable	ν	{5, 10, 20, 50}
Number of variables	Observable	n	{2, 3, 5, 10, 15, 20}
Number of points	Observable	N	$\{10n, 20n, 50n, 100n\}$
Sampling algorithm	Control		{SRS, GHalton, MmR}
Basin identification	Control		{NBC, NBC+, TS, NBBC, NBBC+}
Nested factor	Type	Symbol	Levels
Threshold factor	Control	ϕ	{1.1, 1.2, 1.3, 1.4, 1.5, 1.6, 1.8, 2.0, 2.25, 2.5, 2.75, 3.0, 3.5}
Number of neighbors	Control	k	$\{1, 2, \ldots, 14, 16, 18, 20, 25, 30, 35, 40, 50, 70, 90, 110, 130, 150\}$

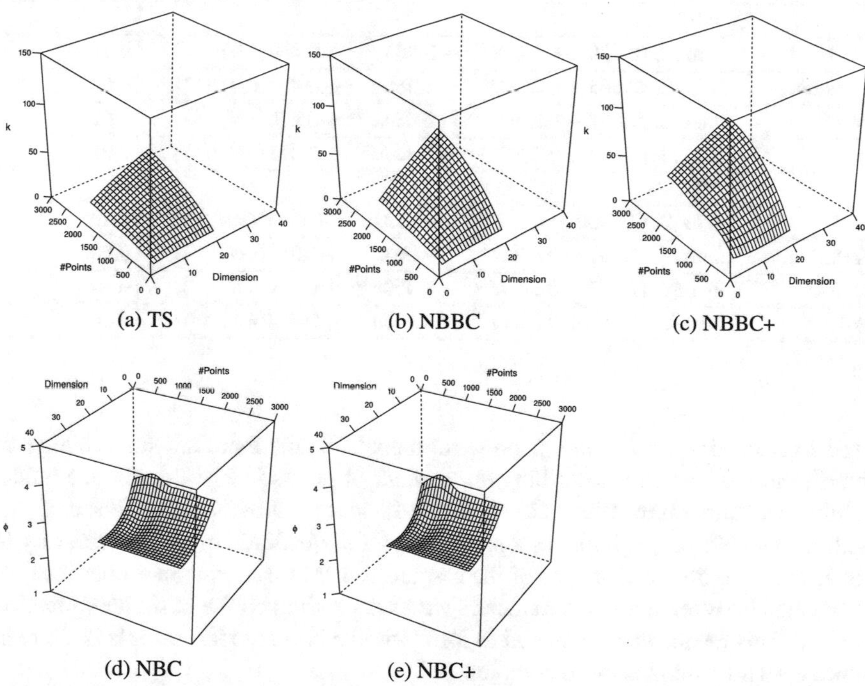

(a) TS (b) NBBC (c) NBBC+

(d) NBC (e) NBC+

Fig. 3 LOESS models for the k and ϕ values obtained from the experimental data

Table 2 The formulas of the parametric regression models

Sampling	Formula	R^2
	NBC	
SRS	$1.275 + 3.927n^{-1}$	0.76
GHalton	$1.312 + 2.774n^{-1}$	0.72
MmR	$1.366 + 0.979n^{-1}$	0.28
All	$1.318 + 2.560n^{-1}$	0.50
	NBC+	
SRS	$1.232 + 3.704n^{-1}$	0.69
GHalton	$1.274 + 2.527n^{-1}$	0.63
MmR	$1.325 + 0.928n^{-1}$	0.27
All	$1.277 + 2.386n^{-1}$	0.46
	TS	
SRS	$\max\{2, \lceil 5.626 + 0.075N^{1/2} - 2.433n^{1/2} + 0.269(nN)^{1/2}\rceil\}$	0.54
GHalton	$\max\{2, \lceil 5.315 - 0.046N^{1/2} - 2.373n^{1/2} + 0.292(nN)^{1/2}\rceil\}$	0.58
MmR	$\max\{2, \lceil 6.618 - 0.129N^{1/2} - 2.459n^{1/2} + 0.310(nN)^{1/2}\rceil\}$	0.56
All	$\max\{2, \lceil 5.853 - 0.033N^{1/2} - 2.421n^{1/2} + 0.290(nN)^{1/2}\rceil\}$	0.56
	NBBC	
SRS	$\max\{2, \lceil 8.850 - 0.152N^{1/2} - 3.905n^{1/2} + 0.466(nN)^{1/2}\rceil\}$	0.61
GHalton	$\max\{2, \lceil 7.505 - 0.266N^{1/2} - 3.568n^{1/2} + 0.483(nN)^{1/2}\rceil\}$	0.63
MmR	$\max\{2, \lceil 7.057 - 0.270N^{1/2} - 2.686n^{1/2} + 0.470(nN)^{1/2}\rceil\}$	0.61
All	$\max\{2, \lceil 7.804 - 0.230N^{1/2} - 3.387n^{1/2} + 0.473(nN)^{1/2}\rceil\}$	0.61
	NBBC+	
SRS	$\max\{2, \lceil 15.300 + 1.262N^{1/2} - 6.593n^{1/2} + 0.252(nN)^{1/2}\rceil\}$	0.49
GHalton	$\max\{2, \lceil 7.197 + 0.639N^{1/2} - 4.003n^{1/2} + 0.371(nN)^{1/2}\rceil\}$	0.58
MmR	$\max\{2, \lceil 5.252 + 0.227N^{1/2} - 2.430n^{1/2} + 0.439(nN)^{1/2}\rceil\}$	0.59
All	$\max\{2, \lceil 9.250 + 0.710N^{1/2} - 4.342n^{1/2} + 0.354(nN)^{1/2}\rceil\}$	0.54

used to determine appropriate parametric models. While there are small bumps in the direction of the axis describing the number of points in Fig. 3d and e, a model solely depending on the dimension seems sufficient for NBC variants. For methods with the number of neighbors as a parameter, a full quadratic model is necessary to cover the three obtained shapes of the response surface (see Fig. 3a–c) because the interaction between the two parameters varies depending on the basin identification method. This means that compared to Ref. [15], the NBC model formula is simpler, while the TS formula is more complex.

Table 2 contains the formulas of the found parametric models and corresponding coefficients of determination R^2 on the training data. In Fig. 4, one can see that for NBBC+, the sampling algorithm has some influence on the regression model, as would be expected due to the involvement of the nearest-better distances in the key

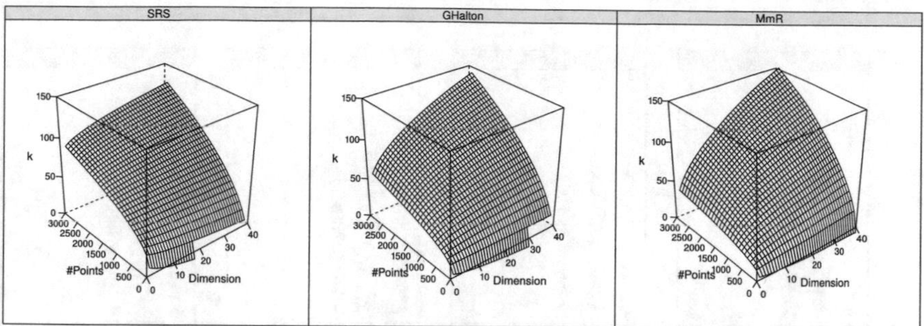

Fig. 4 Parametric regression models for NBBC+ obtained from the experimental data

Table 3 Changed factors for the validation experiment in Sect. 5.2

Factor	Type	Symbol	Levels
Number of local optima	Non-observable	ν	$\{3, 12, 25, 75\}$
Number of variables	Observable	n	$\{2, 4, 8, 16, 24, 32\}$

number computed there. For TS and NBBC, no influence was observed, and hence the corresponding figures are not shown.

Discussion For the NBC models, the R^2 values indicate a better fit of the models to the data for samples with higher irregularity. This may be because, for these samples, it is more important to choose a high enough threshold to avoid predicting false positives. The differences between NBC and NBC+ models are hardly visible (not shown in figures), but the formulas reveal slightly lower thresholds for NBC+, as would be plausible from the otherwise stricter selection criterion.

5.2 *Validation*

To validate the regression models from Table 2, we carry out another experiment using different random numbers and different values for ν and n than before, see Table 3. The optima are not included in the point sets in this experiment to provide a realistic scenario. Therefore, we now calculate the two performance measures from Sect. 5.1 with the number of found basins defined as

$$o = \sum_{i=1}^{|\mathscr{O}|} \min \left\{ 1, \sum_{j=1}^{|\mathscr{P}|} b(\mathbf{x}_j, \mathbf{x}_i^*) \right\},$$

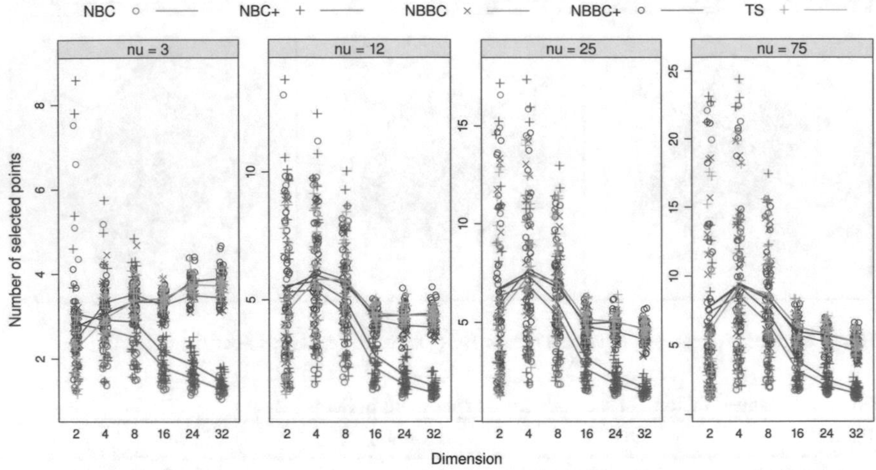

Fig. 5 The number of selected points for the tuned basin identification methods. Lines connect the mean values of the samples

where

$$b(\mathbf{x}, \mathbf{x}^*) = \begin{cases} 1 & \text{if } \mathbf{x} \in \text{basin}(\mathbf{x}^*), \\ 0 & \text{else} \end{cases}$$

is an indicator for membership in the attraction basin of optimum $\mathbf{x}^*$, realized as described in Ref. [13]. The sum $\sum_{j=1}^{|\mathscr{P}|} b(\mathbf{x}_j, \mathbf{x}_i^*)$ counts how many solutions are sampled in basin i, and the minimum term in the outer sum binarizes this value. Dividing by the number of existing basins, we again obtain the ratio of sampled basins as $o/|\mathscr{O}|$. With this setup, we let the methods again try to identify the basins and compare with results regarding precision and recall visually.

Figure 5 shows the number of selected points depending on dimension and number of optima. The two NBC variants select relatively many points in low dimensions n and relatively few for large n. A necessary condition for reaching Pareto-optimality regarding basin ratio and precision is that the number of selected solutions is somewhere between one and $\nu = |\mathscr{O}|$. So, it is already obvious from Fig. 5 that sometimes too many solutions are selected for $n \leq 4$ and $\nu \leq 12$. It is also surprising that NBC+ actually selects more points on the validation set than NBC. This indicates that the NBC+ models obtained in Sect. 5.1 may be chosen too optimistic. Thus, it might be advisable to use the NBC models also for NBC+, to obtain the desired stricter behavior. For the other methods, the effects of the dimension are not as pronounced and on problems with only $\nu = 3$ optima, the number of selected points even increases in higher dimensions. Interestingly, for $\nu > 3$, the behavior of all methods is non-monotonic in the dimension with a maximum of selected points at $n = 4$.

Figure 6 shows the performance measures precision and basin ratio plotted against each other. Higher values are better for both measures. The differences observed in

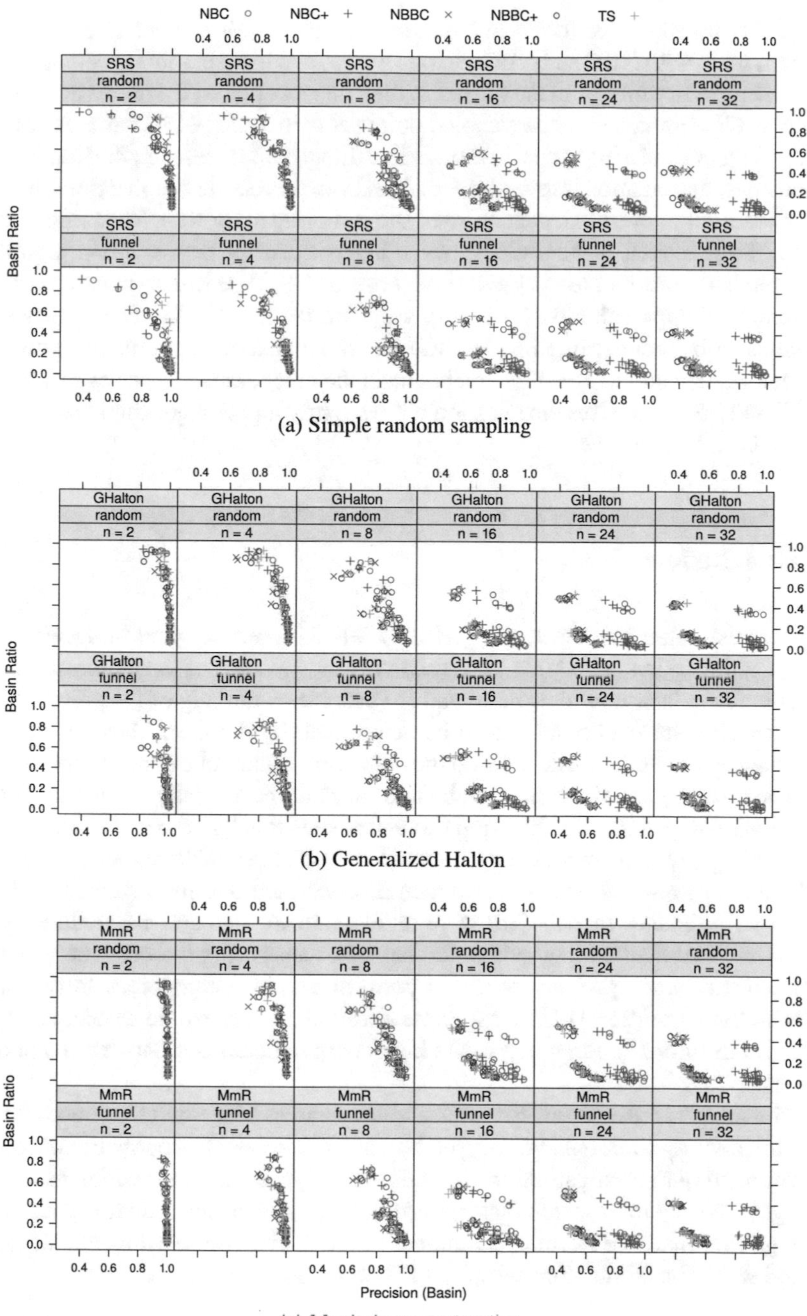

(a) Simple random sampling

(b) Generalized Halton

(c) Maximin reconstruction

Fig. 6 Precision and basin ratio on the validation set

Fig. 5 also transfer over to the two quality indicators. We can see two clusters of methods here, with NBC and NBC+ in one group, and the other three methods in the other one. The two groups mainly differ in their dependency on the dimension and the sampling algorithm. When the sampled point set is irregular, NBC variants obtain a lower precision and a higher basin ratio in low dimensions (see, e. g., Fig. 6a). In high dimensions, the situation is reversed, i. e., NBC variants obtain a higher precision and lower basin ratio, even independently of the sampling algorithm. Most importantly for us, the precision of NBBC is always independent of the sampling algorithm and overall its results are very close to those of TS. A surprise, however, is that the combined approach NBBC+ is also very similar to them. On two-dimensional problems with random sampling, NBBC+ has somewhat of an intermediate position between the two groups (see Fig. 6a), but in all the other cases, it behaves similarly to NBBC and TS. This raises the question if the slightly higher algorithmic complexity really adds a benefit here.

6 Conclusions

With nearest-better ball clustering (NBBC), we achieved our goal of developing a basin identification method that (a) is relatively independent of the uniformity of the points in the database used as input and (b) provides a ranking of the points so that the number of selected points could be determined by the user. This was done by inserting nearest-better distances into the core point notion of density-based clustering. The resulting metric could then be used to rank the available solutions. NBBC obtains almost identical results to topographical selection by Törn and Viitanen [12]. Parametric regression models can be used to predict suitable values for the one configuration parameter that all basin identification methods investigated here have. Previously published models [15] were refined and we also derived models for the new basin identification methods tested here. For answering the question of which approach and which parameters would perform best as a component in two-stage optimization algorithms [14], an experiment testing just that would be necessary. We leave that to future research due to the high computational cost incurred by adding local search algorithms.

Combining nearest-better distances and the nearest-better neighborhood did not seem to offer any additional benefit yet, but many more variants could be conceived. We restricted our investigations to methods that aggregate all information in a single key figure, but future research may also define multidimensional feature spaces and apply general machine learning techniques to them. For example, any classification method would be eligible for the subset selection task investigated here.

References

1. Braun, M., Heling, L., Shukla, P., Schmeck, H.: Multimodal scalarized preferences in multi-objective optimization. In: Proceedings of the Genetic and Evolutionary Computation Conference, GECCO '17, pp. 545–552. ACM, New York (2017)
2. Conway, J.H., Sloane, N.J.A.: Sphere Packings, Lattices and Groups, Grundlehren der mathematischen Wissenschaften, vol. 290. Springer, Berlin (1988)
3. De Rainville, F.M., Gagné, C., Teytaud, O., Laurendeau, D.: Evolutionary optimization of low-discrepancy sequences. ACM Trans. Model. Comput. Simul. **22**(2), 9:1-9:25 (2012)
4. Ding, C., Peng, H.: Minimum redundancy feature selection from microarray gene expression data. J. Bioinform. Comput. Biol. **3**(2), 185–205 (2005)
5. Ester, M., Kriegel, H.P., Sander, J., Xu, X.: A density-based algorithm for discovering clusters in large spatial databases with noise. In: Proceedings of the Second International Conference on Knowledge Discovery and Data Mining, KDD'96, pp 226–231. AAAI Press, Menlo Park (1996)
6. Manning, C.D., Raghavan, P., Schütze, H.: Introduction to Information Retrieval. Cambridge University Press, New York (2008)
7. Maree, S.C., Alderliesten, T., Thierens, D., Bosman, P.A.N.: Niching an estimation-of-distribution algorithm by hierarchical gaussian mixture learning. In: Proceedings of the Genetic and Evolutionary Computation Conference, GECCO'17, pp. 713–720. ACM, New York (2017)
8. Meinl, T., Ostermann, C., Berthold, M.R.: Maximum-score diversity selection for early drug discovery. J. Chem. Inf. Model. **51**(2), 237–247 (2011)
9. Preuss, M.: Multimodal Optimization by Means of Evolutionary Algorithms. Springer, Berlin (2015)
10. Preuss M, Schönemann, L., Emmerich, M.: Counteracting genetic drift and disruptive recombination in (μ +/, λ)-EA on multimodal fitness landscapes. In: Proceedings of the 2005 Conference on Genetic and Evolutionary Computation, GECCO'05, pp. 865–872. ACM, New York (2005)
11. Thomsen, R.: Multimodal optimization using crowding-based differential evolution. IEEE Congress Evol. Comput. (CEC) **2**, 1382–1389 (2004)
12. Törn, A., Viitanen, S.: Topographical global optimization. In: Floudas, C.A., Pardalos, P.M. (eds.) Recent Advances in Global Optimization. Princeton Series in Computer Sciences, pp. 384–398. Princeton University Press, Princeton, NJ (1992)
13. Wessing, S.: The multiple peaks model 2. Algorithm Engineering Report TR15-2-001, Technische Universität Dortmund. https://ls11-www.cs.uni-dortmund.de/_media/techreports/tr15-01.pdf (2015a)
14. Wessing, S.: Two-stage methods for multimodal optimization. Ph.D. thesis, Technische Universität Dortmund (2015b)
15. Wessing, S., Rudolph, G., Preuss, M.: Assessing basin identification methods for locating multiple optima. In: Pardalos, P.M., Zhigljavsky, A., Žilinskas, J. (eds.) Advances in Stochastic and Deterministic Global Optimization, pp. 53–70. Springer, Berlin (2016)

Deflection and Stretching Techniques for Detection of Multiple Minimizers in Multimodal Optimization Problems

Konstantinos E. Parsopoulos and Michael N. Vrahatis

Abstract Multimodal optimization refers to problems where the detection of many local or global minimizers is desirable. A number of methodologies have been developed in the past decades for this purpose. Deflection and stretching are two techniques that can be integrated with any optimization algorithm in order to detect multiple minimizers by properly transforming the objective function. Requiring only a minimal number of control parameters, both techniques have been used with metaheuristics as well as gradient-based optimization algorithms, enhancing their performance while demanding only minor implementation effort. Up until now their applications span various scientific fields, ranging from game theory and numerical optimization to astrophysics, computational intelligence, and medical informatics. The present chapter offers a comprehensive presentation of the two techniques and demonstrates their use through simple examples. Also, their latest developments and applications of the past two decades are concisely reviewed.

1 Introduction

The ongoing scientific and technological developments produce challenging problems that often involve the optimization of multimodal functions with numerous local and global optimizers. Without loss of generality we will henceforth refer only to minimization cases, while maximization can be straightforwardly addressed by changing the sign of the objective function. Even in relatively simple cases, the multimodal minimization problem can become NP-hard [8]. In several applications of

K. E. Parsopoulos (✉)
Department of Computer Science and Engineering, University of Ioannina,
GR-45110 Ioannina, Greece
e-mail: kostasp@cse.uoi.gr

M. N. Vrahatis
Computational Intelligence Laboratory (CI Lab), Department of Mathematics,
University of Patras, GR-26110 Patras, Greece
e-mail: vrahatis@math.upatras.gr

M. Preuss et al. (eds.), *Metaheuristics for Finding Multiple Solutions*,
Natural Computing Series,
https://doi.org/10.1007/978-3-030-79553-5_6

this type, it is required to detect some or all global and/or local minimizers. A typical example coming from game theory is the detection of Nash equilibria that correspond to steady-state solutions of a game [21]. Different Nash equilibria correspond to different outcomes of the game. Thus, the outcome of the game can be predicted if all Nash equilibria are known. Another interesting application is the detection of periodic orbits of nonlinear mappings, which are used to model conservative or dissipative dynamical systems. Fixed points, i.e., points that remain invariant by the mapping, concentrate scientific interest. Developing techniques for detecting multiple fixed points has been an area of ongoing research for many decades [27].

The past decades have witnessed the development of various efficient and effective optimization algorithms. The introduction of methods aiming at the detection of multiple minimizers either through specialized ad hoc procedures (operators) or by using external techniques integrated with the corresponding algorithms is of utmost importance. *Niching* is a term used in metaheuristics literature to describe methods that maintain multiple solutions in multimodal domains, in contrast to existing evolutionary and swarm intelligence optimization techniques that have been designed to detect single solutions. An excellent review of niching techniques is offered by Li et al. in [10].

The present chapter is devoted to the *deflection* and *stretching* techniques [20], which were introduced as mathematical tools for addressing two different problems in multimodal optimization. Specifically, deflection is designed to facilitate the detection of many (local or global) minimizers, while stretching aims at concurrently alleviating many local minimizers. Both techniques are based on the filled functions approach where the original objective function is transformed after the detection of a new minimizer. The applied transformation aims at eliminating the targeted minimizer (and possibly some other minimizers) by transforming it to a maximizer.

Requiring only a minimal number of control parameters and minor implementation effort, both deflection and stretching have been successfully used with metaheuristics as well as gradient-based optimization algorithms. Up until now their applications span various scientific fields, ranging from game theory and numerical optimization to astrophysics, computational intelligence, and medical informatics.

The rest of the chapter is organized as follows: Sect. 2 is devoted to deflection and its applications, while Sect. 3 analyzes stretching and presents a number of recent variants and applications. Indicative experimental results are offered for both techniques in Sect. 4. The paper closes with conclusions in Sect. 5.

2 Deflection Technique

In the following paragraphs, the basic deflection scheme is presented and demonstrated on a well-known test function, followed by a review of recent applications.

2.1 Basic Scheme

The deflection technique has been studied in Magoulas et al. [13] for the alleviation of local minima in artificial neural networks training with the backpropagation algorithm. Later it was adopted for the detection of multiple minimizers with the particle swarm optimization algorithm by Parsopoulos and Vrahatis [19, 20]. It belongs to the category of filled functions techniques [5, 6], where the original objective function is transformed into a new one where the already detected minimizers are eliminated. This way, optimization algorithms are driven away from the detected minimizers, prohibiting their repetitive detection in subsequent runs.

Putting it formally, let:

$$f : S \subset \mathbb{R}^n \rightarrow \mathbb{R}, \tag{1}$$

be the original objective function under consideration. Let us make the necessary assumption that $f(\mathbf{x})$ is bounded from below, and let:

$$M = \left\{ \mathbf{x}_1^*, \mathbf{x}_2^*, \ldots, \mathbf{x}_m^* \right\},$$

be a set comprising already detected minimizers. Then, deflection replaces $f(\mathbf{x})$ with a transformation of the following type:

$$F_M(\mathbf{x}) = \frac{f(\mathbf{x})}{T_1(\mathbf{x}; \mathbf{x}_1^*, \lambda_1) T_2(\mathbf{x}; \mathbf{x}_2^*, \lambda_2) \cdots T_m(\mathbf{x}; \mathbf{x}_m^*, \lambda_m)}, \tag{2}$$

where $\mathbf{x}_i^* \in M$ for all $i = 1, 2, \ldots, m$, and $T_i(\mathbf{x}; \mathbf{x}_i^*, \lambda_i)$ are proper functions with control parameters $\lambda_i \in \mathbb{R}$, respectively. These functions shall be ideally selected such that $F_M(\mathbf{x})$ has exactly the same minimizers as $f(\mathbf{x})$ except for the already detected ones in M. This property is called the *deflection property* [13] and dictates that any sequence of points converging to a detected minimizer:

$$\lim_{k \to \infty} \mathbf{x}_k = \mathbf{x}_i^*, \quad \mathbf{x}_i^* \in M,$$

shall not produce a minimum of $F_M(\mathbf{x})$ at $\mathbf{x} = \mathbf{x}_i^*$. It shall be noticed that the product in the denominator of Eq. (2) consists of one term for each detected minimizer. Thus, the optimization algorithm initially starts with the set $M = \emptyset$, and the original function $f(\mathbf{x})$. Then, as soon as a minimizer is detected, the corresponding term is multiplied with the current denominator, producing a new transformation that is subsequently used. Thus, the form of the deflection transformation dynamically changes during the run of the algorithm, depending on the detected minimizers.

A critical point is the magnitude of the transformation induced by each function $T_i(\mathbf{x}; \mathbf{x}_i^*, \lambda_i)$. Although it is desirable to eliminate $\mathbf{x}_i^*$ from the set of minimizers of $F_M(\mathbf{x})$, it is equally important to confine the effect of the transformation only to a neighborhood of $\mathbf{x}_i^*$ that ideally covers only its basin of attraction. For this reason, the use of tunable functions $T_i(\mathbf{x}; \mathbf{x}_i^*, \lambda_i)$ is necessary.

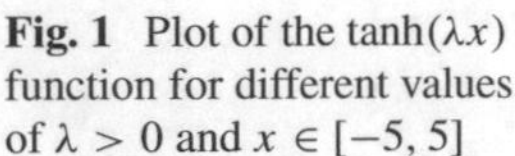

Fig. 1 Plot of the $\tanh(\lambda x)$ function for different values of $\lambda > 0$ and $x \in [-5, 5]$

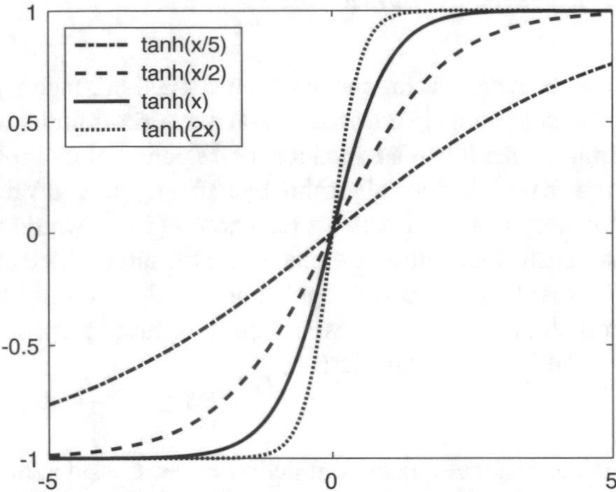

As proposed in [19], the hyperbolic functions:

$$T_i(\mathbf{x}; \mathbf{x}_i^*, \lambda_i) = \tanh(\lambda_i \|\mathbf{x} - \mathbf{x}_i^*\|), \quad \lambda_i > 0, \tag{3}$$

fulfill the aforementioned properties, with $\| \cdot \|$ denoting a distance norm (typically the Euclidean ℓ_2-norm). Figure 1 illustrates the shape of the $\tanh(\lambda x)$ function for different values of its parameter. Focusing on the nonnegative values of x, we can clearly see that the function is equal to zero for $x = 0$, while it asymptotically converges to 1 for higher values. Therefore, the effect of $T_i(\mathbf{x}; \mathbf{x}_i^*, \lambda_i)$ of Eq. (3) in the deflection transformation $F_M(\mathbf{x})$ of Eq. (2) is expected to be huge when $\mathbf{x}$ lies close to $\mathbf{x}_i^*$ because, as $\|\mathbf{x} - \mathbf{x}_i^*\|$ approximates zero, it produces a singularity point of $F_M(\mathbf{x})$ at $\mathbf{x}_i^*$. On the other hand, the effect becomes milder as $\mathbf{x}$ moves away from $\mathbf{x}_i^*$. The magnitude of the transformation is essentially controlled by the parameter λ.

It can be easily derived from Eq. (2) that the sign of $f(\mathbf{x})$ plays crucial role. Indeed, the proposed deflection approach can work properly only if $f(\mathbf{x}) > 0$ for all $\mathbf{x} \in S$. For this reason, we will henceforth assume that this property holds by definition for the problem under consideration. In cases where this property does not hold or it is unclear whether the objective function is strictly positive everywhere in the search space, the user can enforce it by simply using a shift-up transformation of $f(\mathbf{x})$ as follows:

$$f_\alpha(\mathbf{x}) = f(\mathbf{x}) + \alpha, \tag{4}$$

where $\alpha > 0$ is a sufficiently large positive constant. Possible estimation (e.g., through Monte Carlo sampling) of a lower bound of $f(\mathbf{x})$ can be very useful to this end. Otherwise, arbitrarily large positive values of α can be used.

Let us now demonstrate the deflection technique on a well-known 2-dimensional test problem, namely, the Levy no. 5 test function, defined as:

$$f(\mathbf{x}) = \zeta_1(\mathbf{x})\, \zeta_2(\mathbf{x}) + (x_1 + 1.42513)^2 + (x_2 + 0.80032)^2, \tag{5}$$

where:

$$\zeta_1(\mathbf{x}) = \sum_{i=1}^{5} i \cos((i-1)x_1 + i), \qquad \zeta_2(\mathbf{x}) = \sum_{j=1}^{5} j \cos((j+1)x_2 + j).$$

In the range $[-10, 10]^2$ the Levy no. 5 test function has 760 local minima. We now focus on the range $[-2, 2]^2$, which is illustrated in Fig. 2. The global minimizer of the function in this range is $\mathbf{x}^* = (-1.3069, -1.4248)^T$, denoted with a black star in the lower left basin of the contour plot of Fig. 2, with $f(\mathbf{x}^*) = -176.1376$. Since this objective function is not always positive, we transformed it according to Eq. (4) with $\alpha = 180$. Also, we shall note that its local minima become deeper as we approximate the global minimizer.

In order to demonstrate the deflection technique, we selected a number of minimizers to apply the deflection transformation. Figure 3 illustrates the deflection transformation applied on the local minimizer $\mathbf{x}_1 = (-0.3521, -0.8003)^T$ with $f(\mathbf{x}_1) = -144.3250$ for $\lambda = 1$ (up) and $\lambda = 2$ (down). Apparently, the effect of the deflection transformation is ameliorated as λ increases, while smaller values have wider influence affecting also neighboring maxima. Figure 4 illustrates the concurrent application of deflection on the three local minimizers $\mathbf{x}_1 = (-0.3521, -0.8003)^T$, $\mathbf{x}_2 = (-1.3069, -0.1957)^T$, and $\mathbf{x}_3 = (-0.3557, 0.3330)^T$, for $\lambda = 2$ (up) and $\lambda = 1$ (down). Deflection applied on multiple minimizers becomes increasingly beneficial because it replaces the detected minimizers with maximizers that can repel the minimization algorithms toward undiscovered minimizers.

A critical observation is that higher values of λ may have only local effect but they tend to introduce new local minima in the neighborhood of the deflected point. This is the well-known *mexican hat effect*; addressing it remains an open problem when the basins of attraction of the minimizers are unknown. Nevertheless, the newly introduced local minima are always higher than the deflected ones and, thus, they can be explicitly neglected by a stochastic algorithm. This can be achieved by using the *repulsion* technique proposed and integrated with deflection in [19] for the particle swarm optimization algorithm.

According to this approach, whenever a particle $\mathbf{x}$ moves into the repulsion area of a deflected minimizer $\mathbf{x}_i^*$, which is defined as a sphere of radius $\rho_i > 0$ around the minimizer, it is immediately repelled away by assuming the new position:

$$\mathbf{x} := \mathbf{x} + \rho_i \frac{\mathbf{x} - \mathbf{x}_i^*}{\|\mathbf{x} - \mathbf{x}_i^*\|}.$$

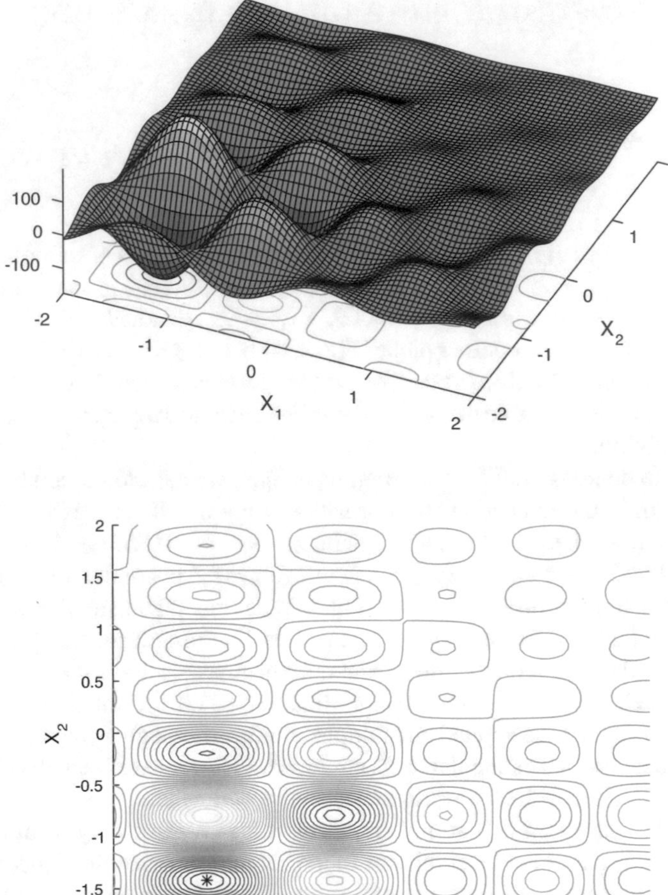

Fig. 2 Surface plot (up) and contour plot (down) of the Levy no. 5 test function of Eq. (5) in the range $[-2, 2]^2$

Again, the problem is the determination of ρ_i, which shall be large enough to prohibit the algorithm from converging close to $\mathbf{x}_i^*$ but, at the same time, avoiding enclosing other minimizers in its repulsion area. Possible knowledge regarding the distribution of the minimizers in the search space may be beneficial for a proper setting. Otherwise, small fixed values of ρ_i are preferable.

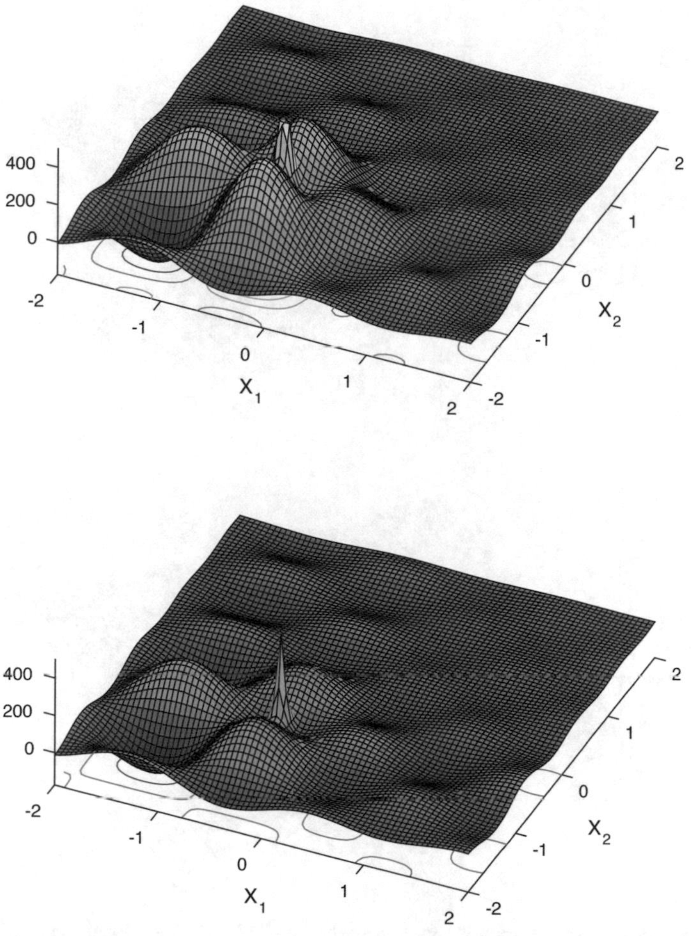

Fig. 3 Deflection transformation applied on the local minimizer $\mathbf{x}_1 = (-0.3521, -0.8003)^T$ for $\lambda = 1$ (up) and $\lambda = 2$ (down)

2.2 *Variants and Applications*

The deflection technique proved to be beneficial for various algorithms in diverse scientific fields. Important applications have appeared in computational intelligence, game theory, and astrophysics, including optimization, partitioning, Nash equilibria, and periodic orbit problems.

The alleviation of local minima in artificial neural network training was the inaugural application of deflection in Magoulas et al. [13] in 1997. A few years later in 2004, Parsopoulos and Vrahatis [19] employed deflection for the enhancement of the particle swarm optimization algorithm. Specifically, it was used for the detection of all minimizers in various problems, including multimodal optimization test problems,

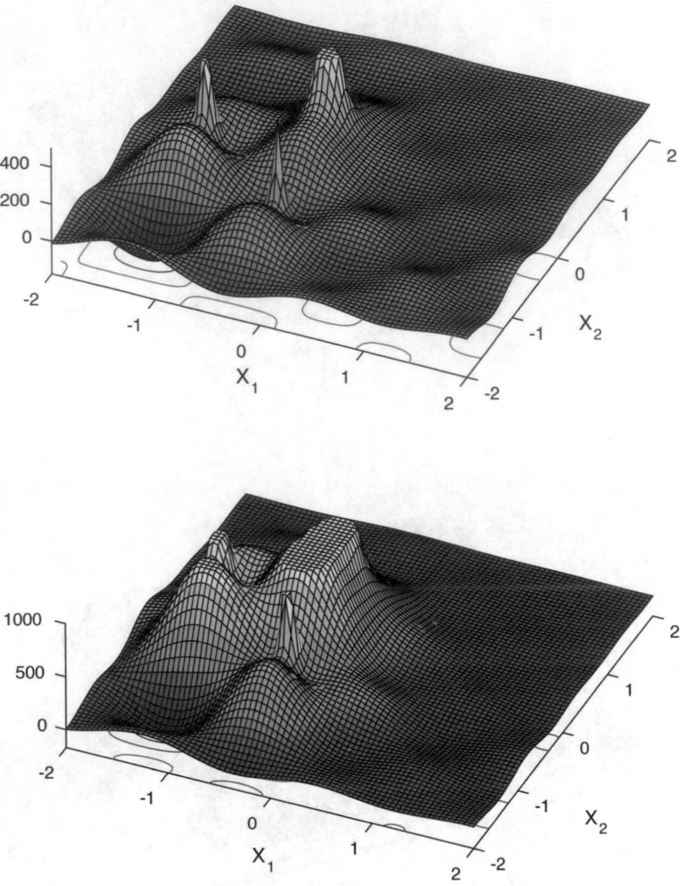

Fig. 4 Deflection transformation concurrently applied on the local minimizers $\mathbf{x}_1 = (-0.3521, -0.8003)^T$, $\mathbf{x}_2 = (-1.3069, -0.1957)^T$, and $\mathbf{x}_3 = (-0.3557, 0.3330)^T$, for $\lambda = 2$ (up) and $\lambda = 1$ (down)

and periodic orbits of nonlinear mappings. The obtained results were impressive for two variants of the studied algorithm.

Motivated by the success of deflection, Pavlidis et al. [21] proposed in 2005 the computation of multiple Nash equilibria in finite strategic games from the field of game theory through computational intelligence methods. Covariance matrix adaptation evolution strategies, differential evolution, and particle swarm optimization equipped with deflection were tested on formal games of 3 up to 5 players with very promising results.

In 2006, Gao and Tong [4] proposed UEAS, a space contraction technique that employed deflection (and stretching as well) in a general iterative optimization scheme that was successfully combined with the differential evolution algorithm.

The experimental setup followed closely that of Parsopoulos and Vrahatis [19], considering similar test problems. In the same year, Zhang and Liu [36] followed the pioneering work in [21]. Specifically, n-players strategic games were considered and deflection was used for learning all Nash equilibria by deflecting the players' payoff function. Experimental results for games with 2 and 3 players revealed the promising behavior of the proposed approach.

Tong et al. extended in 2008 the application range of the UEAS approach to the hardware/software partitioning problem [32]. Deflection was adopted as a crucial part of the search procedure for the detection of multiple partitioning configurations aiming at the minimization of implementation time and cost. In the same year, Zhang and Fan [35] extended the previous work of [36] for learning Nash equilibria through an adaptive policy gradient employing deflection as the mechanism for detecting multiple equilibria.

The ability of deflection in detecting global minimizers was further verified in 2009 by Li et al. [9]. In this case, the chaotic ant swarm was used as the main search algorithm. The dynamics of the algorithm were analyzed on several test problems, showing high ability of detecting all minimizers. In the same year, the previous applications of deflection-equipped algorithms on the detection of Nash equilibria were extended by Liu and Dumitrescu [12]. In their work, a new metaheuristic, namely, roaming optimization, was considered along with differential evolution and particle swarm optimization. Experimental results verified that deflection was beneficial also for the new algorithm.

Chaos control and detection of periodic orbits of chaotic systems were investigated by Gao et al. [3] in 2012. This work followed the basic analysis of the pioneering work in [19] and considered also different chaotic systems. Also, a zooming procedure proved to be beneficial when combined with deflection for the specific application.

Finally, deflection recently constituted an essential mechanism for the production of training samples in a computationally efficient approach based on Gaussian process regression to assess the accessibility in the main-belt asteroids, as proposed in 2017 by Shang and Liu [31]. Numerical simulations demonstrated that the proposed method could achieve significant results with minimal time requirements.

3 Stretching Technique

In the following paragraphs, the basic stretching scheme is presented followed by a number of applications.

3.1 Basic Scheme

The stretching technique was originally introduced by Parsopoulos et al. [15] as a tool for the alleviation of local minima in metaheuristic optimization. It is based on the same essential properties of filled functions as the deflection technique, although its use is recommended in different cases. Specifically, stretching shall be used in environments with large number of local minimizers because it can eliminate in one application a given local minimizer as well as all higher minimizers, while leaving lower minimizers unaffected.

Making our description more concrete, let us assume again the problem of Eq. (1) with the accompanying assumption of a lower bounded objective function $f(\mathbf{x})$. Then, given a detected local minimizer $\mathbf{x}^*$, stretching consists of the following two transformations:

$$G(\mathbf{x}) = f(\mathbf{x}) + \gamma_1 \|\mathbf{x} - \mathbf{x}^*\| \left(\operatorname{sign}\left(f(\mathbf{x}) - f(\mathbf{x}^*)\right) + 1\right), \tag{6}$$

$$H(\mathbf{x}) = G(\mathbf{x}) + \gamma_2 \frac{\operatorname{sign}\left(f(\mathbf{x}) - f(\mathbf{x}^*)\right) + 1}{\tanh\left(\lambda\left(G(\mathbf{x}) - G(\mathbf{x}^*)\right)\right)}, \tag{7}$$

where $\gamma_1, \gamma_2 > 0$, are control parameters of the transformations; $\lambda > 0$ is the control parameter of the hyperbolic tangent function; and $\operatorname{sign}(\cdot)$ is the well-known three-valued sign function defined as:

$$\operatorname{sign}(x) = \begin{cases} -1, & x < 0, \\ 0, & x = 0, \\ 1, & x > 0. \end{cases}$$

The transformation $G(\mathbf{x})$ of Eq. (6) is responsible for "stretching" the objective function upwards, while leaving unaffected all points of lower function values than the detected minimizer. This way, all local minima that are higher than the detected one are eliminated. Then, the transformation $H(\mathbf{x})$ of Eq. (7) transforms the detected minimizer to a maximizer, similarly to the deflection approach. Thus, there is double gain of eliminating not only the detected minimizer but also all minimizers with higher values as well.

A direct consequence of this property is that, contrary to the previously presented deflection technique, stretching shall be applied only on the best local minimizer found so far. This property also renders stretching inappropriate for detecting all minimizers of the objective function because it eliminates multiple local minima in one application. Moreover, if it is applied on a global minimizer it eliminates also all the other global minimizers.

The mexican hat effect is present also in stretching, introducing artificial local minimizers around the targeted one. However, their depth can be controlled through the control parameters of the technique. Let us consider again the Levy no. 5 function of Eq. (5). The transformation $G(\mathbf{x})$ of Eq. (6) applied on the local minimizer $\mathbf{x} = (-0.3572, 1.3166)^T$ is depicted in the upper part of Fig. 5 for $\gamma_1 = 200$. We can clearly observe that the parts of the objective function with values equal or greater than the stretched minimizer are widely affected, while lower minima remain unaffected appearing as holes on the stretched landscape. Also, we can see that the targeted minimizer remains a minimizer after this transformation.

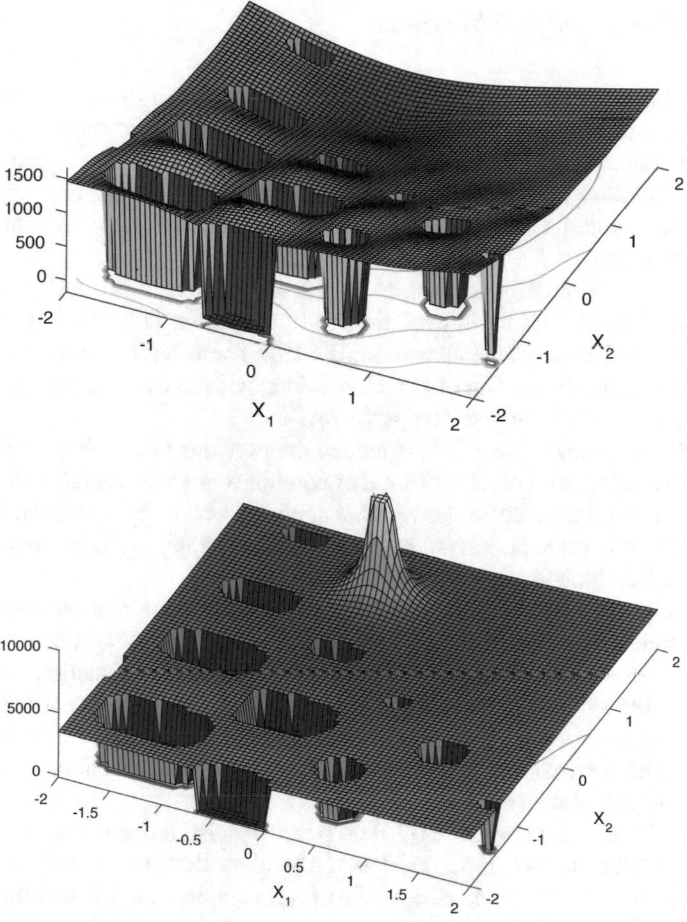

Fig. 5 Stretching transformation $G(x)$ (up) and $H(x)$ (down) applied on the local minimizer $\mathbf{x} = (-0.3572, 1.3166)^T$ for $\gamma_1 = \gamma_2 = 50$ and $\lambda = 0.05$

Retaining $\gamma_1 = 200$, the transformation $H(\mathbf{x})$ of Eq. (7) applied on the same local minimizer is depicted in the lower part of Fig. 5 for $\gamma_2 = 1000$ and $\lambda = 0.005$. Evidently, this transformation turns the targeted minimizer into a maximizer while leaving all the deeper minima unaffected. The parameters γ_2 and λ have a direct effect on the shape of the transformed objective function around the minimizer. Smaller values of γ_2 may produce the mexican hat effect. On the other hand, λ has the opposite effect, namely smaller values raise the maximizer higher, thereby ameliorating the mexican hat effect. Taking into consideration the interesting interplay of the two parameters, an inexperienced user may find it easier to keep the one parameter constant to a reasonable value and tune the other one.

3.2 Variants and Applications

Stretching has been combined with both gradient-based and stochastic optimization algorithms, including trajectory-based and population-based methods. Its flexibility and effectiveness has established it as a viable approach in the relevant literature. So far applications have appeared in software/hardware partitioning, global optimization, semi-infinite programming, red blood cell motion, clustering, and process control problems.

Stretching was initially introduced by Parsopoulos et al. in 2001 [15] for the alleviation of local minima in evolutionary optimization [14, 16, 17], and it was further extended in subsequent works [18, 19]. These works provided a general framework for the application of stretching along with an experimental justification based on a number of standard test problems.

In 2007, Wang and Zhang [33] extended the previous pioneering works by considering stretching with an algorithm that combines gradient search and simulated annealing. Their experiments showed that applying stretching on a local minimizer detected through gradient descent and, subsequently, applying simulated annealing on the stretched function can be highly beneficial.

Later, in 2008, Tong et al. successfully applied a scheme that combined deflection and stretching on the hardware/software partitioning problem in 2008 [32]. In the same year, Wang and Zhang [34] proposed an auxiliary function method that implements the stretching transformation such that it remains descending wherever the function values are higher than the stretched minimum. This was a step toward addressing the mexican hat effect described in the previous sections. This method was later integrated with memetic algorithms in [2].

In 2009, Pereira and Fernandes [26] proposed a global reduction method for solving semi-infinite programming problems. The algorithm was based on simulated annealing, which incorporated a sequence of local applications of stretching to compute the solutions of the lower-level problem. A penalty method was also used for the approximation of the solution of the finite reduced optimization problem, along with a globalization procedure based on line search to guarantee a sufficient decrease of the underlying merit function.

The same authors presented in [22] and later extended in [23] the aforementioned method of simulated annealing with stretching to solve constrained multi-global optimization problems, where all global solutions are sequentially computed by applying stretching with an adaptive simulated annealing variant, while constraint-handling is carried out through a non-differentiable penalty function.

The stretched simulated annealing approach was used also by Pinho et al. [28] for the characterization of the behavior of red blood cell motion through a glass microchannel. Specifically, the radial displacement of forty red blood cells was considered and different functions were used to approximate their displacement by means of global optimization.

Ribeiro et al. [29] proposed the first attempt of applying stretching in a parallel computing environment. Specifically, they considered the problem of solving multilocal programming problems through a parallel stretched simulated annealing, which is based on partitioning the search space. This way, the resolution of the search domains is increased, facilitating the discovery of new solutions while retaining time efficiency. This approach was further analyzed in [24] and extended for the constrained multi-local optimization problem in [25, 30].

In 2013, Lu et al. [11] employed a gradient-based update procedure for the parameters of the kernel-based fuzzy c-means algorithm and tackled the local minima problem by using stretching. Experiments on both artificial and real-world datasets showed that the stretched algorithm with optimized kernel parameters was superior to other competing algorithms.

He et al. [7] proposed in 2014 a combination of stretching with the simulated annealing algorithm, which is iteratively applied on the stretched function, while its trial points generation scheme is especially designed to promote diversity. Numerical results from a large number of test problems suggest that the hybrid method is effective for global optimization.

In 2015, Drąg and Styczeń [1] applied stretching for the control and optimization of multistage technological processes. In their study, the multistage differential-algebraic constraints with unknown consistent initial conditions were considered. The infinite-dimensional optimal control problem was transformed into a finite-dimensional optimization task through the direct shooting method, while simulated annealing with stretching was successfully used to solve the corresponding constrained optimization problem.

4 Experimental Evaluation

Table 1 reports experimental results from the application of deflection with the constriction coefficient version of particle swarm optimization on several test problems, as well as on the problem of detecting periodic orbits of the Hénnon nonlinear map-

Table 1 Numerical results from the application of deflection on various test problems reported in [19]

Problem	Range	Minimizers	Average iterations total/Per minimizer
Egg-holder (TP_{UO-21} in [20])	$[-5, 5]^2$	12/12	315 / 26.2
Levy no. 3	$[-10, 10]^2$	10	2913 / 291.3
Levy no. 5	$[-10, 10]^2$	2	3026.5 / 1036.5
Hénnon mapping ($\cos a = 0.24$)	$[-1, 1]^2$	11/11	249 / 22.6
Hénnon mapping ($\cos a = 0.8$)	$[-1, 1]^2$	2/2	21 / 10.5
Battle of sexes game		3	52 / 17.3

ping and the detection of Nash equilibria for the "battle of sexes" game, as they appear in [19]. For the complete definition of each problem and the exact experimental configuration the reader is referred to [19] and citations therein.

For the case of the egg-holder function and the two versions of the Hénnon mapping, the target was the detection of all global minimizers. Therefore, deflection was applied in combination with the repulsion strategy proposed in [19] as a simple mean to ameliorate the consequences of the mexican hat effect by repelling the particles from the deflected minimizers.

For the rest of the problems, stretching was used to eliminate the multitude of local minima in the ranges of interest. It shall be noted that the two Levy-family problems contain hundreds of local minimizers. Stretching was applied after the detection of the first local minimizer, allowing the algorithm to eventually converge to the global minimizer. Similar performance was achieved for the game theory problems. The reader is referred to [19, 20] for a complete analysis of the results.

5 Conclusions

Since their introduction almost 20 years ago, deflection and stretching have served as the means for locating multiple minimizers and alleviating local minimizers in various applications. During the past years they have been combined with a multitude of algorithms including evolutionary and swarm intelligence methods such as particle swarm optimization, differential evolution, ant colony optimization, and covariance matrix adaptation evolution strategies, as well as trajectory-based stochastic optimization approaches such as simulated annealing, and gradient-based algorithms.

Besides the benchmarking on a plethora of standard test problems, the ongoing scientific activity has resulted in a number of interesting applications from various fields. These include the detection of Nash equilibria from game theory, the detection of periodic orbits and chaos control in nonlinear mappings and chaotic systems, the study of the accessibility in main-belt asteroids from astrophysics, semi-infinite programming, the study of red blood cells motion from medical informatics, the kernel-based fuzzy c-means algorithm from clustering, as well as control and optimization problems.

Deflection and stretching have attracted the interest of the scientific community for many years. It is the authors' belief that the ongoing necessity for multimodal optimization algorithms will add further merit to these approaches, especially in terms of new applications, integration with other algorithms, and theoretical investigations.

Acknowledgements The authors would like to thank the editors for their invitation to contribute in this volume as well as the reviewers for their constructive comments.

References

1. Drąg, P., Styczeń, K.: Simulated annealing with constraints aggregation for control of the multistage processes. In: Proceedings of the Federated Conference on Computer Science and Information Systems, pp. 461–469. IEEE (2015)
2. Fan, L., Wang, Y.: A minimum-elimination-escape memetic algorithm for global optimization: MEEM. Int. J. Innov. Comput. Inf. Control **8**(5(B)), 3689–3703 (2012)
3. Gao, F., Qi, Y., Yin, Q., Xiao, J.: Solving problems in chaos control through an differential evolution algorithm with region zooming. Appl. Mech. Mater. **110–116**, 5048–5056 (2012)
4. Gao, F., Tong, H.: UEAS: a novel united evolutionary algorithm scheme. In: King, et al. (eds.) Lecture Notes in Computer Science (LNCS), vol. 4234, pp. 772–780. Springer (2006)
5. Ge, R.P.: A filled function method for finding a global minimizer of a function of several variables. Math. Program. **46**, 191–204 (1990)
6. Ge, R.P., Qin, Y.F.: A class of filled functions for finding global minimizers of a function of several variables. J. Optim. Theory Appl. **54**(2), 241–252 (1987)
7. He, J., Dai, H., Song, X.: The combination stretching function technique with simulated annealing algorithm for global optimization. Optim. Methods Softw. **29**(3), 629–645 (2014)
8. Horst, R., Tuy, H.: Global Optimization - Deterministic Approaches. Springer, New York, USA (1996)
9. Li, L., Yang, Y., Peng, H.: Computation of multiple global optima through chaotic ant swarm. Chaos, Solitons Fractals **40**, 1399–1407 (2009)
10. Li, X., Epitropakis, M.G., Deb, K., Engelbrecht, A.: Seeking multiple solutions: an updated survey on niching methods and their applications. IEEE Trans. Evol. Comput. **21**(4), 518–538 (2017)
11. Lu, C., Zhu, Z., Gu, X.: Kernel parameter optimization in stretched kernel-based fuzzy clustering. In: Zhou, Z.-H., Schwenker, F. (eds.) Lecture Notes in Artificial Intelligence (LNAI), vol. 8183, pp. 49–57. Springer (2013)
12. Lung, R.I., Dumitrescu, D.: Evolutionary multimodal optimization for Nash equilibria detection. In: Krasnogor, N. et al. (eds.) Studies in Computational Intelligence: Nature Inspired Cooperative Strategies for Optimization, vol. 236, pp. 227–237. Springer (2009)
13. Magoulas, G.D., Vrahatis, M.N., Androulakis, G.S.: On the alleviation of the problem of local minima in back-propagation. Nonlinear Anal. Theory Methods Appl. **30**(7), 4545–4550 (1997)
14. Parsopoulos, K.E., Plagianakos, V.P., Magoulas, G.D., Vrahatis, M.N.: Improving particle swarm optimizer by function "stretching". In: Hadjisavvas, N., Pardalos, P.M. (eds.) Advances in Convex Analysis and Global Optimization, vol. 54 of Nonconvex Optimization and Applications, chap. 28, pp. 445–457. Kluwer Academic Publishers, The Netherlands (2001)
15. Parsopoulos, K.E., Plagianakos, V.P., Magoulas, G.D., Vrahatis, M.N.: Objective function "stretching" to alleviate convergence to local minima. Nonlinear Anal. Theory, Methods Appl. **47**(5), 3419–3424 (2001)
16. Parsopoulos, K.E., Plagianakos, V.P., Magoulas, G.D., Vrahatis, M.N.: Stretching technique for obtaining global minimizers through particle swarm optimization. In: Proceedings of the Particle Swarm Optimization Workshop, pp. 22–29. Indianapolis (IN), USA (2001)
17. Parsopoulos, K.E., Vrahatis, M.N.: Modification of the particle swarm optimizer for locating all the global minima. In: Kurkova, V., Steele, N.C., Neruda, R., Karny, M. (eds.) Artificial Neural Networks and Genetic Algorithms. Computer Science, pp. 324–327. Springer, Wien, Austria (2001)
18. Parsopoulos, K.E., Vrahatis, M.N.: Recent approaches to global optimization problems through particle swarm optimization. Nat. Comput. **1**(2–3), 235–306 (2002)
19. Parsopoulos, K.E., Vrahatis, M.N.: On the computation of all global minimizers through particle swarm optimization. IEEE Trans. Evol. Comput. **8**(3), 211–224 (2004)
20. Parsopoulos, K.E., Vrahatis, M.N.: Particle Swarm Optimization and Intelligence: Advances and Applications. Information Science Publishing (IGI Global), Hershey (PA), USA (2010)
21. Pavlidis, N.G., Parsopoulos, K.E., Vrahatis, M.N.: Computing Nash equilibria through computational intelligence methods. J. Comput. Appl. Math. **175**, 113–136 (2005)

22. Pereira, A.I., Fernandes, E.M.G.P.: Constrained multi-global optimization using a penalty stretched simulated annealing framework. In: AIP Conference Proceedings 1168, vol. 2, pp. 1354–1357. AIP (2009)
23. Pereira, A.I., Ferreira, O., Pinho, S.P., Fernandes, E.M.G.P.: Multilocal programming and applications. In: Zelinka, I. et al. (eds.) Handbook of Optimization, ISRL 38, pp. 157–186. Springer (2013)
24. Pereira, A.I., Rufino, J.: Solving multilocal optimization problems with a recursive parallel search of the feasible region. In: Murgante, B., et al. (ed.) Lecture Notes in Computer Science (LNCS), vol. 8580, pp. 154–168. Springer (2014)
25. Pereira, A.I., Rufino, J.: Solving constrained multilocal optimization problems with parallel stretched simulated annealing. In: Gervasi, O., et al. (ed.) Lecture Notes in Computer Science (LNCS), vol. 9156, pp. 534–548. Springer (2015)
26. Pereira, A.I.P.N., Fernandes, E.M.G.P.: A reduction method for semi-infinite programming by means of a global stochastic approach. Optimization **58**(6), 713–726 (2009)
27. Petalas, Y.G., Parsopoulos, K.E., Vrahatis, M.N.: Stochastic optimization for detecting periodic orbits of nonlinear mappings. Nonlinear Phenom. Complex Syst. **11**(2), 285–291 (2008)
28. Pinho, D., Pereira, A.I., Lima, R.: Motion of red blood cells in a glass microchannel: a global optimization approach. In: AIP Conference Proceedings 1168, vol. 2, pp. 1362–1365. AIP (2009)
29. Ribeiro, T., Rufino, J., Pereira, A.I.: PSSA: parallel stretched simulated annealing. In: AIP Conference Proceedings 1389, pp. 783–786. AIP (2011)
30. Rufino, J., Pereira, A.I., Pidanic, J.: coPSSA: constrained parallel stretched simulated annealing. In: 25th International Conference Radioelektronika. IEEE (2015)
31. Shang, H., Liu, Y.: Assessing accessibility of main-belt asteroids based on Gaussian process regression. J. Guid. Control Dyn. **40**(5), 1144–1154 (2017)
32. Tong, Q., Zou, X., Tong, H., Gao, F., Zhang, Q.: Hardware/software partitioning in embedded system based on novel united evolutionary algorithm scheme. In: Proceedings of the International Conference on Computer and Electrical Engineering (ICCEE 2008), Phuket, Thailand. IEEE (2008)
33. Wang, Y.-J., Zhang, J.-S.: An efficient algorithm for large scale global optimization of continuous functions. J. Comput. Appl. Math. **206**, 1015–1026 (2007)
34. Wang, Y.-J., Zhang, J.-S.: A new constructing auxiliary function method for global optimization. Math. Comput. Model. **47**, 1396–1410 (2008)
35. Zhang, H., Fan, Y.: An adaptive policy gradient in learning Nash equilibria. Neurocomputing **72**, 533–538 (2008)
36. Zhang, H., Liu, P.: A momentum-based approach to learning Nash equilibria. In: Shi, Z., Sadananda, R. (eds.) Lecture Notes in Artificial Intelligence (LNAI), vol. 4088, pp. 528–533. Springer (2006)

Multimodal Optimization by Evolution Strategies with Repelling Subpopulations

Ali Ahrari and Kalyanmoy Deb

Abstract This work presents a niching method based on the concept of repelling subpopulations for multimodal optimization. It utilizes several existing concepts and techniques in order to develop a new multimodal optimization algorithm that does not make any of specific assumptions on the shape, size, and distribution of minima. In the proposed method, several subpopulations explore the search space in parallel. Offspring of weaker subpopulations are forced to maintain a distance from the fitter subpopulations and the previously identified niches. This defines taboo regions to hinder the exploration of the same regions of the search space and previously identified niches. The size of each taboo region is adapted independently so that the method can handle challenges of minima with dissimilar basin sizes and irregular distribution. The local shape of a basin is approximated by the distribution of the subpopulation members converging to that basin. The proposed niching strategy is incorporated into the state-of-the-art evolution strategies, and the resulting method is compared with some of the most successful multimodal optimization methods on composite test problems. A comparison of numerical results demonstrates the superiority of our proposed method.

1 Introduction

[1]In recent decades, optimization of practical problems by using evolutionary algorithms has gained much interest [7], where the problem landscape may exhibit chal-

[1]Some materials in Sects. 1 and 2 of this work have been excerpted from our previously published paper: *"Ali Ahrari, Kalyanmoy Deb, and Mike Preuss, 'Multimodal Optimization by Covariance Matrix Self-Adaptation Evolution Strategy with Repelling Subpopulations', Evolutionary Computation, 25:3 (Fall, 2017), pp. 439–471. ©2017 by the Massachusetts Institute of Technology, published by the MIT Press"*. Permission to excerpt the materials from our previous work has been granted by The MIT Press.

A. Ahrari (✉) · K. Deb
Michigan State University, 48824 East Lansing, MI, USA

K. Deb
e-mail: kdeb@egr.msu.edu
URL: https://www.coin-lab.org

M. Preuss et al. (eds.), *Metaheuristics for Finding Multiple Solutions*, Natural Computing Series,
https://doi.org/10.1007/978-3-030-79553-5_7

lenging features such as multimodality, discontinuity, ill-condition, and correlation [16]. On the other hand, practical considerations have defined new notions in optimization in order to tackle problems with conflicting objectives (multi-objective optimization) [7], uncertainties (robust optimization) [4], costly evaluations (surrogate-assisted optimization) [19], or when distinct good solutions are desired (multimodal optimization) [8]. Finding multiple good and distinct solutions, instead of a single one, can provide a variety of distinct good solutions for the decision-maker. This is particularly useful in engineering problems, in which the decision-maker can consider other factors that were possibly overlooked during the mathematical modeling of the problem (e.g., aesthetics), to select the final solution. Moreover, finding all the high-fitness optimal solutions might be inherently critical. A typical example is finding all the resonance frequencies that lead to high vibration amplitude [8].

Multimodal optimization is usually achieved by a diversity preservation strategy, called niching, incorporated into a global optimization method, which we call the core algorithm, to enable parallel convergence to different minima. Early niching methods were proposed for genetic algorithms (GAs), including crowding [9] and fitness sharing [14]. Several subsequent studies analyzed and developed the niching methods in the realm of GAs [22, 23]. Similar or new niching strategies were also incorporated to other metaheuristics such as particle swarm optimization (PSO) [12, 26], differential evolution (DE) [3, 6], and covariance matrix adaptation evolution strategy (CMA-ES) [24, 30]. A comprehensive review of niching methods for parameter optimization can be found in the work of Singh and Deb [32], and a few recent review papers [8, 21].

Regardless of the core algorithm, niching strategies can be classified into two groups: radius-based and non-radius-based [33]. Radius-based niching methods are those that rely on a distance threshold, generally referred to as the niche radius, to check whether two individuals share the same niche. The oldest technique in this group is fitness sharing, which reduces the fitness of similar individuals that share a niche. Radius-free niching methods, in contrast, do not depend upon the definition of the threshold distance, and hence can be considered more robust and desirable than radius-based methods. The oldest method in this group is crowding, in which a descendant may replace the most similar parent.

For multimodal optimization, algorithms with minimal assumptions on properties of the fitness landscape are highly desirable; hence, radius-free niching methods show a significant advantage over those that depend on a user-tuned niche radius parameter. In most radius-based techniques with fixed niche radius, the niche radius should be tuned by the user, or set to a default value (e.g., see the recommended value by Deb and Goldberg [10]). This implicitly assumes the basins are almost uniformly distributed in the entire search space. In general, even fine-tuning of the niche radius might be of little use, especially when the distribution of minima is not uniform, or basins are of different shapes and sizes.

Another common shortcoming of many niching methods is their reliance on the Euclidean distance to cluster individuals to different niches, according to which closer individuals are more likely to share the same basins, or equivalently, farther individuals belong to different niches. This strategy implicitly presumes that niches are roughly spherical. Considering that ill-conditioned problems are one of the main challenges in real-parameter optimization [16], exploitation of the Euclidean distance metric may considerably degrade the results. As an alternative, some studies [30] utilized the Mahalanobis distance metric, which may handle this problem to some extent, even though in general, basins might have any arbitrary shape.

This study develops a novel niching method for multimodal optimization, which combines the concept of repelling subpopulations and state-of-the-art evolution strategies. The resulting method, called evolution strategy with repelling subpopulations (RS-ES) can learn the relative size and possibly the shape of the basins, and thus, handle the challenge of non-circular basins. Limitations of RS-ES are only those imposed by the core algorithm. No assumption on the distribution of global minima, the shape, or the size of the basins is required. RS-CMSA has no niche radius parameters, and for all other parameters, the default values are shown to be robust. On the secondary level, this study suggests a new metric to quantify the usefulness of an arbitrary test problem. This will particularly help improve standard test suites for comparison of niching methods, such as CEC'2013 [20] test suite for multimodal optimization.

The rest of this article is organized as follows. Section 2 elaborates different components of RS-ES in detail. In Sect. 3, three variants of RS-ES are compared with some of the state-of-the-art niching methods on a set of 15 composite functions. Finally, conclusions are drawn in Sect. 4.

2 Niching with Repelling Subpopulations

Different components of the proposed method are explained in this section.

2.1 Core Algorithm

Our niching strategy can be used with any evolution strategy; however, we select covariance matrix adaptation evolution strategy (CMA-ES) [18] and covariance matrix self-adaptation evolution strategy (CMSA-ES) [5], the state-of-the-art evolution strategies for continuous parameter optimization. Both methods adapt the full covariance matrix of multivariate normal distribution. In CMA-ES, the global mutation strength, the so-called step-size (σ_{mean}), is adapted based on the concept of cumulation, and a few other heuristics are employed to update the covariance matrix ($\mathbf{C}$). By adapting the covariance matrix, distribution of sampled solutions in CMA-ES may gradually conform to the global/local shape of the fitness landscape in order to handle rotated ill-conditioned problems efficiently [17]. One may speculate that this property can be particularly utilized to learn the shape of the basin to which the algorithm is converging in order to cope with the challenge of non-spherical basins, as investigated in a former study [30].

In CMSA-ES, the mutative step-size control has replaced the cumulative step-size adaptation and a simple procedure adapts the covariance matrix [5]. CMSA-ES was demonstrated to compete with the original CMA-ES in some multimodal problems; however, in rotated ill-conditioned problems, where efficient adaptation of the covariance matrix plays a critical role, it fell behind CMA-ES [5]. Nevertheless, CMSA has a significant advantage from the practical point of view: it is simpler and relies on fewer assumptions on the problem, which makes it more flexible. For example, it can employ intermediate selection schemes, in which a fraction of parents may survive to the next generation, or when the life of each member is limited to a specific number of generations [28]. Employing such strategies in CMA-ES violates the assumptions on the distribution of samples, upon which the concept of cumulative step-size adaptation relies, and thus, a performance decline can be assumed. The proposed niching strategy in this study is tested with both CMA and a modified version of CMSA with an intermediate selection scheme. The latter provides the possibility of an arbitrary trade-off between the non-elite 'comma' selection scheme and the completely elitist 'plus' selection scheme [28].

2.2 Main Niching Ideas

In the proposed niching method, the population size is divided into N_s subpopulations of size λ ($\mathbf{P}_i, i = 1, 2, ..., N_s$), which explore the search space in parallel. Every subpopulation has its own mutation parameters ($\sigma_{\text{mean}}, \mathbf{C}$), center ($\boldsymbol{x}_{\text{mean}}$), elite members ($\mathbf{E}$), and normalized taboo distance ($\hat{d}$). This can be denoted by $\mathbf{P}_i = \left\{\sigma_{\text{mean}}, \mathbf{C}, \boldsymbol{x}_{\text{mean}}, \mathbf{E}, \hat{d}\right\}_i$.

Members of each subpopulation must maintain a sufficient distance from some specific points in the search space, called taboo points. The set of taboo points for $\mathbf{P}_i$ is the union of:

- Previously identified minima ($\boldsymbol{y}_m, m = 1, 2, \ldots, N_{\text{AP}}$), which are stored in an archive denoted by **Archive**, unless $\mathbf{P}_i$ is better than $\boldsymbol{y}_m$. N_{AP} is the number of points in **Archive**. An independent normalized taboo distance ($\hat{d}_m$) is associated with each point in **Archive**.
- The centers of fitter subpopulations. The fitness of a subpopulation is measured by the fitness of its best individual. A single normalized taboo distance ($\hat{d}_0$) is allocated for all the subpopulations.

Remarkably, the set of taboo points is subpopulation-dependent, and a better subpopulation has a smaller taboo set.

A sampled solution is called *taboo-acceptable* if it satisfies the distance criterion for all the taboo points; otherwise, it is discarded without evaluation. This process goes on until λ taboo-acceptable solutions are generated. The overall effect of such rejection is reshaping the distribution of solutions so that subpopulations do not search previously explored regions or those under exploration by other subpopulations.

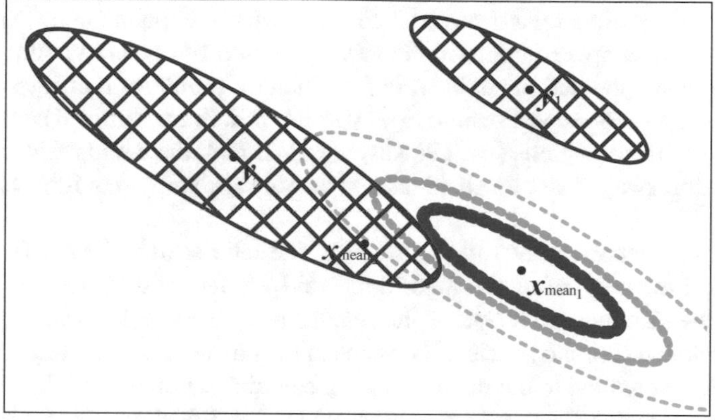

(a) Taboo regions perceived by subpopulation $\mathbf{P}_1$

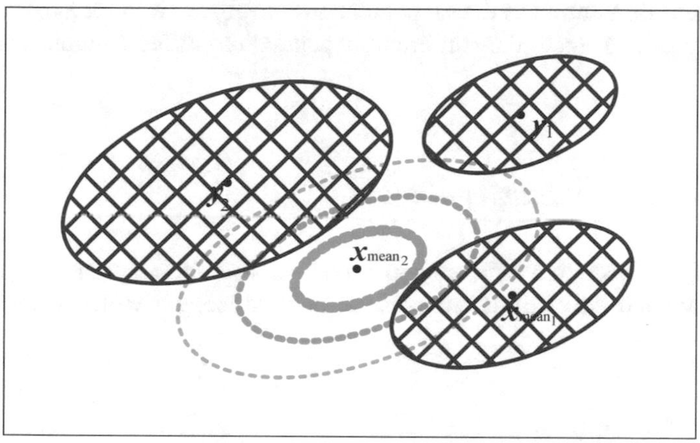

(b) Taboo regions perceived by subpopulation $\mathbf{P}_2$

Fig. 1 Taboo regions perceived by subpopulation $\mathbf{P}_1$ and $\mathbf{P}_2$ ($\mathbf{P}_1$ is fitter than $\mathbf{P}_2$) in presence of archived points $\boldsymbol{y}_1$ and $\boldsymbol{y}_2$, with normalized taboo distance of $\hat{d}_2 \geq \hat{d}_0 \geq \hat{d}_1$. $\boldsymbol{x}_{\text{mean}_1}$ and $\boldsymbol{x}_{\text{mean}_2}$ are the centers of subpopulations $\mathbf{P}_1$ and $\mathbf{P}_2$, respectively

The taboo regions are ellipsoids whose centers lie on the taboo points. Shapes of the taboo regions for $\mathbf{P}_i$ is exclusively determined by the strategy parameters of $\mathbf{P}_i$, while their sizes are affected by the normalized taboo distances of the taboo points as well. Normalized taboo distances of the points in **Archive** are adapted such that a larger basin has a greater $\hat{d}_m$ to enable RS-ES to handle the challenge of unequally sized basins efficiently. Notably, this niching strategy affects only the sampling stage, which means selection, recombination, and adaptation are performed locally; therefore, subpopulations may converge at different times and may have totally different strategy parameters, which allows for the identification of basins with different shapes and sizes.

An exemplary situation is depicted in Fig. 1, in which subpopulations $\mathbf{P}_1$ and $\mathbf{P}_2$ ($\mathbf{P}_1$ is fitter than $\mathbf{P}_2$) sample solutions in the vicinity of archived points $\boldsymbol{y}_1$ and $\boldsymbol{y}_2$. Taboo regions for a subpopulation conform to the mutation profile of that subpopulation, while their sizes are proportional to $\hat{d}_1$ and $\hat{d}_2$ as well. Since sampled solutions that fall inside the taboo regions (red ellipses) are discarded, the taboo regions push the subpopulation away from already identified minima ($\boldsymbol{y}_1$ and $\boldsymbol{y}_2$) or fitter subpopulations.

The idea of taboo regions may remind of the tabu search (TS) [13], in which a short-term memory helps the algorithm avoid the recently visited regions [31]; however, the size and the shape of the regions are preset in TS while RS-ES can learn the relative size and possibly the shape of the basins. The archived solutions in RS-CMSA, in contrast to the memory in TS, contains minima identified during the optimization process, disregarding how recent the identification has occurred. More importantly, taboo regions are coupled to the mutation strength of the subpopulation, which means they shrink as the subpopulation converges. Even for a fixed iteration, the taboo regions perceived by different subpopulations differ in number, shape, and size.

2.3 Evolution of Subpopulations

In RS-ES, the evolution of a subpopulation consists of sampling, selection, and recombination, in which the diversity preservation strategy is applied to the sampling stage.

2.3.1 Distance Metric

A distance metric should be defined to check whether the j-th solution ($\boldsymbol{x}_{ij}$) of $\mathbf{P}_i$ satisfies the distance metric with respect to the k-th taboo point ($\boldsymbol{y}_k$). The shape of a taboo region can be coupled to the covariance matrix by using the Mahalanobis distance instead of the Euclidean distance [30]. The Mahalanobis distance (d^{M}) of the point $\boldsymbol{x}$ from the center $\boldsymbol{x}_{\mathrm{mean}}$ with respect to the covariance matrix $\mathbf{C}$ is defined as follows:

$$d^{\mathrm{M}} = \sqrt{(\boldsymbol{x} - \boldsymbol{x}_{\mathrm{mean}})^T \, \mathbf{C}^{-1} \, (\boldsymbol{x} - \boldsymbol{x}_{\mathrm{mean}})}. \qquad (1)$$

The Mahalanobis distance can be interpreted as the scaled version of the Euclidean distance, where the scaling factors and directions are determined by $\mathbf{C}$. This definition can be employed to define the normalized Mahalanobis distance d^{M}_{ij-k} between sample solution $\boldsymbol{x}_{ij}$ and the taboo point $\boldsymbol{y}_k$ with respect to the parameters of the subpopulation as follows:

$$d_{ij-k}^{\mathrm{M}} = \frac{\sqrt{\left(\boldsymbol{x}_{ij} - \boldsymbol{y}_k\right)^T \mathbf{C}_i^{-1} \left(\boldsymbol{x}_{ij} - \boldsymbol{y}_k\right)}}{\sigma_{\mathrm{mean}_i}}. \quad (2)$$

Contours of points with identical normalized Mahalanobis distance from $\boldsymbol{y}_k$ form concentric ellipsoids whose centers lie on $\boldsymbol{y}_k$, the shape of which is defined by $\mathbf{C}_i$. The normalized distance increases inversely proportional to σ_{mean_i}. This can be beneficial for landscapes with multiple global minima close to each other.

2.3.2 Rejection Rule

The sampled solution $\boldsymbol{x}_{ij}$ will be rejected due to proximity to the taboo point $\boldsymbol{y}_k$ if $d_{ij-k}^{\mathrm{M}} \leq \hat{d}_k$. To be a taboo-acceptable solution, $\boldsymbol{x}_{ij}$ must satisfy the distance criterion for all the taboo points. Only taboo-acceptable solutions are evaluated, and other sampled solutions are rejected without evaluation. This process continues until λ taboo-acceptable solutions are generated.

The covariance matrix of a subpopulation determines the shape of the taboo regions perceived by that subpopulation. It is assumed that the covariance matrix gradually conforms to the shape of the basin to which it is converging. Consequently, the taboo region may gradually conform to the basin shape to overcome the challenge of non-spherical basins, which was highlighted earlier.

2.3.3 Selection and Recombination

Although contemporary evolution strategies, including CMSA-ES and CMA-ES, employ the comma selection scheme [2], preserving the elite member has been preferred for niching with ESs [11]. Shir and Bäck [29] demonstrated the superiority of the plus scheme in multimodal optimization for the special cases of (1+λ)-ES and (1, λ)-ES. A more general elite preservation scheme, in which N_{elt} parents ($0 < N_{\mathrm{elt}} \leq \mu$) can survive to the next generation can thus be advantageous. The surviving parents may participate in subsequent update of the subpopulation parameters ($\boldsymbol{x}_{\mathrm{mean}}, \sigma_{\mathrm{mean}}, \mathbf{C}$). Similar to the concept of finite life span [28], in which the life span of offspring is limited to κ generations, or gradual fitness decay [1], in which the fitness of individuals declines as they grow older, this general case of the selection scheme enables an arbitrary trade-off between advantages and disadvantages of both extreme schemes. Such an intermediate selection scheme can even surpass both extremes in some situations [1].

In RS-ES, the number of parents is more than one and besides, recombination is performed. This makes the generalization of the selection scheme more complicated than the case of (1+λ)-ES. Some parents may be selected from the surviving elites and the rest from the recently generated offspring. Because of the complexity of the adaptation of the mutation profile in CMA-ES, we only consider the elite selection scheme when CMSA-ES is used as the core search algorithm and use the comma selection scheme when CMA-ES is used.

The employed approach for including the elites in the update of the strategy variables, in addition to the decision variables, follows the main goals of elitism. When the traditional concept of mutative strategy parameter control is employed, the underlying goal of elitism is to preserve the successful mutations, those that have resulted in better offspring, so that such success may be repeated in the subsequent generations. This motivation can be extended to successful directions in a similar way: preserving successful directions ($z_{ij} = (\boldsymbol{x}_{ij} - \boldsymbol{x}_{\text{mean}})/\sigma_{ij}$) of the surviving parents, which is employed in this study. At the end of each generation, the solutions (the union of λ recently generated and N_{elt} surviving solutions from the previous iteration) are sorted in increasing order of the objective value. The μ-best solutions are then selected as the parents for the next generation ($\mu = \lceil \lambda/2 \rceil$). Parameters of the subpopulations are then updated similar to CMSA-ES [5], with two exceptions. First, logarithmically decreasing weights (w_j, $j = 1, 2, \ldots, \mu$) [18] are assigned to the parents and second, the update of the global step size is performed as follows:

$$\sigma_{\text{mean}_i} \leftarrow \frac{\sigma_{\text{mean}_i} \left(\prod_{j=1}^{\mu} \sigma_{ij}^{w_j} \right)}{\left(\prod_{j=1}^{\lambda+N_{\text{elt}}} \sigma_{ij}^{\frac{1}{\lambda+N_{\text{elt}}}} \right)}. \tag{3}$$

This revision in the adaption of the global mutation strength was performed to compensate for anisotropic distribution of σ_{ij}'s in the subpopulation, caused by the rejection of certain samples, such that under a random selection of the parents, the expected change of σ_{mean_i} is zero. Ignoring such compensation results in a bias toward increasing σ_{mean_i}. The reason is that a greater σ_{ij} is more likely to produce a sample farther away from $\boldsymbol{x}_{\text{mean}_i}$, which is more likely to lie outside the taboo regions. A larger σ_{mean_i} means larger taboo regions, and preference of greater σ_{ij}'s again. This ever-increasing σ_{mean_i} risks divergence of $\mathbf{P}_i$. Although some statistical criteria can be provided to terminate nonconverging subpopulations, such subpopulations will waste many evaluations.

2.4 Restart Strategy with Increasing Population

We employ the restart strategy with an increasing population size [15]. Since both CMA-EA and CMSA-ES require a large population size for problems with many unwanted local minima, this strategy improves the robustness of the method. For multimodal optimization, there can be situations in which this strategy even surpasses a well-tuned but fixed population size. For example, basins that are easy to identify can be detected in the early restarts, when the population size is small, whereas the subsequent restarts with a larger population size aim at finding harder minima. When all subpopulations in a restart are terminated, λ is multiplied by a user-defined parameter named $PopIncFac$, unless the remaining evaluation budget is not sufficient for the subsequent restart with the increased population size to conclude properly. If so, the population size is increased by a smaller factor, or the number of subpopulations might be decreased to make sure most subpopulations will have a sufficient budget to converge. This process continues until the available budget of function evaluations is depleted.

The criteria for termination of a subpopulation are based on or inspired from those developed for CMA-ES [15]. In RS-ES, a subpopulation is terminated if one of the following situations occurs:

- Condition number criterion: the condition number of $\mathbf{C}$ exceeds 10^{14}.
- Stagnation criterion: the median of the 20 newest values is not smaller than the median of the 20 oldest values, respectively, in the two arrays containing the best recent function values and the median of recent function values of the last $0.2t + 120 + 30D/\lambda$ iterations. The elite solutions are not considered in the calculation of the best and median values.
- No improvement criterion: the range of the best recent function values (excluding elites) during the last $10 + 30D/\lambda$ iterations is smaller than $TolHistFun$.

The terminated subpopulations are analyzed at the end of the restart to check whether they specify new basins. The newly identified basins are added to **Archive** and the normalized taboo distance of each archived solution is updated based on the attraction power of the corresponding basin.

2.5 Adaptation of the Normalized Taboo Distance

The normalized taboo distance should ideally be proportional to the repelling power of the taboo point. This means a great repelling power should be assigned for a basin with a large attraction region, otherwise, it may still attract subpopulations in future. For each arbitrary basin, this parameter is adapted based on the number of subpopulations that have converged to that basin. When a restart concludes, the best solution of every terminated subpopulation is compared to the solutions in **Archive**, to check whether it is desirable. For the case when only global minima are desired, a converged solution is desirable only if its fitness is not worse than the fitness of the best solution in **Archive** minus ϵ_f (target tolerance on the objective function). The adaption procedure of the normalized taboo distances is as follows:

(i) Discard all undesirable solutions. This will reduce the size of remaining converged solutions to $N_{\text{converged}}$.
(ii) For all the remaining solutions ($\boldsymbol{y}_i, i = 1, 2, \ldots, N_{\text{converged}}$), employ the hill-valley function of Ursem [34] to determine whether a solution refers to a new basin. The line search is rendered using Golden Section Search with a maximum of ten function evaluations. If it appears to be a new basin, add this solution to **Archive**. If it is not a new basin, identify the archived point which represents the same basin.

(iii) Use the default value of $\hat{d}_0$ for the normalized taboo distance of the newly archived points. $\hat{d}_0$ is equal to the 25th percentile of normalized taboo distances of the solutions stored in **Archive**. For the zeroth restarts, set $\hat{d}_0 = 1$.

(iv) Count the number of subpopulations that converged to the basin of each archived point (N_{rep_m}). Update the normalized taboo distances using the following equation:

$$\hat{d}_m \leftarrow \begin{cases} \hat{d}_m \left(1 + N_{\text{rep}_m} - \alpha_{\text{new}} \times \dfrac{N_{\text{converged}}}{N_{\text{AP}}}\right)^{\tau_{\hat{d}}}, & \text{if } N_{\text{rep}_m} > \alpha_{\text{new}} \times \dfrac{N_{\text{converged}}}{N_{\text{AP}}}; \\ \hat{d}_m \left(1 - N_{\text{rep}_m} + \alpha_{\text{new}} \times \dfrac{N_{\text{converged}}}{N_{\text{AP}}}\right)^{-\tau_{\hat{d}}}, & \text{otherwise.} \end{cases} \tag{4}$$

If N_{rep_m} is high, Eq. 4 increases the normalized taboo distance of the m-th archived point in order to reduce the probability of convergence to the same basin in the future restarts; otherwise, this equation decreases it since it might be unnecessarily large. $0 \leq \alpha_{\text{new}} \leq 1$ is a user-defined control parameter (by default, $\alpha_{\text{new}} = 0.5$) which specifies the fraction of solutions that can converge to an already identified basin without further increase in the corresponding taboo distance. It can be interpreted as the desired rate of new basin identification. $\tau_{\hat{d}}$ specifies the learning rate of the normalized taboo distance, with the recommended value of $\tau_{\hat{d}} = 1/\sqrt{D}$.

2.6 Boosting Time Efficiency

The main time complexity of the proposed niching strategy originates from the sampling part, in which the distance of each sampled solution should be checked against all the taboo points. It is also likely that most sampled solutions are rejected because of the distance criteria, especially in the early iterations of each restart when the mutation strength is high. Two different strategies are proposed and employed in RS-ES to alleviate this problem: temporary shrinkage of the taboo regions and checking for more critical taboo points earlier. These two strategies are explained in this section.

2.6.1 Temporary Shrinkage of the Taboo Regions

The fraction of the rejected sampled points is directly related to the overall size of the taboo regions. There might be situations in which the taboo regions are so large that almost all the sampled solutions of a subpopulation are rejected. To avoid stagnation, the distance condition is loosened by a temporary reduction of the normalized taboo distance of every taboo point whenever a sampled solution is rejected:

$\hat{d}_k \leftarrow \sqrt[D]{0.99}\hat{d}_k$. This modification is valid only for the current subpopulation and only for the current iteration.

2.6.2 Checking for More Critical Taboo Points Earlier

If the sampled solution $\boldsymbol{x}_{ij}$ violates the distance condition to any taboo point, it is rejected, and checking the distance condition with respect to the other taboo points is not required. This can save a lot of computation if the most critical taboo points, those that are most likely to reject a sampled solution, are known so that the distance condition is checked with respect to them earlier. This means that if the sampled solution is going to be rejected, the algorithm discovers it sooner.

At each iteration and for each subpopulation, we calculate mean rejection probability (*MRP*) with respect to each taboo point, which is the probability that an arbitrary sampled solution by that subpopulation is rejected because of the proximity to that taboo point. *MRP* is employed as a metric to quantify criticality of the taboo point $\boldsymbol{y}_k$ for $\mathbf{P}_i$. Using some simplifications, an upper limit for *MRP* of taboo point $\boldsymbol{y}$ for subpopulation $\mathbf{P}$ can be calculated as follows:

$$MRP(\mathbf{P}_i, \boldsymbol{y}) = \Phi\left(\frac{L + \hat{d}\bar{u}_1\sigma_{\text{mean}}}{u_1\sigma_{\text{mean}}}\right) - \Phi\left(\frac{L - \hat{d}\bar{u}_1\sigma_{\text{mean}}}{u_1\sigma_{\text{mean}}}\right), \tag{5}$$

where Φ computes the cumulative distribution function of standard normal distribution, L is the Euclidean distance between $\boldsymbol{x}_{\text{mean}}$ and $\boldsymbol{y}$, and u_1 is the greatest eigenvalue of the subpopulation covariance matrix $\mathbf{C}$.

A sampled solution is checked against the taboo points in order of their criticality values. Note that the criticality measure is not a function of the sampled solution; therefore, for each subpopulation, it is calculated once per iteration and used to generate all λ taboo-acceptable solutions. Moreover, it can be safely assumed that a taboo point never rejects a sampled solution if its criticality is sufficiently small (less than 0.01 by default). Such a taboo point is called non-critical and is ignored when checking the distance condition.

2.7 *Initialization of Subpopulations*

At the beginning of each restart, N_s subpopulations are initialized such that their centers lie far from each other and the archived solutions. All new subpopulations have similar initial strategy parameters. Starting from a conservatively large σ_{mean}, a candidate point for the center ($\boldsymbol{x}_{\text{mean}_i}$) is randomly sampled in the search space, and its normalized Mahalanobis distance to the archived solutions and the centers of other subpopulations is calculated. If the distance condition is satisfied, the candidate is accepted as the center of $\mathbf{P}_i$, otherwise discarded, and a new candidate is sampled. If multiple consecutive tries (e.g., 100) fail, σ_{mean} is slightly reduced ($\sigma_{\text{mean}} \leftarrow \sqrt[D]{0.99}\,\sigma_{\text{mean}}$), which reduces the overall size of the taboo regions. This process continues until N_{s} subpopulations are generated. Such initialization, although not critically important, helps the method avoid taboo regions more efficiently in early iterations of a restart, especially in lower dimensions.

2.8 Parameter Setting

The default values of the parameters are suggested, except for subpopulation size increase factor ($PopIncFac$) and the default number of subpopulations (N_{s}^0). A smaller N_{s}^0 should be associated with a smaller $PopIncFac$. $N_{\text{s}}^0 = 10$ with $PopIncFac = 1.4$ is a reasonable choice. The value of $TolHistFun$ should be set equal or smaller than the desired tolerance on the objective function.

The flowchart of RS-ES is provided in Fig. 2.

3 Numerical Evaluation

Three variants of RS-ES are considered for benchmarking and comparison:

- RS-CMSA, in which CMSA-ES with elite selection ($N_{\text{elt}} = 1 + [0.15\mu]$) is employed as the core search algorithm.
- RS-CMSA$^{(\text{NoElt})}$, in which CMSA-ES with comma selection ($N_{\text{elt}} = 0$) is employed as the core search algorithm.
- RS-CMA, in which CMA-ES with comma selection is employed as the core search algorithm.

These variants of RS-ES are tested on a set of 15 scalable composite functions proposed in a previous study [25], which are summarized in Table 1. The available code for these test problems can be used only for specific problem dimensions ($D =$ 2, 10, 30, 50). We consider these test problems in a low dimension ($D = 2$) and a moderate dimension ($D = 10$).

Some of the most successful niching methods in literature, including NMMSO [12], LIPS [26], NSDE [27] and PNPCDE [6] are selected and evaluated on this test suite for comparison. Several factors were considered for selection of these methods, including the year of publication, strength of the reported numerical results, minimal tuning effort, and niche radius independence.

The well-known mean peak ratio (MPR) [20] is employed as the comparison metric. We consider the default parameter setting for the tested methods, except that we perform an exhaustive search to find the best value of the most important control parameter of each method. First, this important parameter is identified and defined as function of an independent parameter β (see Table 2). This function is defined simply

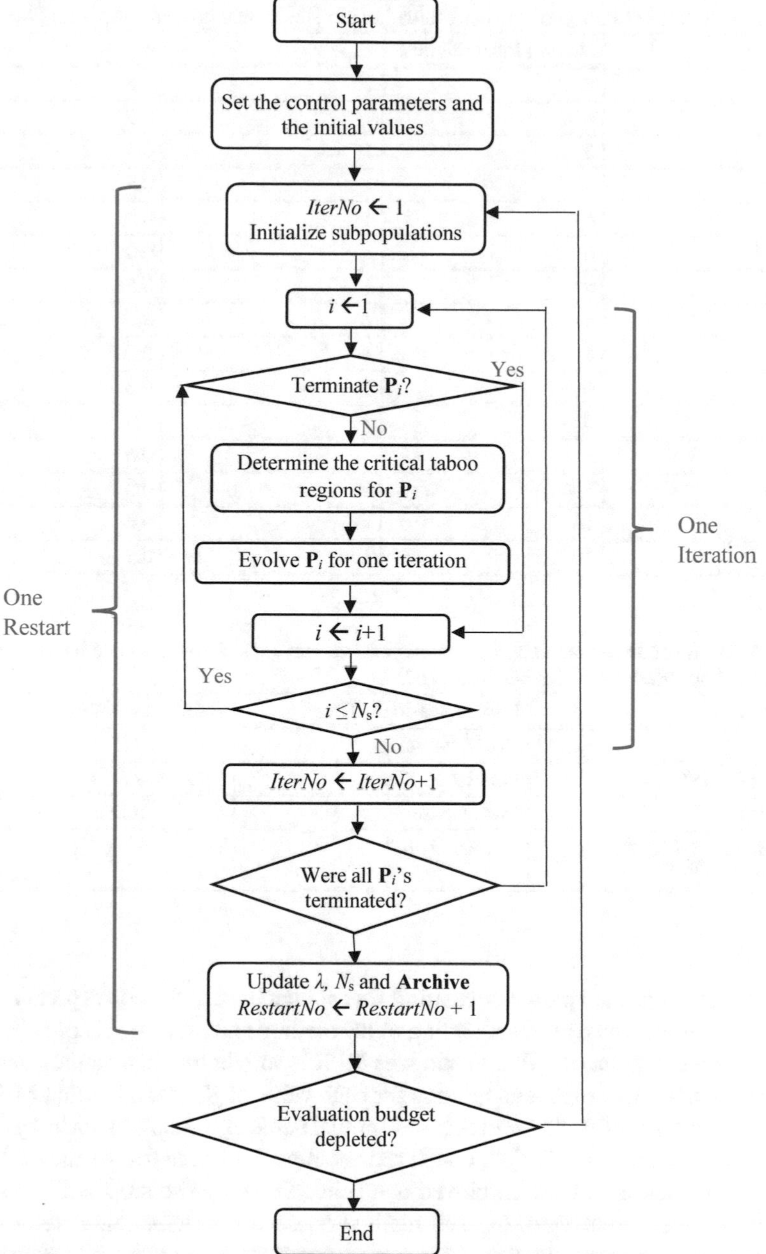

Fig. 2 Flowchart of RS-ES

Table 1 Test problems for performance evaluation of niching methods, adapted from [25]

Function	No. of global minima	r_{nich}	ϵ
CF1	8	1	0.5
CF2	6	1	0.5
CF3	6	1	0.5
CF4	6	1	0.5
CF5	6	1	0.5
CF6	6	1	0.5
CF7	6	1	0.5
CF8	6	1	0.5
CF9	6	1	0.5
CF10	6	1	0.5
CF11	8	1	0.5
CF12	8	1	0.5
CF13	10	1	0.5
CF14	10	1	0.5
CF15	10	1	0.5

Table 2 The most important control parameter of each method is defined as function of β. Integer values for β are tried to find the best value ($\beta^{\star}$)

Method	**Parameter to be Tuned**	**Candidate Values**
LIPS	No. of Particles	$5 \times 2^{\beta} \times \sqrt{D}$
PNPCDE, NSDE	Population Size	$5 \times 2^{\beta} \times \sqrt{D}$
NMMSO	Swarm Size	$5 \times 2^{\beta-1} \times D$
RS-CMA, RS-CMSA, RS-CMSA$^{(\text{NoElt})}$	$PopIncFac$	$2^{2^{-(\beta-1)}}$

based on the suggested parameter setting for each method in the corresponding publication, intuition, or our understanding of the method and multimodal optimization. The best integer value of β that maximizes MPR is sought for each method and each dimension using five independent runs for each value of β, assuming that MPR is a unimodal function of β. After finding the optimal value of β (i.e., $\beta^{\star}$), 25 independent runs are performed with $\beta = \beta^{\star}$, which represent performance for the methods with near-optimal setting on the employed test suite. Table 3 presents $\beta = \beta^{\star}$ and MPR for each method. Peak ratio of each method on each problem, calculated over 25 independent runs, is provided in Table 4. The standard deviation of the performance of the tested methods, calculated over the PR of the tested methods, is provided in the last column for each problem. The obtained results reveal that:

Table 3 Mean Peak Ratio (MPR) and the rank of each method when tested with near-optimal parameter setting ($\beta = \beta^*$)

Method	$D = 2$			$D = 10$			Overall	
	β^*	MPR	Rank	β^*	MPR	Rank	MPR	Rank
LIPS	5	0.78	4	4	0.475	6	0.627	5
NSDE	6	0.761	6	5	0.536	3	0.648	4
PNPCDE	2	0.68	7	3	0.482	5	0.581	6
NMMSO	2	0.856	1	4	0.209	7	0.533	7
RS-CMSA	2	0.841	2	3	0.506	4	0.673	1
RS-CMSA$^{(NoElt)}$	3	0.764	5	4	0.538	2	0.651	3
RS-CMA	5	0.785	3	5	0.543	1	0.664	2

- Overall, RS-CMSA is the most successful method, although there is not a huge gap among the performance of these methods. For $D = 2$, NMMSO, and for $D = 10$, RS-CMA are the most successful methods.
- There is a considerable difference in scalability of the methods. For example, NMMSO is the most successful methods for $D = 2$, but it turned out to be the least successful one for D=10. This will demonstrate the importance of scalable problems to identify pros and cons of each niching strategy or even the core algorithm with respect to problem dimension.
- For RS-ES, the elite preserving strategy is useful in $D = 2$, but disadvantageous for $D = 10$ (compare RS-CMSA with RS-CMSA$^{(NoElt)}$). One may conclude that in low dimensions, the advantages of elite preservation for multimodal optimization are significant; however, in higher dimensions, its disadvantage for global search turns out to be significant. This interesting observation demands further study. A finite life span as introduced in [28] and further studied in [1] can be a potential alternative.
- The standard deviation of PR of different methods for a specific test problem provides valuable information on the usefulness of that test problem. For example, for 10-D CF14 with 10 global minima, almost all the tested methods have $PR = 0.1$. While it cannot be deduced from the provided data, one can speculate that it is one specific global minimum that is identified in all runs. This means other nine global minima are too hard to be found by any method. Such "too easy or too hard" scenario is undesirable in a benchmark function, as it cannot illuminate the advantages and disadvantages of different methods. In contrast, test problems for which SdDev(PR) (calculated over PR of different methods) is high, are good choices to be included in a standard test suite. For example, there is a huge difference in performance of the tested methods for 10-D CF4.

Table 4 Peak Ratio of each method for each function in two dimensions ($D = 2$ and $D = 10$), calculated over 25 independent runs when $\beta = \beta^{\star}$. The last column presents the standard deviation of PRs of the tested method for each problem

Function	D	Method							StDev($\boldsymbol{PR}$)
		LIPS	NSDE	PNPCDE	NMMSO	RS-CMSA	RS-CMSA$^{(NoElt)}$	RS-CMA	
CF1	2	0.88	0.99	0.75	0.99	0.98	0.77	0.89	0.1
CF2	2	1	0.99	0.99	1	1	0.98	0.99	0.01
CF3	2	0.98	1	1	1	0.99	0.87	0.99	0.05
CF4	2	0.99	1	0.67	0.97	1	0.95	0.91	0.12
CF5	2	0.99	0.99	0.97	1	0.99	0.82	0.91	0.07
CF6	2	0.83	0.67	0.67	0.99	0.95	0.89	0.93	0.13
CF7	2	0.67	0.67	0.67	0.81	0.67	0.78	0.74	0.06
CF8	2	0.67	0.69	0.68	0.92	0.86	0.67	0.67	0.11
CF9	2	0.67	0.67	0.67	0.67	0.67	0.67	0.67	0
CF10	2	0.67	0.65	0.6	0.67	0.67	0.67	0.67	0.03
CF11	2	0.96	0.87	0.53	0.96	0.99	0.89	0.91	0.16
CF12	2	0.57	0.55	0.5	0.79	0.78	0.63	0.65	0.11
CF13	2	0.72	0.56	0.42	0.83	0.83	0.7	0.7	0.15
CF14	2	0.51	0.5	0.5	0.53	0.58	0.5	0.5	0.03
CF15	2	0.6	0.62	0.6	0.71	0.67	0.67	0.65	0.04
CF1	10	0.53	0.79	0.57	0.13	0.63	0.66	0.7	0.21
CF2	10	0.57	0.66	0.59	0.43	0.63	0.59	0.66	0.08
CF3	10	0.67	0.93	0.97	0.66	0.67	0.67	0.67	0.14
CF4	10	0.43	0.89	0.81	0.01	0.47	0.67	0.64	0.29
CF5	10	0.57	0.93	0.9	0.39	0.67	0.61	0.67	0.19
CF6	10	0.56	0.5	0.5	0	0.57	0.61	0.62	0.22
CF7	10	0.63	0.33	0.27	0.25	0.67	0.67	0.67	0.2
CF8	10	0.51	0.5	0.5	0.21	0.57	0.58	0.6	0.13
CF9	10	0.55	0.51	0.5	0.29	0.55	0.59	0.59	0.1
CF10	10	0.33	0.31	0.23	0.03	0.33	0.33	0.33	0.11
CF11	10	0.5	0.5	0.35	0.05	0.53	0.61	0.57	0.19
CF12	10	0.39	0.34	0.36	0.26	0.41	0.44	0.43	0.06
CF13	10	0.36	0.36	0.26	0.14	0.31	0.44	0.39	0.1
CF14	10	0.1	0.1	0.1	0.05	0.1	0.1	0.1	0.02
CF15	10	0.43	0.4	0.33	0.25	0.49	0.52	0.5	0.1

4 Summary and Conclusions

This work has developed a novel method for multimodal optimization, which can learn the shapes and sizes of the basins to handle the challenge of the irregular distribution of dissimilar global minima. The method, called evolution strategy with repelling subpopulations (RS-ES), combines the state of the art of evolution strategies as the core search algorithm with the concept of repelling subpopulations, as the diversity preservation strategy. RS-ES with three options has been compared with some of the most successful niching methods on a set of scalable composite test problems on low and moderate dimensions. Different variants of RS-ES have outperformed these methods in general, when the mean peak ratio is considered as the performance metric.

The drastic effect of the problem dimension on the relative performance demonstrates a considerable difference in the scalability of the tested methods. For example, NMMSO has been the most successful methods for low dimensions, but it turned out to be the least successful one for moderate dimensions. This demonstrates the importance of using scalable problems to identify pros and cons of each niching strategy or even the core search algorithm with respect to problem dimension and discourages reliance on low-dimensional non-scalable classical niching problems in standard test suites.

A comparison of RS-ES with and without elite preservation has demonstrated that elitism can be useful for multimodal optimization in low dimensions. In higher dimensions, the non-elite selection is advantageous, possibly because of the negative effect of elitism on global search capability of the core search algorithm. This demands further study on possible alternatives to handle this dilemma for multimodal optimization.

For a fixed test problem, the distribution of performance metric among different methods (standard deviation of peak ratio in our case) can clarify the merit of that test problem in benchmarking niching methods. If all the methods show similar performance on that test problem, especially if only particular global minima are identified in all runs, that test problem may not highlight the pros and cons of the tested methods. This criterion can guide the selection of test problems for a standard test suite for multimodal optimization, which plays a critical role in the development of this field.

Acknowledgements Permission to excerpt some parts of this article from the following publication was granted by MIT Press: Ali Ahrari, Kalyanmoy Deb, and Mike Preuss, 'Multimodal Optimization by Covariance Matrix Self-Adaptation Evolution Strategy with Repelling Subpopulations', Evolutionary Computation, 25:3 (Fall, 2017), pp. 439–471. ©2017 by the Massachusetts Institute of Technology, published by the MIT Press.

Computational work in support of this research was performed at Michigan State University's High-Performance Computing Facility. The source codes for NSDE and LIPS and the tested composite functions were provided by Ponnuthurai N. Suganthan. The source code of PNPCDE was provided by Subhodip Biswas. The source code of NMMSO was provided by Jonathan Fieldsend. The authors would like to thank them for sharing their source codes and providing guidelines to tune the control parameters. The source code of RS-CMSA-ES (in MATLAB) can be found at the COIN lab website: https://coin-lab.org and https://www.researchgate.net/profile/Ali_Ahrari/research.

References

1. Ahrari, A., Kramer, O.: Finite life span for improving the selection scheme in evolution strategies. Soft Comput. 1–13 (2015)
2. Bäck, T., Foussette, C., Krause, P.: Contemporary Evolution Strategies. Springer Science & Business Media (2013)

3. Basak, A., Das, S., Tan, K.C.: Multimodal optimization using a biobjective differential evolution algorithm enhanced with mean distance-based selection. IEEE Trans. Evolut. Comput. **17**(5), 666–685 (2013)
4. Beyer, H.G., Sendhoff, B.: Robust optimization-a comprehensive survey. Comput. Methods Appl. Mech. Eng. **196**(33), 3190–3218 (2007)
5. Beyer, H.G., Sendhoff, B.: Covariance matrix adaptation revisited–the CMSA evolution strategy–. In: Parallel Problem Solving from Nature–PPSN X, pp. 123–132. Springer (2008)
6. Biswas, S., Kundu, S., Das, S.: An improved parent-centric mutation with normalized neighborhoods for inducing niching behavior in differential evolution. IEEE Trans. Cybernet **44**(10), 1726–1737 (2014)
7. Coello, C.A.C., Lamont, G.B.: Applications of Multi-objective Evolutionary Algorithms, vol. 1. World Scientific (2004)
8. Das, S., Maity, S., Qu, B.Y., Suganthan, P.N.: Real-parameter evolutionary multimodal optimization - a survey of the state-of-the-art. Swarm Evolut. Comput. **1**(2), 71–88 (2011)
9. De Jong, K.A.: Analysis of the behavior of a class of genetic adaptive systems (1975)
10. Deb, K., Goldberg, D.E.: In: An investigation of niche and species formation in genetic function optimization, pp. 42–50. Morgan Kaufmann Publishers Inc. (1989)
11. Debski, R., Dreżewski, R., Kisiel-Dorohinicki, M.: Maintaining population diversity in evolution strategy for engineering problems. In: New Frontiers in Applied Artificial Intelligence, pp. 379–387. Springer (2008)
12. Fieldsend, J.E.: Running up those hills: Multi-modal search with the niching migratory multi-swarm optimiser. In: 2014 IEEE Congress on Evolutionary Computation (CEC), pp. 2593–2600. IEEE (2014)
13. Glover, F.: Future paths for integer programming and links to artificial intelligence. Comput. Oper. Res. **13**(5), 533–549 (1986)
14. Goldberg, D.E., Richardson, J.: Genetic algorithms with sharing for multimodal function optimization. In: Genetic algorithms and their applications: Proceedings of the Second International Conference on Genetic Algorithms, pp. 41–49. Lawrence Erlbaum, Hillsdale, NJ (1987)
15. Hansen, N.: Benchmarking a bi-population CMA-ES on the BBOB-2009 function testbed. In: Proceedings of the 11th Annual Conference Companion on Genetic and Evolutionary Computation Conference: Late Breaking Papers, pp. pp. 2389–2396. ACM (2009)
16. Hansen, N., Auger, A., Finck, S., Ros, R.: Real-parameter black-box optimization benchmarking 2010: Experimental setup. Technical Report RR-7215, INRIA (2010)
17. Hansen, N., Kern, S.: Evaluating the cma evolution strategy on multimodal test functions. In: Parallel Problem Solving from Nature-PPSN VIII. pp. 282–291. Springer (2004)
18. Hansen, N., Ostermeier, A.: Completely derandomized self-adaptation in evolution strategies. Evolut. Comput. **9**(2), 159–195 (2001)
19. Jin, Y.: Surrogate-assisted evolutionary computation: Recent advances and future challenges. Swarm Evolut. Comput. **1**(2), 61–70 (2011)
20. Li, X., Engelbrecht, A., Epitropakis, M.G.: Benchmark functions for CEC'2013 special session and competition on niching methods for multimodal function optimization. Evolutionary Computation and Machine Learning Group, RMIT University, Technical report (2013)
21. Li, X., Epitropakis, M.G., Deb, K., Engelbrecht, A.: Seeking multiple solutions: an updated survey on niching methods and their applications. IEEE Trans. Evolut. Comput. **21**(4), 518–538 (2017)
22. Mahfoud, S.W.: Niching methods for genetic algorithms. Urbana **51**(95001), 62–94 (1995)
23. Mengshoel, O.J., Goldberg, D.E.: The crowding approach to niching in genetic algorithms. Evolut. Comput. **16**(3), 315–354 (2008)
24. Preuss, M.: Niching the CMA-ES via nearest-better clustering. In: Proceedings of the 12th annual conference companion on Genetic and evolutionary computation, pp. 1711–1718. ACM (2010)
25. Qu, B.Y., Suganthan, P.N.: Novel multimodal problems and differential evolution with ensemble of restricted tournament selection. In: 2010 IEEE Congress on Evolutionary Computation (CEC), pp. 1–7. IEEE (2010)

26. Qu, B.Y., Suganthan, P.N., Das, S.: A distance-based locally informed particle swarm model for multimodal optimization. IEEE Trans. Evolut. Comput. **17**(3), 387–402 (2013)
27. Qu, B.Y., Suganthan, P.N., Liang, J.J.: Differential evolution with neighborhood mutation for multimodal optimization. IEEE Trans. Evolut. Comput. **16**(5), 601–614 (2012)
28. Schwefel, H.P., Rudolph, G.: Contemporary Evolution Strategies. Springer (1995)
29. Shir, O.M., Bäck, T.: Niching with derandomized evolution strategies in artificial and real-world landscapes. Nat. Comput. **8**(1), 171–196 (2009)
30. Shir, O.M., Emmerich, M., Bäck, T.: Adaptive niche radii and niche shapes approaches for niching with the CMA-ES. Evolut. Comput. **18**(1), 97–126 (2010)
31. Siarry, P., Berthiau, G.: Fitting of tabu search to optimize functions of continuous variables. Int. J. Numer. Methods Eng. **40**(13), 2449–2457 (1997)
32. Singh, G., Deb, K.: Comparison of multi-modal optimization algorithms based on evolutionary algorithms. In: Proceedings of the Genetic and Evolutionary Computation Conference (GECCO-2006), New York. pp. 1305–1312. Proceedings of the Genetic and Evolutionary Computation Conference (GECCO-2006), New York (2006)
33. Stoean, C., Preuss, M., Stoean, R., Dumitrescu, D.: Multimodal optimization by means of a topological species conservation algorithm. EEE Trans. Evolut. Comput. **14**(6), 842–864 (2010)
34. Ursem, R.K.: Multinational evolutionary algorithms. In: Proceedings of the 1999 Congress on Evolutionary Computation, 1999. CEC 99. vol. 3. IEEE (1999)

Two-Phase Real-Valued Multimodal Optimization with the Hill-Valley Evolutionary Algorithm

S. C. Maree, D. Thierens, T. Alderliesten, and P. A. N. Bosman

Abstract The aim of multimodal optimization (MMO) is to obtain all global optima of an optimization problem. In this chapter, we introduce a general framework for two-phase MMO evolutionary algorithms (EAs), in which different high-fitness regions (niches) are located in the first phase via clustering, and each of the located niches is separately optimized with a core search algorithm in the second phase. One such two-phase MMO EA is the Hill-Valley Evolutionary Algorithm (HillVallEA). In HillVallEA, the remarkably simple hill-valley clustering method is used. The idea behind hill-valley clustering is that two solutions belong to the same niche (valley) when there is no hill in between them, which can be easily tested by performing additional function evaluations. We compare hill-valley clustering to two other recently introduced fitness-informed clustering methods: nearest-better clustering and hierarchical Gaussian mixture learning. We show how these clustering methods, as well as different core search algorithms, influence the resulting optimization performance of the two-phase MMO framework on the commonly used CEC 2013 niching benchmark suite. Our results show that HillVallEA, equipped with the core search algorithm Adapted Maximum-Likelihood Gaussian Model Univariate (AMu) as core search algorithm, outperforms all other MMO EAs, both within the limited benchmark budget, and in the long run. HillVallEA-AMu was the winner of the GECCO niching competition in 2018 and 2019, and is currently, to the best of our knowledge, the best performing algorithm on this benchmark suite.

S. C. Maree (✉) · T. Alderliesten
Amsterdam UMC, University of Amsterdam, Amsterdam, The Netherlands
e-mail: s.c.maree@amsterdamumc.nl

T. Alderliesten
e-mail: t.alderliesten@amsterdamumc.nl

D. Thierens
Utrecht University, Utrecht, The Netherlands
e-mail: d.thierens@uu.nl

P. A. N. Bosman
Centrum Wiskunde & Informatica, Amsterdam, The Netherlands
e-mail: peter.bosman@cwi.nl

M. Preuss et al. (eds.), *Metaheuristics for Finding Multiple Solutions*,
Natural Computing Series,
https://doi.org/10.1007/978-3-030-79553-5_8

1 Introduction

Model-based evolutionary algorithms (EAs) adapt an underlying search model based on certain features of the fitness landscape [21]. Classically, these features are related to linkage, or dependence, of problem variables. The performance of many EAs deteriorates when the fitness landscape is multimodal, as high-quality solutions can be found in different parts of the search space, which may prevent narrowing down the search to a specific region. Being able to explicitly deal with multimodality may therefore be beneficial to many EAs. In addition, exploring multiple niches can provide additional insight into the structure of the problem at hand. Real-world problems are often not unimodal, and by providing the decision-maker with multiple high-quality solutions, the final solution can be chosen based on external factors that are best considered once a set of interesting alternatives is known [24].

In optimization, *niching* refers to a method of obtaining and maintaining solutions in multiple niches, i.e., high-fitness parts of the search space. Niching methods originated as a tool for preserving population diversity in EAs, but are now generally designed for multimodal optimization (MMO) [15]. MMO is applicable to optimization problems with all types of problem variables but is mainly applied to problems with real-valued variables, as we do in this chapter, because of a natural notion of distance and locality in real-valued fitness landscapes.

One of the difficulties with niching methods is that they often introduce additional parameters such as a minimal niche size or the (expected) number of niches [15]. This is however generally unknown a priori, especially in a black-box setting, and niching methods applied in such more general optimization settings should therefore make as few assumptions as possible on the size, shape, or number of niches.

A naive approach to MMO is restarting a *core search algorithm* randomly in different parts of the search space. In case of real-valued optimization, commonly used core search algorithms are the covariance matrix self-adaptation evolution strategy (CMSA) [4] and variants of the adapted maximum-likelihood Gaussian model iterated density-estimation evolutionary algorithm (AMaLGaM) [6]. A downside of this naive serial approach is that optima within larger niches are obtained many times before a core search algorithm optimizes a smaller niche. This makes this approach often expensive, and it is difficult to determine when to stop restarting. A parallel—diversity-preserving—search of the search space based on crowding [8] or fitness-sharing [9] can overcome this, however, these methods do not explicitly separate solutions over multiple niches.

A recent MMO EA is the repelling-subpopulations method (RS-CMSA) [2], in which multiple instances of the core search algorithm are randomly initialized in the search space. However, to prevent multiple instances from converging to the same optimum, rejection sampling is used to push them away from each other, and also from previously obtained optima.

As opposed to random initialization of core search algorithms within the search space, two-phase MMO EAs aim in the first phase at locating different niches in the search space. In the second phase of these MMO EAs, core search algorithms are specifically initialized in these niches. The Nearest-better EA (NEA2+) [22, 23] is one such two-phase MMO EA, which uses nearest-better clustering (NBC) in the first phase to cluster an initial set of solutions. NBC is a fitness-informed clustering method that uses a distance measure based on the concept of a nearest-better solution. The idea behind NBC is that local optima can be detected by the observation that there are no *nearby* solutions with a better fitness. However, the distinction between *nearby* and *far away* is rather difficult to make as it depends on both the number of solutions that is clustered, and the problem dimensionality.

In this chapter, we introduce a general framework for two-phase MMO EAs that can be equipped with different clustering methods and core search algorithms. Besides NBC, we consider hierarchical Gaussian mixture learning (HGML) [19], which is based on the same nearest-better concept as in NBC, but the correlation between the search space and fitness values is exploited to determine a cluster set that best captures the structure of the fitness landscape. The third fitness-informed clustering method that we consider is hill-valley clustering (HVC). HVC is based on the hill-valley test, which states that when there is a hill between two solutions, they belong to a different niche (valley). To test this in practice, intermediate test solutions are sampled and evaluated. The hill-valley test was first used in the multinational EA [26], where it was used to test every newly sampled solution against a population of solutions. The same test was also used by RS-CMSA, but only in a post-processing step, to determine whether two presumed optima are distinct. As core search algorithms, we consider CMSA and different variants of AMaLGaM.

The remainder of this chapter is organized as follows. In Sect. 2, we introduce the framework for two-phase MMO EAs. We discuss the three fitness-informed clustering-based niching methods NBC, HGML, and HVC in detail in Sect. 3. In Sect. 4, we describe the different core search algorithms. In Sect. 5, we experimentally compare the ability of the three clustering methods to locate multiple niches. We furthermore evaluate the performance of the different clustering methods and core search algorithms within the framework on the test problems of the CEC 2013 niching benchmark suite [14], and compare the best performing two-phase MMO EAs of the framework to other MMO EAs, both when allowing a smaller and a larger computational budget. We finally conclude in Sect. 6.

2 Framework for Two-Phase MMO EAs

In this chapter, we consider a general framework for two-phase MMO EAs (see Algorithm 1), that is given by,

$$\mathcal{E} = \text{TwoPhaseMMOEA}(f, \text{ClusterMethod}, \text{CoreSearchAlgorithm}, \varepsilon), \quad (1)$$

where $f : X \to \mathbb{R}$ is the to-be-minimized fitness function, with $X \subseteq \mathbb{R}^d$ the d-dimensional search space. The aim of an MMO EA is to obtain all global optima of f, which are collected in the elitist archive $\mathcal{E} \subset X$. The user specifies the used

Algorithm 1: Framework for Two-phase MMO EAs

```
function: [ℰ] = TwoPhaseMMOEA(f, ClusterMethod, CoreSearchAlgorithm, ε)
input   : Fitness function f : X → ℝ, with X ⊆ ℝ^d;        // To be minimized
          ClusterMethod;                                   // NBC, HGML, or HVC
          CoreSearchAlgorithm;                             // See Table 1
          ε ≥ 0;                                // Threshold for global optima
output  : Set of presumed global optima ℰ ⊂ X;
N = 64;                                           // Initial population size
N_c = 0.8 · N_c^rec(CoreSearchAlgorithm);          // Cluster size (Table 1)
τ = τ^rec(CoreSearchAlgorithm);               // Selection pressure (Table 1)
ℰ = {}, 𝒫 = {}, 𝒦 = {};     // Elitist archive, Population, Cluster set
while budget remaining do
    // First phase - locating niches
    𝒫 = sample(N, 𝒫, 𝒦, d + 1, 2.0);              // Also evaluates solutions
    𝒮 = truncation_selection(𝒫, τ);
    ρ = 0.01 · (|X|/|𝒮|)^(1/d);        // For initializing clusters with |C| = 1
    if ClusterMethod = HVC then 𝒮 = 𝒮 ∪ ℰ;       // Add elites to selection
    𝒦 = ClusterMethod(𝒮);                          // Cluster the selection
    // Second phase - niche optimization
    foreach C ∈ 𝒦 do
        if ClusterMethod = HVC and x_(0) ∈ C is an elite e ∈ ℰ then
            continue;        // Skip already-optimized niches (HVC only)
        x = CoreSearchAlgorithm(C; N_c, ρ, ℰ);          // Run till convergence
        if x is a presumed distinct global optimum then
            ℰ = ℰ ∪ {x};                          // Add x to elitist archive
            foreach e ∈ ℰ that satisfies f(x) + ε < f(e) do
                ℰ = ℰ\{e};        // Remove low-fitness elites from archive
    if no new solution was added to ℰ then
        N ← 2.0 · N;                       // Increase initial population size
        N_c ← 1.1 · N_c;                              // Increase cluster size
```

clustering method and *core search algorithm*, which are discussed in respectively Sects. 3 and 4. A fundamental difficulty of real-valued multimodal optimization is to distinguish between (high-quality) local optima and global optima. Therefore, the user specifies an acceptable fitness tolerance $\varepsilon \in \mathbb{R}_{\geq 0}$. All solutions with a fitness value less than ε worse than the best obtained fitness value are considered to be a global optimum. This tolerance can be set *large* when the aim is to also locate (high-quality) local optima.

In the first phase, an initial population $\mathcal{P} \subset X$ of $|\mathcal{P}| = N$ solutions is sampled. Then, truncation selection is performed by selecting the $\lfloor \tau N \rfloor$ fittest solutions, and this set of selected solutions $\mathcal{S} \subset \mathcal{P}$ is clustered, i.e., $\mathcal{K} = \text{ClusterMethod}(\mathcal{S})$, with $\mathcal{K} = \{C_0, C_1, \ldots, C_{K-1}\}$ being the set of K clusters $C_i \subset X$, for which we will discuss three fitness-informed clustering methods in Sect. 3. By clustering only the selected solutions, less effort is spent on low-fitness regions in the search space. On the other hand, small niches could be accidentally discarded when the selection pressure is too high.

In the second phase, from each cluster $C \in \mathcal{K}$, a core search algorithm is initialized and run sequentially until convergence. Core search algorithms return a presumed optimal solution, i.e., $\mathbf{x} = \text{CoreSearchAlgorithm}(C; N_c, \rho, \mathcal{E})$. Three additional parameters need to be specified to initialize a core search algorithm. N_c is the population size used by the core search algorithm internally in the subsequent optimization of that niche. The initial search distribution multiplier $\rho \in \mathbb{R}_{>0}$ is used to initialize a cluster when it contains only one solution. In that case, the covariance matrix used by the core search algorithm is initialized by the identity matrix, multiplied by ρ. The aim is to set ρ small enough so that clusters are initialized within a single niche. In case of a bounded search space $X \subset \mathbb{R}^d$, when scattering N solutions equidistantly, the expected distance between two solutions is $\sqrt[d]{|X|/N}$, where $|X|$ is the volume of X. We therefore set $\rho = 0.01 \cdot \sqrt[d]{|X|/N}$, so that newly sampled solutions are expected to be closer-by than the previous set of selected solutions. Alternatively, when it is not possible to (easily) determine $|X|$, the average distance between solutions can instead be used to set ρ, as we will see in Sect. 3.3.1. Finally, the elitist archive $\mathcal{E}$ is provided to the core search algorithms which is used only to determine whether to terminate, as discussed in detail later in Sect. 4.1. Core search algorithms are run sequentially in the order of the fitness value of the best solution of the cluster from which they are initialized. This order prioritizes clusters that are more likely to result in a global optimum, which is important when optimization is performed with a limited budget, as is the case in this chapter.

Each run of a core search algorithm results in a presumably (locally) optimal solution $\mathbf{x}$. The user-defined fitness tolerance ε specifies whether a solution is added to the elitist archive, or whether existing solutions need to be removed from the elitist archive $\mathcal{E}$, similar to the post-processing step in [2]. Additionally, even though clusters are aimed to be initialized in different niches, it could be that specific global optima are obtained multiple times. To prevent this, without having to specify a niche-radius that determines whether two solutions are distinct, we use the hill-valley test [26] to determine whether two presumed optima belong to different niches. In the hill-valley test (see Algorithm 2), a straight line is drawn between two solutions in the search space, and the fitness is evaluated on $N_t = 5$ equidistantly located test solutions along this line. If the fitness of any of the test solutions is worse than the fitness of both solutions, there is a *hill* between the two solutions, and the two solutions presumably belong to different *valleys* (niches). As we will see in Sect. 3.3, the hill-valley test is also an essential part of HVC.

If there is budget remaining (in terms of function evaluations or computation time) after all core search algorithms are terminated, a restart is performed with a larger population size. Schemes like these that increase the population size of an EA over time, i.e., with restarts, are common to overcome setting these parameters [3, 7, 12].

Here, the initial population size is set to $N = 64$ (independent of the problem dimensionality d). The initial cluster size is set to $N_c = 0.8 \cdot N_c^{\text{rec}}$, where N_c^{rec} is the recommended population size of the corresponding core search algorithm, which we

Algorithm 2: Hill-Valley Test [20, 26]

function: $[B]$ =Hill-Valley($\mathbf{x}$, $\mathbf{y}$, N_t, f)
input : Solutions $\mathbf{x}, \mathbf{y} \in X$;
Number of test solutions $N_t \geq 0$;
Fitness function $f : X \to \mathbb{R}$; `// To be minimized`
output : $\mathbf{x}$ and $\mathbf{y}$ belong to the same niche?; `// Boolean`

for $k = 0, \ldots, N_t - 1$ **do**
  $\mathbf{x}_{\text{test}} = \mathbf{y} + \frac{k+1}{N_t+1}(\mathbf{x} - \mathbf{y})$;
  if $\max(f(\mathbf{x}), f(\mathbf{y})) < f(\mathbf{x}_{\text{test}})$ **then**
    return false;

return true;

will discuss in Sect. 4. Core search algorithms are initialized within a niche, which aims to make the problem landscape locally unimodal, and therefore, a cluster size is used that is smaller than recommended. At the end of a run, only if there were no new elites obtained in that run, we increase N by a factor of 2.0 and N_c by a factor of 1.1. This is to aid with locating smaller niches and optimizing more difficult niches over time. These values were based on previously obtained empirical results [17, 20].

2.1 *Initial Population Sampling*

It has been shown that better spreading the initially sampled population, compared to uniform random sampling, can improve the performance of MMO EAs [27]. Here, we use a combination of greedy scattered subset selection and rejection sampling (see Algorithm 3). Greedy scattered subset selection aims to select a diverse set of solutions from a population [25]. It starts with the solution that has the largest parameter value in a randomly chosen dimension. Then, iteratively, the solution that is furthest away from all previously selected solutions is added, until the desired number of solutions in the subset is reached. When the two-phase MMO EA is initialized, $\psi \cdot N$ solutions are sampled, which is reduced to N solutions using greedy scattered subset selection, where we use $\psi = 2$, empirically determined. After all core search algorithms have been terminated, and there is budget remaining, a new initial population is sampled. By rejection sampling, we aim to reduce the number of solutions in niches that are already explored. For this, let us denote the ith nearest neighbor (nn_i) of a solution $\mathbf{x} \in X$ within a set of solutions $\mathcal{P} \subset X$ by,

$$\text{nn}_i(\mathbf{x}; \mathcal{P}) = \underset{\mathbf{y} \in \mathcal{P} \setminus \{\mathbf{x}\}}{\arg\min} \left(\|\mathbf{x} - \mathbf{y}\| : \mathbf{y} \notin \{\text{nn}_j(\mathbf{x}; \mathcal{P})\}_{j=0}^{i-1} \right), \tag{2}$$

Algorithm 3: Initial Population Sampling [17]

```
function: [P] = sample(N, P^prev, K^prev, N_min, ψ)
input    : Desired number of solutions N ≥ 1;
           Previous population P^prev ⊂ X ⊆ R^d;
           Previous cluster set K^prev = {C_0^prev, C_1^prev, ...}, with C_i^prev ⊂ X;
           Number of nearest neighbors N_min ≥ 1;
           Sample ratio ψ ≥ 1.0;
output   : Population P ⊂ X of size |P| = N;

// Sample ψ · N solutions with rejection sampling
for i = 0, ..., ψ · N − 1 do
    x_i = sample_solution_uniformly();
    if |P^prev| > 0 then
        if nn(x_i; P^prev) ∉ C^prev for all C^prev ∈ K^prev then
            continue;                                                    // accept
        if C^prev_⟨nn(x_i;P^prev)⟩ ≠ C^prev_⟨nn_j(x_i;P^prev)⟩ for any j = 1, ..., N_min − 1 then
            if UniformRandom_[0,1] < 0.9 then set i = i − 1;            // reject

// Select final N solutions with greedy scattered subset
   selection
P = {x with maximum parameter value in random dimension j = 1, ..., d};
for i = 1, ..., N − 1 do
    P = P ∪ {arg max_x(||y − x|| : x ∉ P, y ∈ P)};        // add furthest solution
return P = {x_i, ..., x_N};
```

where $\|\cdot\|$ is the Euclidean distance, and we simply write $\text{nn}(\mathbf{x}; \mathcal{P}) = \text{nn}_0(\mathbf{x}; \mathcal{P}) = \arg\min_{\mathbf{y} \in \mathcal{P} \setminus \{\mathbf{x}\}} (\|\mathbf{x} - \mathbf{y}\|)$ for the nearest neighbor of $\mathbf{x}$ within $\mathcal{P}$. We store for each solution $\mathbf{x}$ of the previous initial population $\mathcal{P}^{\text{prev}}$ to which cluster it belonged. For this, we use the notation $C_{\langle \mathbf{x} \rangle}$ to the denote the cluster that contains $\mathbf{x}$. Then, again, $\psi \cdot N$ solutions are sampled, but now based on rejection sampling, where a sample is rejected with probability $p = 0.9$ if its nearest $N_{\text{min}} = d + 1$ solutions of the previous initial population belonged to the same cluster, i.e., when,

$$C_{\langle \text{nn}(\mathbf{x};\mathcal{P}^{\text{prev}}) \rangle} = C_{\langle \text{nn}_1(\mathbf{x};\mathcal{P}^{\text{prev}}) \rangle} = \cdots = C_{\langle \text{nn}_{N_{\text{min}}-1}(\mathbf{x};\mathcal{P}^{\text{prev}}) \rangle}. \tag{3}$$

Note that only the selection of fittest solutions is clustered, and not the entire population $\mathcal{P}^{\text{prev}}$. In case the nearest solution from the previous generation does not belong to any cluster, the newly sampled solution is always accepted, as this means that a high-fitness solution was obtained in a region that was considered of low fitness in the previous generation.

For only the resulting N solutions, the fitness value is computed. Note that this is however a relatively expensive approach with a computational complexity of $O(N^2 d)$, and when computation time is limited, standard uniform sampling may be preferred at the cost of a slight performance reduction.

3 Fitness-Informed Clustering

We discuss three fitness-informed clustering methods that aim to cluster a set of solutions, such that each cluster resides in a single niche. These clustering methods are fitness-informed in the sense that they are not only based on distances in the search space, but also incorporate corresponding fitness values.

3.1 Nearest-Better Clustering

In Nearest-Better Clustering (NBC) [22–24] (see Algorithm 4), a spanning tree of solutions is constructed by connecting each solution in the search space to the nearest solution that has better fitness, i.e., its *nearest-better solution*. We can formally define the nearest better (nb) of a solution $\mathbf{x} \in X$ within a set of solutions $\mathcal{S} \subset X$ by first defining,

$$\mathcal{S}^+(\mathbf{x}) = \{\mathbf{y} \in \mathcal{S} \backslash \{\mathbf{x}\} : f(\mathbf{y}) \leq f(\mathbf{x})\}, \tag{4}$$

as the subset of $\mathcal{S}$ of all solutions with equal or better fitness than $\mathbf{x}$. Then, let,

$$\text{nb}(\mathbf{x}) = \text{nn}(\mathbf{x}; \mathcal{S}^+(\mathbf{x})) = \arg\min_{\mathbf{y} \in \mathcal{S}^+(\mathbf{x})}(\|\mathbf{x} - \mathbf{y}\|), \quad \text{and,} \quad \delta(\mathbf{x}) = \|\mathbf{x} - \text{nb}(\mathbf{x})\|, \tag{5}$$

where we refer to $\delta(\mathbf{x})$ as the nearest-better distance of solution $\mathbf{x}$ (within $\mathcal{S}$). Ranking the solutions based on fitness allows for an efficient construction of the nearest-better tree. Denote the solution with rank i by $\mathbf{x}_{(i)}$, with $\mathbf{x}_{(0)}$ being the best solution, then $\mathcal{S}^+(\mathbf{x}_{(i)}) = \{\mathbf{x}_{(0)}, \ldots, \mathbf{x}_{(i-1)}\}$. Note that the best solution in the population, $\mathbf{x}_{(0)}$, has no nearest-better solution.

The rationale behind the *nearest-better tree* is that the best solutions found so far in the niche of a local optimum have no nearby solutions with better fitness, and therefore have a relatively long outgoing edge, i.e., a large nearest-better distance. By removing these edges, a number of disconnected sub-trees remains, and each sub-tree forms a cluster. To detect long edges, we compute the mean edge length,

$$\mu_\delta = \frac{1}{|\mathcal{S}|} \sum_{i=1}^{|\mathcal{S}|-1} \delta(\mathbf{x}_{(i)}), \tag{6}$$

and remove edges with length $\delta(\mathbf{x}) \geq \phi \cdot \mu_\delta$. It is not straightforward to calibrate these cutting rules, as distances between solutions depend both on the problem dimensionality and the number of solutions [24]. We use $\phi = 2.0$ in this work, adhering to literature on NBC [22]. In terms of computational complexity, generating the NBC can be performed in $\mathcal{O}(|\mathcal{S}|^2 d)$.

Algorithm 4: Nearest-Better Clustering (NBC) [22]

function: $[\mathcal{K}] = \text{NBC}(\mathcal{S})$
input : Set of solutions $\mathcal{S} \subset X$;
output : Set of clusters $\mathcal{K} = \{C_0, C_1, \ldots, C_{K-1}\}$, with $C_i \subset X$;

Rank solutions $\mathbf{x} \in \mathcal{S}$ by fitness value, such that $\mathbf{x}_{(0)}$ is the best solution;
$C_0 := \{\mathbf{x}_{(0)}\}$; $K = 1$;
$\phi = 2.0$;
$\mu_\delta = \frac{1}{|\mathcal{S}|} \sum_{i=1}^{|\mathcal{S}|-1} \delta(\mathbf{x}_{(i)})$; // Note that $\delta(\mathbf{x}_{(0)})$ does not exist
for $\mathbf{x} \in \mathcal{S}, \mathbf{x} \neq \mathbf{x}_{(0)}$ **do**
  if $\delta(\mathbf{x}) < \phi \cdot \mu_\delta$ **then**
    Add $\mathbf{x}$ to cluster $C_{\langle \text{nb}(\mathbf{x}) \rangle}$ of $\text{nb}(\mathbf{x})$;
  else
    New cluster $C_K := \{\mathbf{x}\}$; $K = K + 1$;
return $\mathcal{K} = \{C_0, \ldots, C_{K-1}\}$;

3.2 *Hierarchical Gaussian Mixture Learning*

In hierarchical Gaussian mixture learning (HGML) [19] (see Algorithm 5), solutions are clustered such that, when fitting a Gaussian mixture model (GMM) to the clusters, it correlates well with the fitness landscape. In that way, when sampling from the GMM, high-fitness solutions are expected. A GMM was used because in the second phase, niches will be explored with core search algorithms that are also Gaussian-based.

3.2.1 Gaussian Mixture Model

Let $\mathcal{N}_\theta$ be a d-dimensional Gaussian distribution, parameterized by $\theta = (\boldsymbol{\mu}, \Sigma)$, where $\boldsymbol{\mu} \in \mathbb{R}^d$ is the distribution mean and $\Sigma \in \mathbb{R}^{d \times d}$ the covariance matrix. We denote its probability distribution function for $\mathbf{x} \in \mathbb{R}^d$ by $p(\mathbf{x}; \theta)$. To fit a Gaussian distribution to a set of data points, one could use the Maximum-Likelihood Estimator (MLE), which yields a closed form solution [10, 13].

A natural extension of the Gaussian distribution is the GMM,

$$p_K(\mathbf{x}; \Theta) = \sum_{k=0}^{K-1} w_k p(\mathbf{x}; \theta_k), \tag{7}$$

with K mixture components, where $\Theta = \{(w_k, \theta_k)\}_{k=0,\ldots,K-1}$ is the set of distribution parameters and w_k are the positive mixing weights, summing up to one. For a GMM, there is no closed form MLE, but estimates can for instance be found via expectation-maximization [10]. This is however a computationally expensive algorithm, even for a fixed number of components K. It is computationally cheaper to first cluster the

Algorithm 5: Hierarchical Gaussian Mixture Learning (HGML) [19]

```
function: [𝒦] = HGML(𝒮)
input   : Set of solutions 𝒮 ⊂ X;
output  : Set of clusters 𝒦 = {C_0, C_1, ..., C_{K-1}}, with C_i ⊂ X;
Rank solutions x ∈ 𝒮 by fitness value, such that x_(0) is the best solution;
C_0 = {x_(0)};
E = {};                      // List of edges of the nearest-better tree
for i = 1, ..., |𝒮| - 1 do
    C_i = {x_(i)};
    E = E ∪ {(C_i, C_⟨nb(x_(i))⟩)};        // Edge from C_i to cluster of
    nearest-better
𝒦_0 := {C_0, ..., C_{|𝒮|-1}};
for n = 1, ..., N - 1 do
    Find the shortest edge (C_f, C_t) ∈ E;
    Create new cluster by merging C = C_f ∪ C_t;
    Initialize new cluster set 𝒦_n := 𝒦_{n-1} \ {C_f, C_t} ∪ C;
    Delete edge (C_f, C_t) from E;
    Replace all occurrences of C_f and C_t in E by the new C;
n⁺ = arg max_n (Round (DFC(𝒦_n), 0.05));
if DFC(𝒦_{n⁺}) < 0 then
    n⁺ = N - 1;
return 𝒦 = 𝒦_{n⁺};
```

data into K clusters. As each solution then belongs to only one cluster, we can subsequently use the MLE to estimate a mixture component on each of the clusters, and set the weights inversely proportional to the number of solutions in each cluster. By using this approach, a set of clusters can be directly and uniquely associated with a GMM. Due to this simplification, overlapping mixture components are no longer possible, which is desirable for niching approaches, as niches are also non-overlapping.

A fundamental difficulty of mixture models is how to determine the number of mixture components K, as more mixture components will inherently result in a better fit of the data [13]. In our case, rather than fitting the data points, the primary target is to fit a GMM that maximizes the probability of sampling high-fitness solutions. We will use this idea to determine a suitable number of mixture components.

3.2.2 Density-Fitness Correlation

Density-Fitness Correlation (DFC) was introduced as a tool to determine whether the probability distribution of an estimation-of-distribution algorithm needs to be adapted for a better match with the fitness landscape [5]. Here, we use the DFC to test, in a greedy way, which GMM is the best match with the structure of the problem at the current state of the search.

Given a GMM, parameterized by Θ, as in Eq. (7), and a set of solutions $\mathcal{S} \subset X$, we compute the DFC as follows. For each solution $\mathbf{x} \in \mathcal{S}$, let $D(\mathbf{x}; \Theta, \mathcal{S}) = \text{Rank}(p_K(\mathbf{x}; \Theta)|\mathcal{S})$ be the density rank of $\mathbf{x}$ under the GMM parameterized by Θ, such that the solution with the highest probability density has rank 0. Furthermore, let $F(\mathbf{x}; \mathcal{S}) = \text{Rank}(f(\mathbf{x})|\mathcal{S})$ be the fitness rank of $\mathbf{x}$, such that the best solution has rank 0. Then, the DFC is given by the Spearman rank correlation between the density and fitness ranks,

$$\text{DFC}(\Theta, \mathcal{S}) = 1 - \frac{6 \sum_{i=0}^{|\mathcal{S}|-1} \left[F(\mathbf{x}_i; \mathcal{S}) - D(\mathbf{x}_i; \Theta, \mathcal{S})\right]^2}{|\mathcal{S}|(|\mathcal{S}|^2 - 1)}, \tag{8}$$

and takes values in $[-1, 1]$. The larger the DFC, the higher the probability that high-fitness solutions are sampled. In our case, as the solutions are clustered first, the cost of computing the DFC can be reduced by computing the DFC per cluster, i.e., per Gaussian $\mathcal{N}_{\theta_k}$. The DFC of the GMM is then the average of the clusters DFC values, weighted by the number of solutions in each cluster, i.e., for a cluster set $\mathcal{K} = \{C_k\}_{k=0}^{K-1}$, with $\theta_k = (\boldsymbol{\mu}_k, \Sigma_k)$ corresponding to cluster C_k, we obtain,

$$\text{DFC}(\mathcal{K}) = \sum_{k=0}^{K-1} \frac{1}{|C_k|} \text{DFC}(\theta_k, C_k). \tag{9}$$

Using the DFC, we can select the best GMM out of a set of candidate GMMs.

3.2.3 Hierarchical Clustering

To generate a set of candidate GMMs, a bottom-up hierarchical clustering approach is used where, initially, each solution is considered to be a separate cluster. Then, iteratively, clusters are merged until one large cluster remains that contains all solutions. After each merge, a GMM is fit to the current cluster set, and the corresponding DFC is computed using Eq. (9). When clustering $|\mathcal{S}|$ solutions, this results in $|\mathcal{S}|$ cluster sets, each corresponding to a GMM. The cluster set with the best DFC is selected as final cluster set.

The merge order is however important, as solely merging based on distance in the search space might result in solutions of two different niches being merged only because they are close in the search space while they may in fact belong to different local optima. To reduce this effect, we make use of the nearest-better tree that we discussed in Sect. 3.1. Edge lengths in the nearest-better tree were only defined as the (Euclidean) distance between two solutions. We naturally extend this definition by defining the distance between two clusters as the Euclidean distance between the cluster means in the search space. We then iteratively merge the two clusters corresponding to the shortest edge in the nearest-better tree. We recall from Sect. 3.1 solutions with long outgoing edges are expected to be (local) optima, and by iteratively merging the shortest edge in the tree, these solutions get merged latest.

Note that, when fitting a Gaussian using MLE, as least $N_{\min} = d + 1$ solutions are required for each cluster in the cluster set to estimate a stable $d \times d$ covariance matrix. Therefore, we ignore all cluster sets that contain any cluster with less than $N_{\min}$ solutions by setting the corresponding DFC to -1, i.e., the worst attainable value.

GMM estimates are prone to statistical noise, especially in smaller clusters, thus if two clusters have a similar DFC, we prefer the larger cluster. To realize this, the DFCs are rounded with an accuracy of 0.05, which was empirically determined [19]. If two rounded DFCs have the same value, the larger cluster is chosen. If all the DFCs are negative, which implies that all of the estimated GMMs correlate poorly with the problem structure, we fall back to using a single cluster that contains all solutions.

When using a lower bound of $N_{\min} = d + 1$ solutions per cluster, HGML has a computational complexity of $\mathcal{O}(|\mathcal{S}|d^2 + |\mathcal{S}|^2 d)$ [19].

3.3 Hill-Valley Clustering

The final clustering method that we discuss in this chapter is hill-valley clustering (HVC) [18, 20] (see Algorithm 6). As discussed previously, in NBC, solutions were clustered together with their nearest-better solution, as long as these are *nearby*. It might however be that even nearby solutions belong to a different niche. In HVC, we explicitly check for this using the hill-valley test (see Algorithm 2), which is also used to determine whether two obtained elites are distinct global optima. In HVC, if a solution does not belong to its nearest-better solution, it is checked whether it belongs to its next-nearest better. For this, we define the ith nearest-better solution of a solution $\mathbf{x} \in X$ within a set of solutions $\mathcal{S} \subset X$ as $\text{nb}_i(\mathbf{x}) = \text{nn}_i(\mathbf{x}; \mathcal{S}^+(\mathbf{x}))$, with $\mathcal{S}^+(\mathbf{x})$ defined as in Eq. (4).

To cluster a set of solutions $\mathcal{S} \subset X$, HVC is initialized by forming the first cluster $C_0 = \{\mathbf{x}_{(0)}\}$ from the best solution $\mathbf{x}_{(0)} \in \mathcal{S}$. Then, we consider the second-best solution $\mathbf{x}_{(1)} \in \mathcal{S}$, and test whether it belongs to the same niche as its nearest-better solution, $\text{nb}(\mathbf{x}_{(1)}) = \mathbf{x}_{(0)}$, using the hill-valley test. When it does, $\mathbf{x}_{(1)}$ is added to the cluster of $\mathbf{x}_{(0)}$. Otherwise, as there are no other better solutions, a new cluster is formed, i.e., $C_1 = \{\mathbf{x}_{(1)}\}$. Next, the third-best solution $\mathbf{x}_{(2)} \in \mathcal{S}$ is tested against the nearest solution that has better fitness, which can either be $\mathbf{x}_{(0)}$ or $\mathbf{x}_{(1)}$, depending on which one is nearer. If $\mathbf{x}_{(2)}$ does not belong to the same niche as its nearest-better solution $\text{nb}(\mathbf{x}_{(2)})$, we test it against $\text{nb}_1(\mathbf{x}_{(2)})$. If it also does not belong to that niche, we create a new cluster from $\mathbf{x}_{(2)}$. For each solution, we check at most its $N_{\min} = d + 1$ nearest-better solutions. In this way, even for the one-dimensional problems, at least two neighboring solutions are considered. This procedure is then repeated for all solutions.

Algorithm 6: Hill-Valley Clustering (HVC) [20]

```
function: [𝒦] = HVC(𝒮, f)
input   : Set of solutions 𝒮 ⊂ X;
          Fitness function f : X → ℝ, with X ⊆ ℝ^d;        // To be minimized
output  : Set of clusters 𝒦 = {C_0, C_1, ..., C_{K-1}}, with C_i ⊂ X;
Rank solutions x ∈ 𝒮 by fitness value, such that x_(0) is the best solution;
C_0 := {x_(0)}; K = 1;
for x ∈ 𝒮 do
    for j = 0, ..., d − 1 do
        y = nb_j(x);   // Break if there are no more better solutions
        if already checked cluster of y then
            continue;
        N_t = 1 + ⌊ ||x − y|| / (|X|/|𝒮|)^(1/d) ⌋;      // Or N_t = 1 + ⌊||x − y||/μ_δ⌋
        if Hill-Valley(x, y, N_t, f) or (N_t = 1 and Rank(f(x)|𝒮) ≥ 0.5 · |𝒮|) then
            Add x to cluster of y;
            break;
    if x was not added to any cluster then
        C_K := {x}; K = K + 1;
return 𝒦 = {C_0, ..., C_{K-1}}
```

3.3.1 Number of Test Solutions N_t

In the post-processing step to filter out global optima that are obtained multiple times, the hill-valley test was used with $N_t = 5$ test solutions. We would like to use fewer test solutions during the clustering process, however, but to have more test solutions when two solutions are further apart. We, therefore, increase the number of test solutions with the distance between the two solutions, scaled by the expected distance between two solutions in the selection, similar to how ρ was set in Sect. 2, i.e.,

$$N_t = 1 + \left\lfloor \frac{\|\mathbf{x} - \mathbf{y}\|}{\sqrt[d]{|X|/|\mathcal{S}|}} \right\rfloor. \tag{10}$$

When the search space is unbounded, or its volume $|X|$ cannot be determined easily, we can replace the expected distance $\sqrt[d]{|X|/|\mathcal{S}|}$ by the mean distance between nearest-better solutions μ_δ as given in Eq. (6), i.e., by setting $N_t = 1 + \lfloor \|\mathbf{x} - \mathbf{y}\|/\mu_\delta \rfloor$.

3.3.2 Reducing Function Evaluations within HVC

Two extensions are added to hill-valley clustering to reduce the number of function evaluations. First, if the nearest-better solution and a next-nearest-better solution belong to the same cluster, we only perform the hill-valley test on the nearest-better solution, and if it is rejected, we reject the next-nearest-better solution without performing the hill-valley test.

Second, it is generally not required to accurately model the low-fitness search space. Therefore, we automatically cluster two solutions $\mathbf{x}, \mathbf{y} \in \mathbb{R}$, with $f(\mathbf{y}) \leq f(\mathbf{x})$ together when they are close to each other and both belong to the worst 50% of the set $\mathcal{S} \subset X$ that is being clustered, i.e., when $N_t = 1$ and $\text{Rank}(f(\mathbf{x})|\mathcal{S}) \geq 0.5 \cdot |\mathcal{S}|$, without performing the hill-valley test.

Additionally, when the hill-valley test has been performed, and when two solutions are clustered together, all test solutions generated by the hill-valley test are added to the cluster as well. When the hill-valley test rejected a solution pair, the test solutions that have been evaluated so far are added to the cluster of the worst, which is the endpoint of the line from which sampling starts. However, the solution that violated the hill-valley test, and thus has worse fitness than the two solutions, is discarded.

3.3.3 Preventing Re-optimization of Niches

To prevent niches from being re-explored after a restart of the framework, all elites are added back to the population before clustering. After clustering, when the best solution in a cluster is an elite, no core search algorithm is initialized from that cluster. This approach can however only be applied when HVC is used as clustering method, as HVC tests whether solutions actually belong to the same niche as an elite, while the other clustering methods simply cluster them to any nearby solution.

4 Core Search Algorithms

We consider the core search algorithms listed in Table 1, which are CMSA [4] and different versions of AMaLGaM [6]. AMaLGaM is an estimation-of-distribution algorithm, where a Gaussian distribution is fitted to selected solutions with maximum likelihood and adapted based on whether improvements are found near or far from the mean. In AMaLGaM, a full covariance matrix is estimated, while a univariate covariance matrix is estimated in AMaLGaM-Univariate. The latter is therefore cheaper in terms of memory and computational effort, as only the diagonal of the covariance matrix needs to be computed and stored. This is especially relevant for problems with a larger problem dimensionality d. More relevant in our case is however that estimating an accurate full-rank covariance matrix requires a larger population size, which implies that fewer generations of AMaLGaM can be run compared to AMaLGaM-Univariate for the same computational budget. In iAMaLGaM (-Univariate), the covariance matrix is estimated incrementally over multiple generations, which allows for a smaller population size.

CMSA was used instead of the more common CMA-ES [11], as it was suggested to perform better when adding elitism [2]. CMSA fits a Gaussian using the population mean of the previous generation, essentially modeling the direction of improvements.

Table 1 Different core search algorithms with corresponding recommended settings as taken from literature: selection pressure $\tau \in (0, 1]$, population size N_c^{rec} for problem dimensionality d

Core search algorithm	Abbreviation	τ^{rec}	N_c^{rec}
CMSA [4]	CMSA	0.50	$3 \log d$
AMaLGaM [6]	AM	0.35	$17 + 3 \cdot d\sqrt{d}$
AMaLGaM-Univariate [6]	AMu	0.35	$10\sqrt{d}$
iAMaLGaM [6]	iAM	0.35	$10\sqrt{d}$
iAMaLGaM-Univariate[6]	iAMu	0.35	$4\sqrt{d}$

4.1 *Termination Criteria for Core Search Algorithms*

As core search algorithms are run in serial, additional termination criteria are required besides the budget defined in terms of function evaluations. AMaLGaM-variant core search algorithms are terminated if the maximum standard deviation of solutions in the search space is too low (10^{-12}), or if the standard deviation of fitness values is too low (10^{-12}) [6]. CMSA is terminated using the recommended criteria [11], and parameters set as in RS-CMSA [2], that is, if the improvement in fitness value over the last $10 + \lfloor 30d/N_c \rfloor$ generations is less than 10^{-5}. Furthermore, fail-safe termination criteria are added to terminate CMSA if the standard deviation of solutions in the search space reaches machine accuracy (10^{-15}) and if the condition number of the covariance matrix is larger than 10^{14}.

4.1.1 Termination When Converging to Elite

In an attempt to prevent core search algorithms from re-exploring niches, in every fifth generation of a core search algorithm the best obtained solution is compared to the nearest-better solution in the elitist archive using the hill-valley test (with $N_t = 5$ test solutions). When it is found that these two solutions belong to the same niche, the core search algorithm is terminated.

4.1.2 Termination When Converging to Local Optimum

The final stopping criterion that we introduce is used to detect whether a core search algorithm is converging to a local minimum. As a proxy for the global minimum, let b be the fitness value of the best solution in the elitist archive. Let a_g be the average fitness of the selected solutions in generation g and $\Delta_g := a_g - (b + \varepsilon)$ the fitness difference with the best elite, incorporating the user-defined tolerance ε of whether a solution can be considered a global optimum. If $\Delta_g \leq 0$, the current selection is better than the so-far obtained elites, and we do not terminate. To predict the value of Δ_{g+1}, we use the fact that AMaLGaM was shown to have exponential convergence on smooth unimodal functions such as the sphere function [6]. Under this assumption, we can write $\Delta_{g+1} = \Delta_g(1-r)$, where r is the rate of convergence. We estimate r by r_n over the previous n generations by,

$$r \approx r_n = 1 - \left(1 - \frac{\Delta_{g-n} - \Delta_g}{\Delta_{g-n}}\right)^{1/n}, \tag{11}$$

with $n = 5$ to reduce statistical noise. To prevent premature termination, this criterion is only applied when Δ_g decreased in the most recent $n = 5$ generations. Finally, we estimate the *time to optimum* (t_o) in order to achieve $\Delta_{g+t_o} = \Delta_g(1-r)^{t_o} = 10^{-12}$, again under the assumption of exponential convergence. Rewriting this in terms of t_o gives,

$$t_o = \frac{\log\left(10^{-12}/\Delta_g\right)}{\log(1-r)} \approx \frac{\log\left(10^{-12}/\Delta_g\right)}{\log(1-r_n)} = \frac{\log\left(10^{-12}/\Delta_g\right)}{\frac{1}{n}\log\left(\Delta_g/\Delta_{g-n}\right)}. \tag{12}$$

The idea is that t_o increases rapidly when converging to a local optimum. Therefore, a core search algorithm is terminated if $g + t_o$ exceeds 50 times the maximum number of generations it took to find any elite in the elitist archive.

5 Experiments

We compare the discussed methods on the test problems of the CEC 2013 niching benchmark suite [14]. This benchmark consists of 20 problems, to be solved within the specified budget in terms of function evaluations (see Table 2). All benchmark problems are defined on a bounded domain. For each of the benchmark problems, the location of the optima and the corresponding fitness values are known, however, these are only used to measure performance, and are not used during optimization.

One key performance measure for MMO EAs is the commonly used peak ratio (PR), defined as the fraction of obtained distinct global optima over the total number of global optima for the problem at hand. We use the benchmark guidelines to determine whether a global optimum is attained [14], at an accuracy level of $\varepsilon = 10^{-5}$. All experiments in this chapter are run with this tolerance level.

Besides the peak ratio, a secondary performance measure that is commonly used is defined as the fraction of obtained *distinct* global optima within the provided solution set. This measure aims to overcome a scenario in which an MMO EA simply returns all obtained solutions, without giving insight whether solutions actually belong to different modes. In our MMO EA framework, the elitist archive is post-processed to remove local optima and duplicate global optima, in a similar fashion as in RS-CMSA. Because of the post-processing step, the perfect score for this measure is always achieved, and it is therefore left out of this chapter.

Table 2 Niching benchmark suite from the CEC 2013 special session on multimodal optimization [14]. For each problem, the function name, problem dimensionality d, number of global optima $\#gopt$, local optima $\#lopt$, and budget in terms of function evaluations are given

P	Function name	d	#gopt	#lopt	Budget (K)
1	Five-Uneven-Peak Trap	1	2	3	50
2	Equal Maxima	1	5	0	50
3	Uneven Decreasing Maxima	1	1	4	50
4	Himmelblau	2	4	0	50
5	Six-Hump Camel Back	2	2	5	50
6	Shubert	2	18	many	200
7	Vincent	2	36	0	200
8	Shubert	3	81	many	400
9	Vincent	3	216	0	400
10	Modified Rastrigin	2	12	0	200
11	Composition Function 1	2	6	many	200
12	Composition Function 2	2	8	many	200
13	Composition Function 3	2	6	many	200
14	Composition Function 3	3	6	many	400
15	Composition Function 4	3	8	many	400
16	Composition Function 3	5	6	many	400
17	Composition Function 4	5	8	many	400
18	Composition Function 3	10	6	many	400
19	Composition Function 4	10	8	many	400
20	Composition Function 4	20	8	many	400

Unless mentioned otherwise, all experiments are repeated 50 times, and resulting performance measures are averaged over all repetitions. Results are tested for statistical significance using the one-sided unpaired Wilcoxon rank-sum test at $\alpha = 0.05$. Additionally, the Bonferroni correction is applied by dividing α by the number of tests performed in each experiment. Source code of HillVallEA and the two-phased MMO framework used for the experiments in this chapter is available online [16].

5.1 Experiment 1: Clustering Comparison

Setup. We investigate how many unique niches (corresponding to global optima) can be located when clustering a population of a given population size with the clustering methods in Sect. 3, without performing subsequent optimization of the located niches. We define a niche to be located when the best solution of a cluster is in that niche. For smooth functions (i.e., benchmark problems 1–10), this is easily

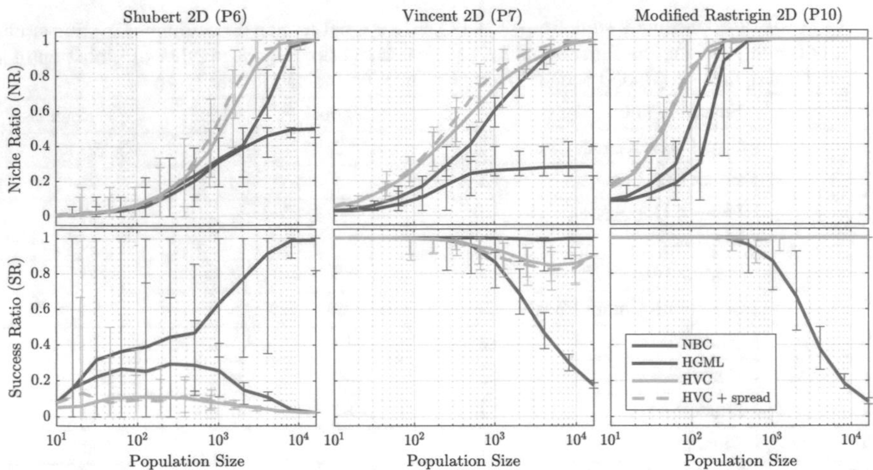

Fig. 1 Niche Ratio (NR) and Success Ratio (SR) for the three clustering methods discussed in Sect. 3, for different population size. Error bars represent min/max scores. *HVC + spread* is HVC applied on a better spread of the initial solutions in the population as described in Sect. 2.1

tested for with the hill-valley test as the locations of their optima are known. The niche ratio (NR) is then given by the number of located niches as a fraction of the total number of niches of the problem. Furthermore, we define the success ratio (SR) as the number of different niches located as a fraction of the number of clusters. Both NR and SR should be maximized and when $NR = SR = 1$, all niches are obtained exactly once.

We compare the clustering methods on three problems with $d = 2$ variables that exhibit different landscape features. The Modified Rastrigin function (P10) that has no local optima and for which all niches are the same shape and size. The Vincent function (P7) that also has only global optima, but of different shape and size, and especially has a few very small niches. Finally, the Shubert function (P6) that has many local optima and global optima with small niches.

All methods cluster the same set of initial solutions, and when the population size is increased, more solutions are added instead of re-sampling the entire population, to reduce randomness. The extra function evaluations used by hill-valley clustering are included in the results. Additionally, we show the improvement in performance for HVC when using a better spread of the sampled population as described in Sect. 2.1.

Results. Results are shown in Fig. 1. Overall, the Modified Rastrigin function is rather simple, as it can be clustered with only few solutions, and both HGML and HVC obtain the perfect score, i.e., $NR = SR = 1$. NBC also obtains a perfect score for a population size of around 10^3 solutions, but as the population size increases, it tends to overestimate the number of clusters. This same behavior is observed for all three test problems. This is because of the cutting rule, as described in Sect. 3.1, that has a fixed threshold. The results clearly indicate that it should be adapted to the population size, as this influences the distribution of distances between neighboring solutions.

For all problems, HVC achieves the best NR for the three test problems, which includes the additional function evaluations, and shows that it is worth spending these additional evaluations in order to locate more niches versus simply increasing the population size of NBC. Additionally, better spreading the initial sampled population shows to further improve HVC.

However, HVC aims to locate all niches, including the ones corresponding to local optima. This can be observed for the Shubert function, for which HVC obtains a large number of clusters. The Vincent and Modified Rastrigin problem have no local optima, therefore, the SR of HVC is closer to 1.0 for these problems. For the Vincent function, the SR deteriorates slightly when the population size increases. This happens because HVC only checks a limited number of nearby solutions ($N_{\text{min}} = d + 1 = 3$ solutions). For the very stretched niches of the Vincent function, and when the population size is not too large, it might be that all those solutions belong to a different niche. Therefore, multiple clusters are (incorrectly) initialized in these stretched niches, and checking more nearby solutions reduces this effect at the cost of additional function evaluations.

HGML is not able to distinguish peaks that are close to each other for both the Shubert and Vincent problem. This is a result of the hierarchical clustering process that is based on the nearest-better distances. For the Shubert function, peaks are divided over nine groups of two optima. HGML obtains all 9 groups correctly, but, as the two peaks in each group are very close to each other, a single Gaussian fits rather similar to these two peaks than two separate ones. Additionally, the hierarchical clustering order is based on nearest-better distances, and since these two peaks are very close to each other, these solutions get wrongly merged into the same cluster early in the merging process. This is also the reason why NBC performs worse on the Shubert function than on the Vincent function compared to HVC.

On the other hand, HGML performs best in terms of the SR, as it obtains $SR = 1$ when the population size is large enough, as each Gaussian it fits accurately captures one (or multiple) modes. To improve the niche ratio of HGML, clustering can be applied every generation, as in [19].

5.2 *Experiment 2: Core Search Algorithms and Clustering Methods*

Setup. We equip the framework for two-phase MMO EAs with each of the five core search algorithms in Table 1 and each of the three clustering methods of Sect. 3. The resulting 15 different two-phase MMO EAs are run on the 20 benchmark problems.

Results. Table 3 shows the obtained average peak ratios for the 15 different two-phase MMO EAs. We refer to the MMO EAs equipped with HVC as the Hill-

Valley Evolutionary Algorithm (HillVallEA). The best average peak ratio (0.89) was obtained by HillVallEA with AMu as core search algorithm (HillVallEA-AMu).

HVC performed generally best, closely followed by NBC, and with quite an improvement over HGML, which is in line with the results obtained in the previous experiment. In terms of core search algorithms, AMu and iAM have the same recommended population size, which allows for the same number of restarts, and the resulting performance is very similar in all cases. A reason could be that the used cluster size, $N_c = 0.8 \cdot N_C^{\text{rec}}$, is too small for iAMu, and therefore deteriorates performance. CMSA sometimes *jumps* out of its own niche shortly after initialization. Because of that, even when small niches are located in the clustering process, they might get lost during optimization, resulting in a lower peak ratio when using HVC or NBC. HGML however finds fewer clusters upon initialization, and jumping out of the located niche seems therefore beneficial, resulting in the best performance of HGML when combining it with CMSA.

In Table 4, the peak ratio per problem obtained with two-phase MMO EAs equipped with AMu and the three clustering methods is shown. For problems 8 and 14, NBC performs slightly better than HVC. However, for problems 9 and 20, HVC greatly outperforms the other two clustering methods. The restart scheme largely overcame that HGML was unable to locate all peaks for problems 6 and 7, but this approach still resulted in a lower peak ratio for almost all problems.

5.3 Experiment 3: MMO EA Comparison

Setup. We compare HillVallEA-AMu, i.e., the two-phase MMO EA equipped with HVC and AMu, to other MMO EAs that performed well in the niching competitions held at the GECCO and CEC conferences in 2016, 2017, and 2018. The included algorithms are NEA2+ [23], RLSIS [27], and RS-CMSA [1].

Results. Table 5 shows the peak ratio per problem. HillVallEA-AMu obtains the best peak ratios for all problems, followed by RS-CMSA, NEA2+, and finally RLSIS.

Table 3 Peak ratios (PR) obtained by equipping the framework for two-phase MMO EAs with different core search algorithms and clustering methods, averaged over all 20 benchmark problems

Clustering Method \ Core Search Algorithm	AM	AMu	CMSA	iAM	iAMu
HVC	0.869	0.892	0.855	0.887	0.748
NBC	0.857	0.873	0.868	0.873	0.798
HGML	0.702	0.715	0.727	0.72	0.707

Table 4 Peak Ratios (PR) obtained by equipping the framework for two-phase MMO EAs with AMu and the three clustering methods, per problem, and the average (avg) over all 20 problems. All scores are averaged over 50 runs. Bold scores are the best obtained per problem, and those not significantly different from it. For problems 1–6 and 10, all instances achieves a perfect score of 1, and are therefore not shown

MMO EA	Avg. PR	Benchmark problem												
		7	8	9	11	12	13	14	15	16	17	18	19	20
AMu+HVC	**0.892**	**1.00**	0.97	**0.97**	**1.00**	**1.00**	**1.00**	0.92	**0.75**	**0.72**	**0.75**	**0.67**	**0.59**	**0.48**
AMu+NBC	0.873	**1.00**	**0.98**	0.71	**1.00**	**1.00**	**1.00**	**0.98**	**0.75**	**0.77**	**0.75**	**0.67**	**0.61**	0.24
AMu+HGML	0.715	0.67	0.92	0.37	0.95	0.90	0.64	0.78	**0.73**	0.53	0.42	0.19	0.08	0.12

Problems 1–5 and 10 are fully solved by all methods (except for two runs of NEA2+). These problems are too simple to give insight in the performance of different algorithms. None of the MMO EAs can obtain the final two Weierstrass peaks of Composition Function 4 for any dimensionality (problem 15, 17, 19, and 20) within the given computational budget. Specifically, for problems 15, 17, and 18, the standard deviation of the obtained peak ratio for HillVallEA-AMu is zero, which suggests that the final peaks cannot be obtained even when run with a larger budget, which is something we investigate in the next experiment. Again, note that, due to the post-processing step in HillVallEA-AMu and RS-CMSA, both algorithms only obtain distinct global optima, and the number of obtained solutions is therefore equal to the number of located optima.

5.4 Experiment 4: Larger Budget

Setup. To show and study maximum performance, we compare HillVallEA-AMu with the best competitor, RS-CMSA, on a computational budget 100 times larger than the original budget. Due to computational limits, this experiment was repeated only 20 times. We consider the problems on which HillVallEA-AMu was not able to obtain a peak ratio of 1 in the previous experiment, i.e., the Shubert and Vincent problems in 3D, and CF3 and CF4 with different problem dimensionalities.

Results. The result for the larger budget is shown in Fig. 2. HillVallEA-AMu outperforms RS-CMSA in all cases by achieving a higher peak ratio for the same number of evaluations. For problem 8 and 9, both methods obtain all peaks when the computational budget is large enough, which suggests that both algorithms are fundamentally sound. The difference between HillVallEA-AMu and RS-CMSA is most clear for the Vincent function (P9), which has a number of very small niches that are difficult to locate, and lacks local optima. Due to its design, HillVallEA-AMu performs especially well in these two scenarios.

For CF3, the PR curve shows that even for $d = 2$, two of the peaks (corresponding to the Weierstrass function) are difficult to obtain. This effect is enhanced to the extent

Table 5 Peak ratios (PR) per problem (P) for different MMO EAs. Higher scores are better and 1 is the maximum achievable score. Mean peak ratios over 50 runs are shown, together with the standard deviation. Average (avg.) computed over all 20 problems. Bold scores are the best obtained per problem, and those not significantly different from it

Problem	HillVallEA-AMu	RS-CMSA	NEA2+	RLSIS
1	**1.000** ±0.000	**1.000** ±0.000	**1.000** ±0.000	**1.000** ±0.000
2	**1.000** ±0.000	**1.000** ±0.000	**1.000** ±0.000	**1.000** ±0.000
3	**1.000** ±0.000	**1.000** ±0.000	**1.000** ±0.000	**1.000** ±0.000
4	**1.000** ±0.000	**1.000** ±0.000	**0.990** ±0.049	**1.000** ±0.000
5	**1.000** ±0.000	**1.000** ±0.000	**1.000** ±0.000	**1.000** ±0.000
6	**1.000** ±0.000	**0.999** ±0.008	**0.991** ±0.023	0.872 ±0.078
7	**1.000** ±0.000	**0.997** ±0.008	0.810 ±0.046	0.920 ±0.025
8	**0.975** ±0.016	0.871 ±0.032	0.567 ±0.045	0.189 ±0.039
9	**0.972** ±0.011	0.730 ±0.018	0.539 ±0.014	0.584 ±0.016
10	**1.000** ±0.000	**1.000** ±0.000	**0.988** ±0.033	**1.000** ±0.000
11	**1.000** ±0.000	**0.997** ±0.023	0.943 ±0.086	**1.000** ±0.000
12	**1.000** ±0.000	0.948 ±0.062	0.785 ±0.094	0.950 ±0.066
13	**1.000** ±0.000	**0.997** ±0.023	0.937 ±0.094	0.793 ±0.103
14	**0.923** ±0.090	0.810 ±0.082	0.810 ±0.075	0.703 ±0.069
15	**0.750** ±0.000	**0.748** ±0.017	0.718 ±0.060	0.720 ±0.059
16	**0.723** ±0.079	0.667 ±0.000	**0.683** ±0.050	0.670 ±0.023
17	**0.750** ±0.000	0.703 ±0.066	0.723 ±0.057	**0.738** ±0.037
18	**0.667** ±0.000	**0.667** ±0.000	**0.650** ±0.050	**0.667** ±0.000
19	**0.593** ±0.078	0.502 ±0.017	0.505 ±0.075	0.515 ±0.041
20	**0.480** ±0.046	**0.482** ±0.043	0.380 ±0.093	0.422 ±0.075
Avg.	**0.892** ±0.030	0.856 ±0.024	0.801 ±0.032	0.787 ±0.032

that for problem 18, with $d = 10$, these two peaks are never obtained, even for the case of the extended computational budget. For CF4, the two peaks of the Weierstrass functions are never obtained, not even for $d = 2$. This suggests that these optima

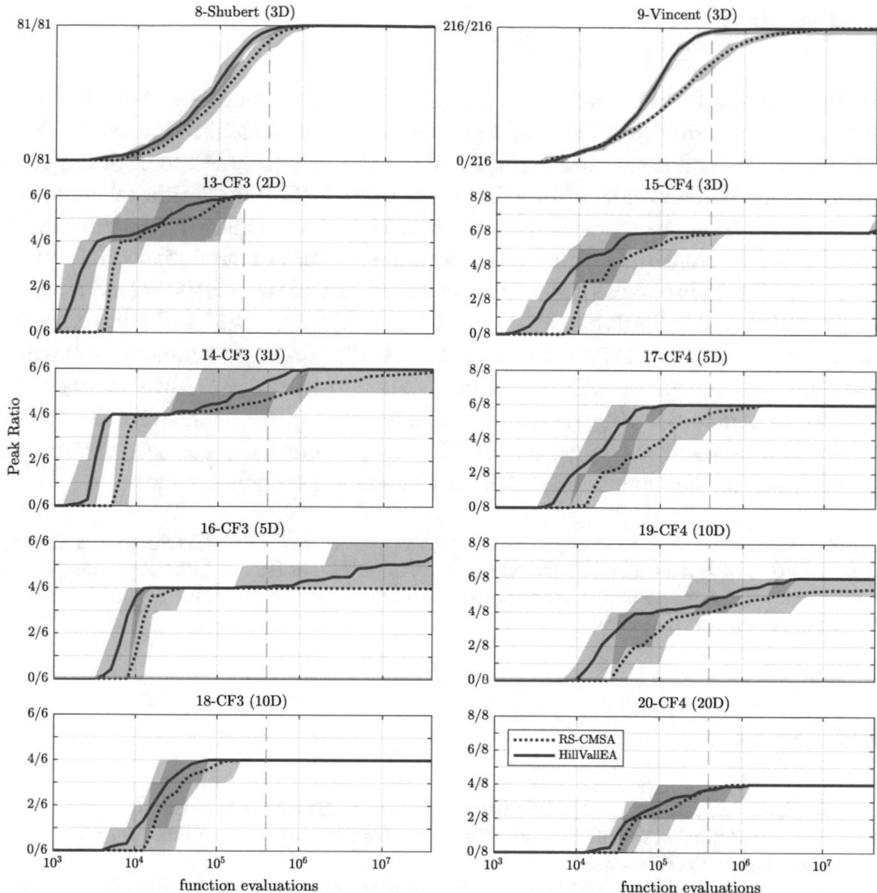

Fig. 2 Peak ratio (PR) shown as a function of the number of function evaluations, for RS-CMSA and HillVallEA-AMu. Shaded areas represent min-max range over multiple repetitions of the experiment. Problems for which HillVallEA did not obtain a peak ratio of 1.0 within the original budget (dashed vertical line) are shown. The allowed budget in terms of function evaluations was 100 times the original budget. Note that the problems are sorted so that all variants of CF3 are in the left column and all variants of CF4 in the right column

might be a needle in a haystack, and therefore fundamentally virtually unobtainable. For increasing problem dimensionality, the two Griewank optima furthermore also become harder to obtain.

6 Conclusion

We introduced a framework for two-phased evolutionary algorithms (EAs) for multimodal optimization (MMO), that equips a clustering method to locate niches in the first phase, and a core search algorithm in the second phase to optimize each niche locally. We showed that hill-valley clustering (HVC) outperforms other fitness-informed clustering methods by locating more niches when allowing the same number of function evaluations. The best performing instance of the framework for two-phase MMO EAs is obtained by equipping the framework with HVC, which we refer to as the Hill-Valley Evolutionary Algorithm (HillVallEA). HillVallEA in particular outperforms all other MMO EAs on the CEC 2013 niching benchmark suite when combined with the core search algorithm AMaLGaM-Univariate (AMu), both within the limited benchmark budget and when allowing many more function evaluations. To the best of our knowledge, HillVallEA-AMu is currently the best performing algorithm on the commonly used CEC 2013 niching benchmark suite.

Acknowledgements This work is part of the research programme IPPSI-TA with project number 628.006.003, which is financed by the Netherlands Organization for Scientific Research (NWO) and Elekta. We also acknowledge financial support of the Nijbakker-Morra Foundation for a high-performance computing system.

References

1. Ahrari, A., Deb, K., Preuss, M.: Benchmarking covariance matrix self adaption evolution strategy with repelling subpopulations for GECCO 2017 competition on multimodal optimization. COIN Report 2017014 (2017)
2. Ahrari, A., Deb, K., Preuss, M.: Multimodal optimization by covariance matrix self-adaptation evolution strategy with repelling subpopulations. Evolut. Comput. **25**(3), 439–471 (2017)
3. Auger, A., Hansen, N.: A restart CMA evolution strategy with increasing population size. In: Proceedings of the IEEE Congress on Evolutionary Computation - CEC 2005, pp. 1769–1776. IEEE Press (2005)
4. Beyer, H.G., Sendhoff, B.: Covariance matrix adaptation revisited - the CMSA Evolution Strategy. In: Parallel Problem Solving from Nature - PPSN 2008, vol. 5199, pp. 123–132. Springer, Berlin (2008)
5. Bosman, P.A.N., Grahl, J.: Matching inductive search bias and problem structure in continuous estimation-of-distribution algorithms. Eur. J. Oper. Res. **185**(3), 1246–1264 (2008)
6. Bosman, P.A.N., Grahl, J., Thierens, D.: Benchmarking parameter-free AMaLGaM on functions with and without noise. Evolut. Comput. **21**(3), 445–469 (2013)
7. Bouter, A., Alderliesten, T., Witteveen, C., Bosman, P.A.N.: Exploiting linkage information in real-valued optimization with the real-valued gene-pool optimal mixing evolutionary algorithm. In: Proceedings of the Genetic and Evolutionary Computation Conference - GECCO 2017, pp. 705–712. ACM Press (2017)
8. De Jong, K.A.: An analysis of the behaviour of a class of genetic algorithms. Ph.D. Thesis University of Michigan, Michigan (2015)
9. Goldberg, D., Richardson, J.: Genetic algorithms with sharing for mulitmodal function optimization. In: Proceedings of the Second International Conference on Genetic Algorithms and Their Application, pp. 41–49. Lauwrence Erlbaum Associates (1987)

10. Gupta, M.R., Chen, Y.: Theory and use of the EM algorithm. Found. Trends Signal Proc. **4**(3), 223–296 (2010)
11. Hansen, N., Ostermeier, A.: Completely derandomized self-adaptation in evolution strategies. IEEE Comput. Intell. Mag. **9**(2), 159–195 (2001)
12. Harik, G.R., Lobo, F.G.: A parameter-less genetic algorithm. In: Proceedings of the Genetic and Evolutionary Computation Conference - GECCO 1999, pp. 258–265 (1999)
13. Hastie, T., Tibshirani, R., Friedman, J.H.: The Elements of Statistical Learning: Data mining, Inference, and Prediction. Springer, New York (2009)
14. Li, X., Engelbrecht, A., Epitropakis, M.G.: Benchmark functions for CEC 2013 special session and competition on niching methods for multimodal function optimization. Technical report, Evolutionary Computation and Machine Learning Group, RMIT University, Australia (2013)
15. Li, X., Epitropakis, M.G., Deb, K., Engelbrecht, A.: Seeking multiple solutions: An updated survey on niching methods and their applications. IEEE Trans. Evolut. Comput. **21**(4), 518–538 (2017)
16. Maree, S.C.: HillVallEA C++ souce code on Github (2019). https://github.com/scmaree/hillvallea
17. Maree, S.C., Alderliesten, T., Bosman, P.A.N.: Benchmarking HillVallEA for the GECCO 2019 competition on multimodal optimization (2019). arXiv:1907.10988
18. Maree, S.C., Alderliesten, T., Bosman, P.A.N.: Real-valued evolutionary multi-modal multi-objective optimization by hill-valley clustering. In: Proceedings of the Genetic and Evolutionary Computation Conference - GECCO 2019, pp. 568–576. ACM Press (2019)
19. Maree, S.C., Alderliesten, T., Thierens, D., Bosman, P.A.N.: Niching an estimation-of-distribution algorithm by hierarchical gaussian mixture learning. In: Proceedings of the Genetic and Evolutionary Computation Conference - GECCO 2017, pp. 713–720. ACM Press (2017)
20. Maree, S.C., Alderliesten, T., Thierens, D., Bosman, P.A.N.: Real-valued evolutionary multi-modal optimization driven by hill-valley clustering. In: Proceedings of the Genetic and Evolutionary Computation Conference - GECCO 2018, pp. 857–864. ACM Press (2018)
21. Pelikan, M., Sastry, K., Cantú-Paz, E. (eds.): Scalable Optimization via Probabilistic Modeling, 1st edn. Springer, Berlin (2006)
22. Preuss, M.: Niching the CMA-ES via nearest-better clustering. In: Proceedings of the Genetic and Evolutionary Computation Conference - GECCO 2010, pp. 1711–1718. ACM Press (2010)
23. Preuss, M.: Improved topological niching for real-valued global optimization. In: Applications of Evolutionary Computation, pp. 386–395 (2012)
24. Preuss, M.: Multimodal Optimization by Means of Evolutionary Algorithms, 1st edn. Springer Publishing Company, Incorporated (2015)
25. Rodrigues, S., Bauer, P., Bosman, P.A.N.: A novel population-based multi-objective CMA-ES and the impact of different constraint handling techniques. In: Proceedings of the Genetic and Evolutionary Computation Conference, GECCO 2014, pp. 991–998. ACM, New York (2014)
26. Ursem, R.K.: Multinational evolutionary algorithms. In: Proceedings of the IEEE Congress on Evolutionary Computation - CEC 1999, pp. vol. 3, 1633–1640. IEEE Press (1999)
27. Wessing, S.: Two-stage methods for multimodal optimization. Ph.D. Thesis, Technische Universität Dortmund (2015)

Probabilistic Multimodal Optimization

Qiang Yang, Wei-Neng Chen, and Jun Zhang

Abstract Multimodal optimization, which aims to discover multiple satisfactory solutions simultaneously, has attracted increasing attention from researchers in the evolutionary computation community. With the aid of niching methods, evolutionary algorithms could simultaneously locate multiple satisfactory solutions in a single run. Although many multimodal evolutionary algorithms have been developed, they are confronted with two limitations: (1) in the niching stage, most niching-based multimodal methods need to compute pairwise Euclidean distances between individuals to separate the population into species, which gives rise to a high computational burden; and (2) in the optimization stage, most existing multimodal algorithms may have limitations in exploring the solution space, due to the utilization of traditional individual-based meta-heuristics, which may easily get trapped in local areas. To resolve the above issues, in this chapter, we introduce probabilistic multimodal optimization algorithms by presenting two probability-based frameworks for multimodal optimization. More specifically, we present a probability-based niching framework to accelerate the niching speed and a probability-based optimization framework to promote the optimization efficiency of multimodal algorithms, respectively. In the former, we utilize locality sensitive hashing to project individuals into buckets with probabilities to divide the population into species. Such a framework can be embedded into different niching methods to accelerate the niching speed. In the latter, we take advantage of the probability distribution of individuals to evolve the population along with a novel adaptive local search method. To instantiate these two

Q. Yang
School of Artificial Intelligence, Nanjing University of Information Science and Technology, Nanjing, China
e-mail: mmmyq@126.com

Q. Yang · W.-N. Chen (✉)
School of Computer Science and Engineering, South China University of Technology, Guangzhou, China
e-mail: cwnraul634@aliyun.com

J. Zhang
Hanyang University, Seoul, Korea
e-mail: junzhang@ieee.org

M. Preuss et al. (eds.), *Metaheuristics for Finding Multiple Solutions*, Natural Computing Series,
https://doi.org/10.1007/978-3-030-79553-5_9

frameworks, we customize them using two distinct approaches, respectively. More concretely, we embed the former into locally informed particle swarm optimization (LIPS) and neighborhood-based crowding differential evolution (NCDE), and customize the latter utilizing an explicit probability-based algorithm, the estimation of distribution algorithm (EDA), and an implicit probability-based algorithm, the continuous ant colony optimization algorithm (ACO), respectively. The efficiency and effectiveness of these two frameworks are carefully examined on a widely used multimodal benchmark set by means of comparing the associated customized algorithms with state-of-the-art multimodal methods. Lastly, the application of the proposed algorithms on multiple pedestrian detection problems is also presented. Experimentally, our approach is seen to perform competitively with traditional methods in this domain.

1 Introduction

MultiModal Optimization Problems (MMOP), which contain multiple global optima, are very common in many areas, e.g., protein structure prediction [64], holographic design [49], data mining [44, 52], tourist itinerary planning [27, 28], and electromagnetic design [15]. Locating all global optima of an MMOP simultaneously is of high significance because it can offer multiple choices to decision makers. Thus, optimization methods seeking multiple satisfactory solutions of an MMOP have attracted increasing attention from researchers in many fields, especially in the evolutionary computation community [36].

Since evolutionary algorithms (EAs) are population-based optimization approaches, they have inherent advantages in preserving multiple solutions in an optimization run, and thus are very suitable for tackling MMOPs [36]. However, originally EAs were designed specifically for locating only one optimum. Therefore, in order to solve MMOPs with high effectiveness and efficiency, special techniques must be designed to equip a traditional EA to enhance its population diversity, so that multiple optima could be simultaneously located. In the literature, the most commonly utilized technique is niching, which divides the population into separate species and then evolves individuals in each species separately to find those optima within the area that the species covers [36]. In this way, different species can locate multiple optima in different areas of the search space. Following this line of thinking, numerous niching methods have been developed, such as crowding [60], speciation [32], and clustering-based niching methods [19].

Nevertheless, most of the existing niching methods suffer from two main drawbacks [77]: (1) the performance of some niching methods is quite sensitive to the setting of the associated niching parameters (e.g., the speciation radius in speciation [32]); (2) the computational cost of most niching methods [19, 23, 32, 48, 51, 60, 75] is high, which results from the indispensable computation of pairwise Euclidean distance between individuals; such high computational cost greatly slows down the execution of EAs since niching is typically performed in each generation.

To alleviate the above-mentioned issues, we have developed a fast probability-based niching framework in [77] via utilizing the locality sensitive hashing method (LSH) [22]. The core idea of this framework is to utilize the Gaussian distribution to randomly sample a hash vector to probabilistically project individuals into different buckets. In this way, the calculation of pairwise Euclidean distance among all individuals in the population can be avoided, and thus the niching speed could be greatly accelerated. In this book chapter, we will elucidate this framework in detail in Sect. 2.

Besides niching methods, another important factor that plays a pivotal role in locating multiple optima of an MMOP is the optimizer, usually the adopted EA which evolves the population. Most existing multimodal algorithms utilize traditional individual-based meta-heuristics, like particle swarm optimization (PSO) [34, 37, 50, 76], differential evolution (DE) [19, 32, 51, 60, 62, 78], and genetic algorithms (GA) [23, 31, 48], as the optimizer to evolve individuals in each species. These meta-heuristics generally have limitations in preserving high search diversity during the evolution. As a result, once the population is trapped into local areas, individuals generally have little chance to jump out of local regions.

With the recent advance of evolutionary computation, some model-based EAs, which randomly sample individuals based on some probability distributions, have been developed to preserve high population diversity. Along this direction, the representative methods include evolutionary strategies (ES) [18], estimation of distribution algorithms (EDA) [25], and ant colony optimization (ACO) [17, 54]. Although these EAs have been extensively researched in recent years, the research mainly focuses on finding only one global optimum. Little effort has been devoted to employing them to deal with multimodal optimization. To fill this gap, we have incorporated niching methods into these EAs to locate multiple optima for multimodal optimization. Particularly, we developed a probability-based optimization framework to evolve the population and then customize it using an explicit probability-based EA (namely the estimation of distribution algorithm, EDA) and an implicit probability-based EA (namely, the ant colony optimization algorithm, ACO), leading to multimodal estimation of distribution algorithms (MEDAs) in [71] and adaptive multimodal ant colony optimization algorithms (AM-ACOs) in [72]. In Sect. 3, we will elaborate this optimization framework and the two customized methods.

In short, this book chapter concentrates on probabilistic multimodal optimization methods, which take advantage of probability distributions to improve the performance of niching and associated optimization methods. By means of the comprehensive introduction of existing probabilistic multimodal optimization methods, we expect to show the advantages of probability distribution-based methods in dealing with multimodal optimization, so as to attract more researchers to contribute to the development of probabilistic multimodal optimization.

This book chapter is organized as follows: Sect. 2 will first review existing representative niching methods and then elaborate the fast probability-based niching framework. In Sect. 3, the probability-based optimization framework will be presented, followed by descriptions of the two customized probability distribution-based optimizers. Section 4 will present the applications of the proposed probabilistic multimodal optimization methods, and Sect. 5 will further discuss the advantages of probabilistic multimodal optimization methods and provide some potential research directions for probability distribution-based EAs on multimodal optimization. At last, Sect. 6 will conclude this chapter.

2 Probability Distribution-Based Niching

In this section, we first review preliminaries on niching methods and local sensitive hashing (LSH). Then, the fast probability-based niching framework based on LSH will be described. At last, verification experiments are presented to demonstrate the effectiveness and efficiency of this niching framework.

2.1 Existing Niching Methods

To simultaneously locate multiple optima of an MMOP, high population diversity preservation is the key to avoiding individuals converging to only one optimum in a single run. To this end, niching is the most commonly used technique to enhance the population diversity of an EA in the literature. In general, niching can be embedded into different operators in an EA, like mutation and selection. In the following, for the convenience of elucidation, we assume *NP* individuals are maintained in a population to locate multiple global optima of a maximization MMOP with dimension size *D*.

As for the selection operator, in most traditional meta-heuristics, such as DE [12, 38], the selection of individuals for the next generation is only performed according to their fitness values. In general, the better the fitness of one individual is, the higher the chance of survival it has. Such selection would easily make the population converge to only one optimum, and thus it is not suitable for multimodal optimization. To prevent individuals from converging to only one area, many researchers have developed niching methods to assist the selection of individuals, so that different individuals can explore different areas of the search space. The most representative methods in this direction are crowding [60], fitness sharing [23], and clearing [48], which are described below, respectively.

Crowding was originally proposed in [14], where each offspring is compared with *CF* (a parameter named crowding factor) individuals randomly selected from the population and the nearest one (measured by Euclidean distance) to this offspring will be replaced if the offspring is better. The time complexity of this crowding scheme is $O(D \times NP \times CF)$. In crowding, the offspring is only compared with the nearest individual in a species formed by randomly selected individuals, and thus the diversity of the population could be enhanced. However, this scheme is very sensitive to the crowding factor CF. To alleviate this predicament, a deterministic crowding method was developed in [45, 60], where CF is set to the population size and every offspring is compared with the nearest individual in the population. Nevertheless, the time complexity of crowding becomes $O(D \times NP^2)$.

Fitness sharing [23] aims to preserve high diversity via reducing the attraction of densely populated regions. Before the selection of individuals, an individual shares its fitness with its nearest neighbors. In particular, the fitness of an individual is divided by the number of its nearest neighbors, and the neighbors are determined by a sharing function based on Euclidean distance. Thus, the denser the area where one individual lies, the larger the number of its nearest neighbors. In this way, the attraction of superior individuals to the dense areas could be alleviated, which is beneficial for avoiding convergence to one optimum. The time complexity of fitness sharing is $O(D \times NP^2)$ as well.

Clearing [48] shares some similarities with fitness sharing in reducing the attraction of individuals to densely populated areas. In particular, during the selection of individuals, the pairwise Euclidean distance between individuals is first calculated and then species are formed based on a dissimilarity threshold (a parameter called clearing radius). In each species, only several best individuals are preserved and others are cleared. Doing this iteratively, similar individuals in the population are eliminated to achieve population diversity. The time complexity of clearing is also $O(D \times NP^2)$.

With respect to the mutation or updating operator, it determines the way to generating offspring and thus plays a key role in diversity maintenance. In traditional meta-heuristics, such as PSO [8, 53, 58, 68, 69] and GA [56], superior individuals in the whole domain have a higher chance to guide the learning or updating of individuals. These learning or updating strategies may attract all individuals into one area and thus are only suitable for locating one global optimum. To adapt existing mutation or updating strategies to multimodal optimization, niching techniques have been incorporated to assist the generation of offspring.

In [33], Li developed a fitness Euclidean distance ratio based PSO, shortened as FER-PSO, via letting each particle learn from its "fittest-and-closest" neighbor instead of the historically best position found by the whole swarm. Such a "fittest-and-closest" neighbor is determined by the fitness Euclidean distance ratio, which is calculated by dividing the fitness difference between the personal best position of the updated particle and that of other particles by their Euclidean distance. Therefore, the niching complexity of FER-PSO is $O(D \times NP^2)$. In [34], a ring topology based PSO (RPSO) was developed by restricting each particle to learn from the personal best positions of its neighbors formed by a small ring topology. Since no Euclidean distance is calculated, the time complexity of RPSO is $O(D \times NP)$. Qu et al. [50] developed a locally informed PSO (LIPS) via updating each particle by utilizing the local best positions of its several nearest neighbors which are determined by Euclidean distance. Thus, the time complexity of LIPS is $O(D \times NP^2)$. Inspired by these works that utilize neighborhood information to update particles in PSO to enhance the diversity, some researchers also adapted this idea to DE and thus developed many mutation variants of DE, such as neighborhood-based crowding DE (NCDE) [51], parent centric normalized neighborhood crowding DE (PNPCDE) [2], and locally informative crowding DE (LoICDE) [3]. In addition, these DE variants also adopt deterministic crowding in the selection of individuals. As a result, the time complexity of these DE variants is $O(D \times NP^2)$.

In addition, niching can also be utilized to form multiple sub-populations. In the literature, multi-population techniques have been widely utilized to enhance diversity to let the population escape from local areas when locating a single global optimum [1, 30, 39]. Even though individuals in each sub-population are evolved separately, the information exchange between sub-populations is designed for locating only one global optimum. To accommodate these techniques into multimodal optimization, niching techniques are especially designed to geometrically separate the whole population into several sub-populations and let each sub-population focus on the optima in the area that this sub-population covers. Along this direction, the most representative methods are speciation [32] and clustering [19].

Speciation aims to divide the whole population into sub-populations by taking both the fitness and geometric distribution of individuals in the search space into account [32]. First, individuals in the population are sorted according to their fitness. Then a species is formed by selecting the best individual (called the seed of the species) and its neighbors whose Euclidean distance to the seed is within a radius (a parameter called species radius). Subsequently, individuals in the formed species are eliminated from the sorted population and another new species is formed in the same way. This procedure continues until the sorted population is empty. After the whole population is separated into different species, each species is independently evolved (including the generation of offspring and the selection of individuals). The time complexity of speciation is $O(D \times NP^2)$.

Although speciation could aid traditional EAs to locate multiple global optima of an MMOP, it is sensitive to the species radius and when the number of individuals in each species is not balanced, it may easily lead to the extinction of some species that contain a very small number of individuals. To alleviate this situation, Gao et al. [19] proposed a clustering method to separate the whole population into sub-populations. Given the number of sub-populations, it partitions the whole population into sub-populations with equal sizes based on pairwise Euclidean distance. The time complexity of clustering is $O(D \times NP^2)$ as well.

Based on speciation and clustering, many researchers have developed a number of variants via incorporating them into different meta-heuristics, such as neighborhood-based speciation DE (NSDE) [51] and locally informative speciation DE (LoISDE) [3]. The time complexity of these variants is also $O(D \times NP^2)$.

As shown in Table 1, we can conclude that most existing niching methods require $O(D \times NP^2)$ in order to locate multiple optima of an MMOP in each generation, which results from the calculation of the pairwise Euclidean distance between individuals in the whole population. Since niching is generally conducted in every generation, such high computational burden can greatly slow down the execution of EAs. To resolve this issue, we have developed a fast probability-based niching framework by means of locality sensitive hashing in [77].

Table 1 Time complexity of existing niching methods

Niching method	Time complexity
Crowding [60]	$O(D \times NP^2)$
Clearing [48]	$O(D \times NP^2)$
Fitness Sharing [23]	$O(D \times NP^2)$
RPSO [34]	$O(D \times NP)$
NCDE [51]	$O(D \times NP^2)$
FER-PSO [33]	$O(D \times NP^2)$
LIPS [50]	$O(D \times NP^2)$
PNPCDE [2]	$O(D \times NP^2)$
LoICDE [3]	$O(D \times NP^2)$
Speciation [32]	$O(D \times NP^2)$
NSDE [51]	$O(D \times NP^2)$
LoISDE [3]	$O(D \times NP^2)$
Clustering [19]	$O(D \times NP^2)$

**NP* is the population size, and *D* is the dimension size

2.2 *Locality Sensitive Hashing (LSH)*

Before elaborating the proposed fast niching framework, we first introduce the locality sensitive hashing method. Originally, LSH [22] was designed for fast approximate nearest neighbor retrieval in high dimensional space. Its key idea is to employ a hash function family containing multiple hash functions to probabilistically project real-valued data into a few bucket sets. In the hash family, each hash function is associated with one bucket set, into which data items are projected using this function.

Specifically, given n data items $(\boldsymbol{x}_1, \boldsymbol{x}_2, \ldots, \boldsymbol{x}_n)$, a hash function projects data into associated buckets as follows:

$$\begin{aligned} h(\boldsymbol{x}_i) &= \left\lfloor \frac{\boldsymbol{x}_i^T \boldsymbol{V} + b}{\theta} \right\rfloor \\ \theta &= \frac{max\{\boldsymbol{x}_i^T \boldsymbol{V} | i \in [1, n]\} - min\{\boldsymbol{x}_i^T \boldsymbol{V} | i \in [1, n]\}}{nb} \end{aligned} \tag{1}$$

where $h(\boldsymbol{x}_i)$ denotes the bucket index of data item $\boldsymbol{x}_i$ after being projected, $\boldsymbol{V}$ is the hash vector that characterizes the associated hash function, b is a random shift generated uniformly within $[0, \theta]$, θ is the real-valued interval between two buckets, nb is the number of buckets, and $\lfloor y \rfloor$ is the floor function which returns the largest integer smaller than y.

As shown in Fig. 1, with the aid of the hash function, each data item is projected into only one bucket. The most advantageous property of hashing is that, it can preserve the similarity between data items in the Euclidean space with a high probability [22]. Specifically, the more similar two data items are in the Euclidean space, the more likely they are projected into the same bucket. Thus, data items located in the same bucket are very similar to each other. In this way, the calculation of pairwise

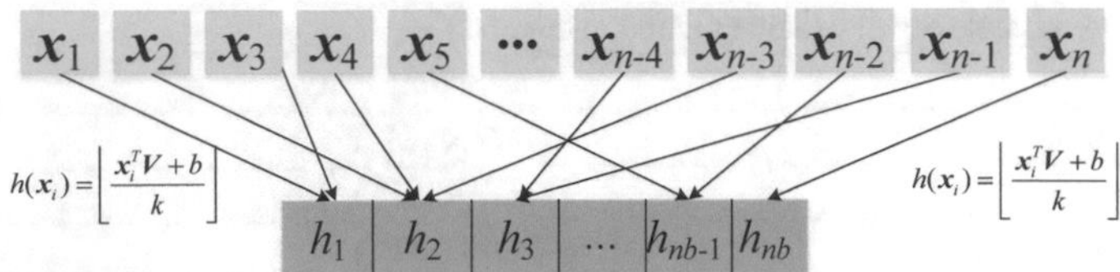

Fig. 1 The working principle of a hash function in LSH

Euclidean distance can be eliminated, and thus we can find the potential neighbors of a data item quickly, especially when the amount of data is very large.

Even though the similarity between data items can be preserved with a high probability via one hash function, it is still possible that some true neighbors of a data may be mistakenly projected into different buckets. Hence, to improve the probability of preserving the true neighbors, LSH generally maintains a hash family which contains multiple hash functions to project data into different bucket sets. In particular, each bucket set consisting of nb buckets is associated with a hash function. By this means, the potential neighbors of a data item can be obtained by the union of the buckets in different bucket sets where this data item lies.

With respect to the generation of the hash vector $\boldsymbol{V}$ in Eq. (1), two concerns need to be taken into consideration: (1) the hash vectors characterizing hash functions should be "locality sensitive" such that similar data items in the Euclidean distance could be hashed into the same bucket with a high probability; and (2) the generation of the hash vectors can be computationally efficient, so that data could be hashed with high efficiency. To this end, the most commonly utilized method to generate hash vectors is random generation based on some p-stable distribution [13]. In the literature, the most well-known and widely adopted p-stable distribution is the Gaussian distribution. Specifically, each element in a hash vector $\boldsymbol{V}$ is randomly generated from a Gaussian distribution and each hash vector is generated independently, so that the projection associated with each hash function is conducted independently.

2.3 Fast Niching

Even though existing niching methods show promising performance in multimodal optimization, they encounter two limitations: (1) most of them need to compute the pairwise Euclidean distance between individuals in the whole population to identify the nearest neighbors, and thus take high computational burden; and (2) the utilization of the true nearest neighbors in each generation may result in the excessive exploitation of neighborhood areas, which may lead to serious loss of the population diversity, and thus weakens the global exploration of the adopted EA.

To resolve the above issues, we have developed a fast niching method via incorporating LSH in [77]. Given that NP individuals are maintained to find all the global

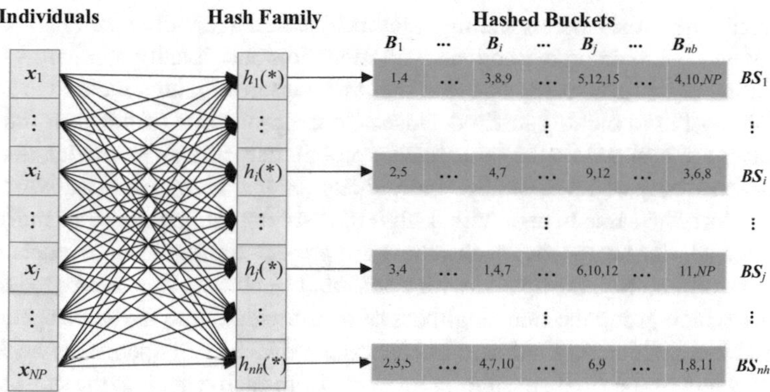

Fig. 2 The working principle of the developed fast niching method

optima of an MMOP, the working principle of the proposed fast niching method is presented in Fig. 2. Specifically, this fast niching method maintains a hash family consisting of nh hash functions as defined in Eq. (1), where each hash function h_i is associated with a bucket set $\boldsymbol{BS}_i$. Subsequently, with the aid of h_i, NP individuals in the current population are projected into nb buckets in $\boldsymbol{BS}_i$. In this way, individuals are separated into different species. In particular, we can consider each bucket set $\boldsymbol{BS}_i$ as a partition of the population into species and each bucket in $\boldsymbol{BS}_i$ as a species. In other words, the developed fast niching method could partition the population into species in different ways, which is implemented by the adopted LSH composed of different hash functions.

When incorporated into EAs to handle MMOPs in each generation, this fast niching method will select randomly a hash function from the hash family to project individuals into the associated bucket set. Thus, it is obvious that in different generations, the partition of the population may be different. However, it should be noticed that since the adopted LSH is locality sensitive, similar individuals would be hashed into the same bucket with a high probability. As a result, although the partition in different generations may be different, the difference is not so significant as similar individuals would be always hashed together with a high probability.

In summary, the proposed fast niching method consists of two major steps:

Step (1) Generating LSH hash family: We take advantage of the Gaussian distribution in [77] to independently generate nh hash vectors to characterize the hash family;

Step (2) Hashing individuals: With the aid of each hash function, individuals are hashed into buckets in the associated bucket set as shown in Fig. 2.

From the above, we can see that the complexity of the developed fast niching is $O(NP \times nh \times D)$. It should be noted that nh is generally much smaller than NP.

Compared with most existing niching methods, which generally take $O(D \times NP^2)$, the proposed fast niching method is much more computationally efficient.

Additionally, it is worth noting that the proposed fast niching method is actually a probability-based niching method. Based on the projection of the hash functions, individuals are hashed into the same bucket probabilistically. In particular, due to the locality sensitivity of LSH, similar individuals (or neighbors of an individual) are projected into the same bucket with a high probability, while dissimilar individuals (or individuals far away from each other) are projected into the same buckets with a very low probability. Compared with traditional Euclidean distance-based niching methods, which group the true neighbors of an individual into the same species, the proposed fast niching method could enhance the diversity of species by accidently projecting individuals that are slightly far away from each other into the same species with a low probability. As a result, we can see that this fast niching method is able to balance the diversity and convergence while searching for multiple optima of an MMOP.

Additionally, it should be also noticed that this approach can be considered as a general framework for niching and can be embedded into existing multimodal algorithms to accelerate the optimization. In [77], we have incorporated this fast niching method into two typical multimodal algorithms: LIPS [50] and NCDE [51], leading to new versions of these two algorithms, which we name as Fast-LIPS and Fast-NCDE, respectively. The flowcharts of these two algorithms are presented in Figs. 3 and 4, respectively. It should be mentioned that in these two algorithms, the

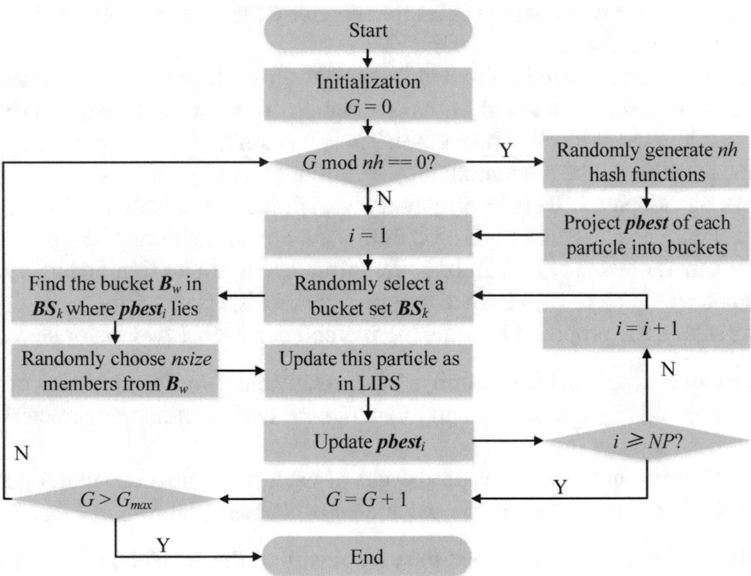

Fig. 3 The flowchart of Fast-LIPS

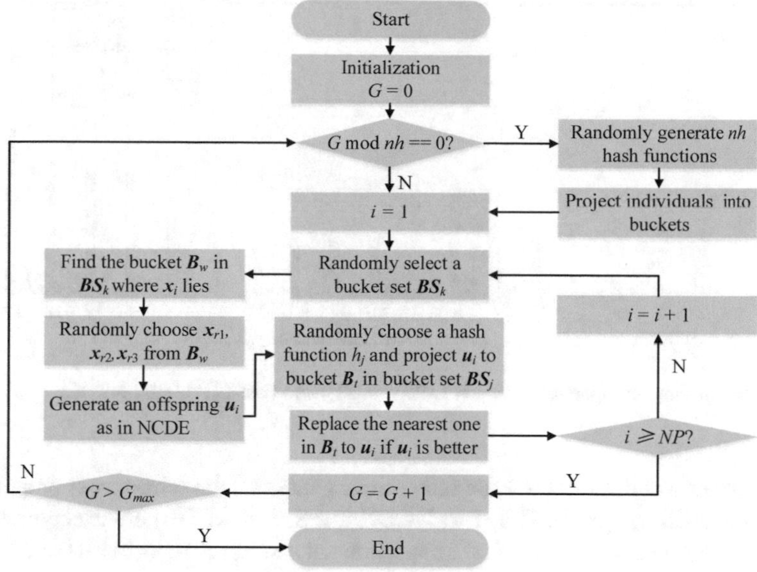

Fig. 4 The flowchart of Fast-NCDE

hash family is randomly regenerated based on the Gaussian distribution every nh generations in order to promote the diversity of the hash functions, and at the same time, the diversity of species sets (namely, the bucket sets $\boldsymbol{BS}$), so that the diversity and the convergence of the adopted EAs can be well balanced.

In the following, these two algorithms are adopted as the representatives of the fast niching framework to verify its effectiveness and efficiency in dealing with MMOPs.

2.4 *Extensive Experiments*

In this subsection, we conduct experiments on the widely used MMOP benchmark set: the CEC'2013 benchmark set [35]. Readers can refer to [35] for details of the benchmark functions, such as the number of the global optima of each function, the dimension size of each function and the maximal number of function evaluations allowed to use for multimodal algorithms to be tested on each function.

In the experiments, we use Fast-LIPS and Fast-NCDE to assess the performance of the proposed fast niching framework and compare the results with those of the original LIPS and NCDE. The population size NP of these two algorithms is set to 100 which is the same setting as in the original LIPS [50] and NCDE [51], the number of buckets nb in each bucket set is set to 20 for Fast-LIPS, and 5 for Fast-NCDE, and the number of hash functions nh is set as 10 for both algorithms. As for the number of fitness evaluations, since it has been already set in [35] for users, we directly

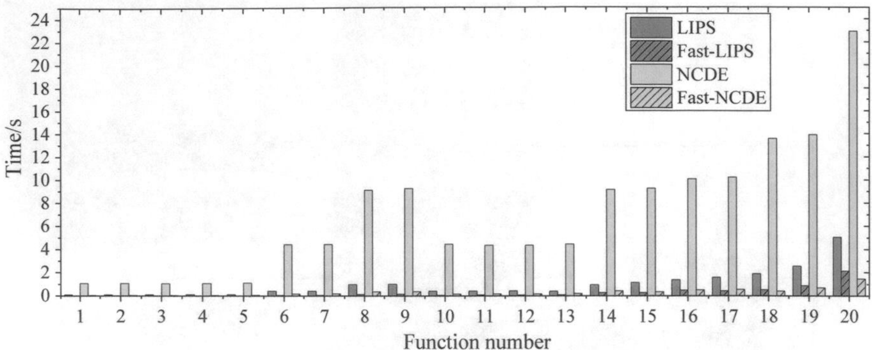

Fig. 5 The runtime comparison between LIPS (NCDE) and Fast-LIPS (Fast-NCDE)

adopt the settings in [35]. For each function, to guarantee the fairness of comparison, the four algorithms (Fast-LIPS, Fast-NCDE, LIPS, and NCDE) are executed for 50 independent runs. In addition, all the experimental results are reported at the accuracy level $\varepsilon = 1.0E - 04$.

Since the main aim of the proposed fast niching method is to accelerate the niching speed and accordingly reduce the execution time of EAs to deal with MMOP, we first focus our attention on the execution time comparison among the four compared algorithms. The comparison results on the 20 CEC'2013 MMOP benchmark functions are presented in Fig. 5.

From Fig. 5, we can obtain the following observations: (1) Comparing Fast-LIPS with LIPS, we can see that Fast-LIPS takes much less time than LIPS. In particular, with the dimension size growing, the execution time of LIPS increases very rapidly, while that of Fast-LIPS increases only mildly. Additionally, the difference between Fast-LIPS and LIPS becomes more and more obvious as the dimension size increases. (2) Similar observations can be made when comparing Fast-NCDE with NCDE. Particularly, the difference between Fast-NCDE and NCDE in terms of execution time is much more significant than that between Fast-LIPS and LIPS. (3) Compared with LIPS, NCDE takes much more time because it needs to conduct pairwise Euclidean distance twice (the first is in the mutation operator and the other is in the selection operator). However, comparing Fast-NCDE with Fast-LIPS, we find that the difference between these two algorithms with respect to the execution time is very small. This benefits from the utilized fast niching method, which greatly reduces the niching time. In short, we can see that the proposed fast niching method could significantly accelerate the niching speed, and thus reduce the execution time of multimodal EAs. This benefit can be attributed to the adopted LSH, which hashes individuals into buckets instead of pairwise Euclidean distance calculation.

Subsequently, we pay attention to the convergence behaviors of the four compared algorithms. Figure 6 displays the comparison results with respect to the change in the number of the found global optima as the optimizers progress.

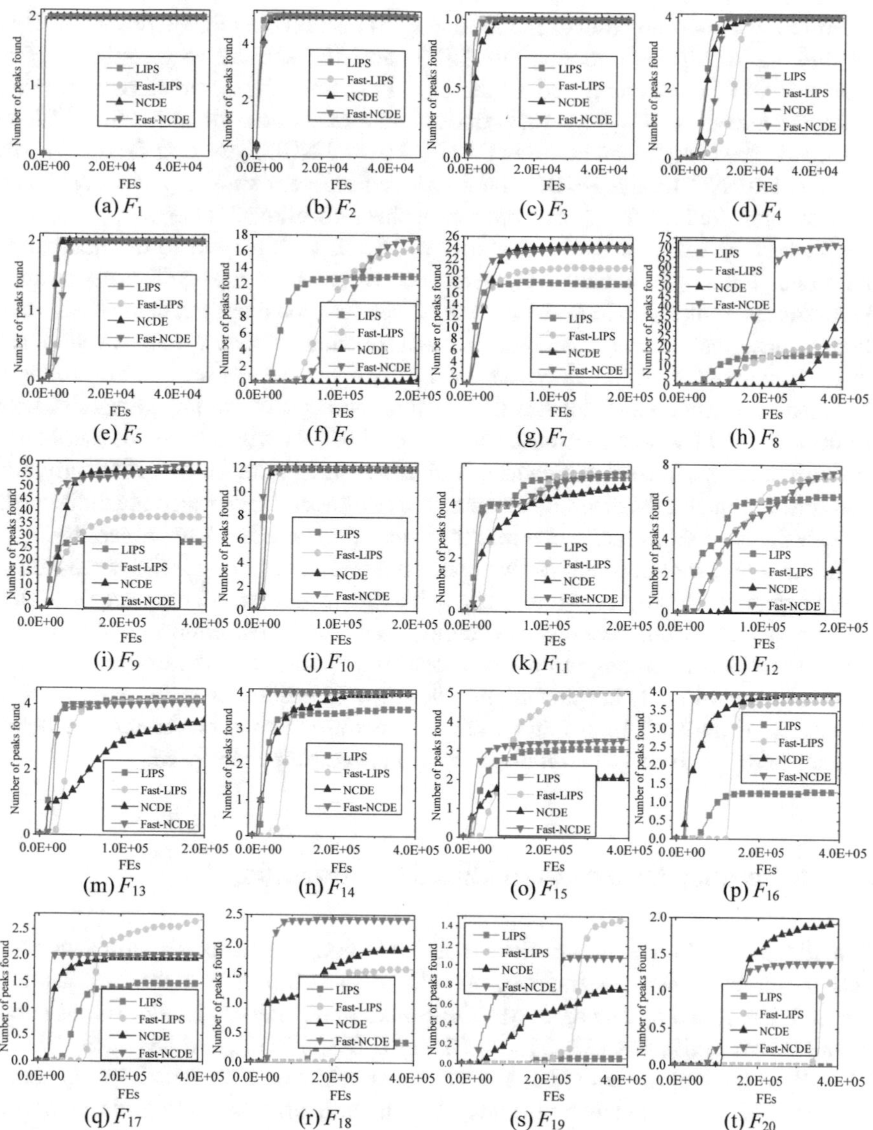

Fig. 6 Comparison results among the four compared algorithms with respect to the number of the found global optima

From Fig. 6, we can observe the following: (1) Comparing Fast-LIPS with LIPS, we find that on F_1-F_5 and F_{10}, Fast-LIPS and LIPS achieve very similar performance. Both of them could find all the global optima of these functions. On other functions, except for F_{13}, Fast-LIPS finds more global optima than LIPS. (2) When it comes to the comparison between Fast-NCDE and NCDE, we find that on F_1-F_5 and F_{10}, Fast-NCDE and NCDE could also achieve very similar performance and both of them find all the global optima of these functions. On F_7, F_{12}, F_{16}, and F_{17}, Fast-NCDE also performs similarly to NCDE with respect to the number of the found global optima. However, on these four functions, Fast-NCDE takes much fewer fitness evaluations to find the same number of global optima than NCDE, and thus it converges faster than NCDE. On other functions, except for F_{20}, Fast-NCDE obtains much better performance than NCDE. In short, we can see that the proposed fast niching method can facilitate EAs to find more global optima compared with traditional Euclidean distance-based niching methods. This benefit can be ascribed to the hash functions in LSH, which hash individuals into buckets probabilistically, and thus further enhance the diversity of each species by accidently projecting individuals slightly far away from each other into the same species with a low probability.

In summary, we can conclude that compared with traditional Euclidean distance-based niching methods, the proposed fast niching method can not only accelerate the niching speed and thus save much execution time of EAs to optimize an MMOP, but also benefit EAs in finding more global optima via enhancing the diversity of each species by occasionally hashing slightly far-away individuals into the same species. By means of the LSH hash family, this fast niching method could obtain a better balance between the population diversity and the convergence speed.

3 Probability Distribution-Based Optimization

In addition to niching methods, the adopted EAs evolving the population also influence the effectiveness and efficiency of multimodal optimization algorithms in locating multiple global optima of an MMOP. So far, most multimodal algorithms adopt meta-heuristics, like GA [23, 31, 48, 57], PSO [33, 34, 50], and DE [2, 3, 19, 32, 51, 60], which employ individuals representing solutions of an MMOP to generate offspring. With the guidance of elites, these meta-heuristics could find the optima quickly, but they encounter great risks in losing high search diversity, thus falling into local areas and miss finding some global optima, which has been demonstrated in [2, 19, 32, 34, 50, 50, 51, 60].

To improve search diversity, researchers have turned to other novel techniques and consequently proposed model-based EAs, such as estimation of distribution algorithms (EDAs) [25], ant colony optimization (ACO) [17] and evolutionary strategies (ES) [18]. These EAs generally utilize probability distribution models instead of individuals to generate offspring, and thus could preserve high population diversity during the evolution. Even though these model-based EAs have been extensively

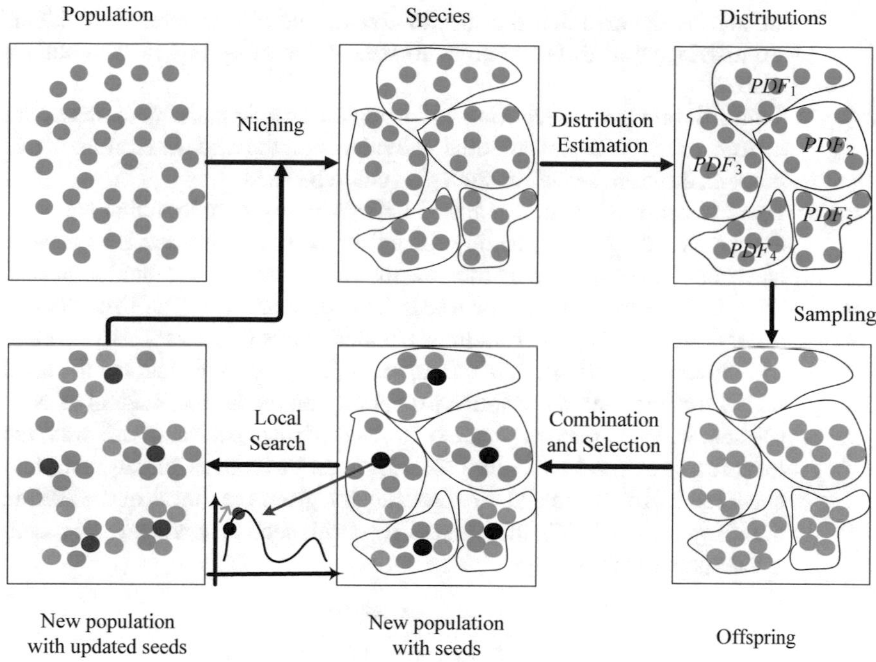

Fig. 7 The general framework of the proposed probability distribution-based optimization

studied in the evolutionary computation community, the focus of the practices is mainly on locating only one global optimum.

In this section, we intend to adapt model-based EAs to deal with MMOPs. In particular, as exhibited in Fig. 7, we propose a general probability distribution-based optimization framework for multimodal optimization. Specifically, this framework consists of the following major steps:

Step (1) **Niching:** In each generation, the population is first divided into several species based on niching methods. Generally speaking, any existing niching method for forming sub-populations as introduced in Sect. 2.1 could be made use of to separate individuals into species.

Step (2) **Distribution Estimation:** After niching, each species is independently treated. Specifically, for each species, we utilize the individuals in that species to estimate its probability distribution based on some specific estimation technique. In this process, it should be noticed that any distribution estimation technique can be utilized, such as the Gaussian distribution and the Cauchy distribution.

Step (3) **Sampling:** Based on the estimated distribution of each species, we randomly generate offspring within that species by taking advantage of some distribution models. In this way, the generated offspring within each species are around the area that the species covers, and thus different individuals are distributed to different areas to seek different optima in the solution space.

Step (4) **Combination and Selection:** The sampled offspring and the parents are then combined to select potential individuals for the next generation. In this process, different selection strategies could be used.

Step (5) **Local Search:** Since model-based EAs generally sample individuals randomly according to a specific probability distribution, these EAs usually have poor local exploitation to refine the accuracy of the obtained solutions. Thus, to remedy this issue, local search methods are widely utilized [79]. In general, any local method could be adopted in this framework. However, to save fitness evaluations, in this framework, we propose to conduct the local search method with an adaptive probability on the best individual of each species, which is generally called the seed of that species. In this way, the local search method can refine solutions in different areas, thereby avoiding convergence to one area alone. Specifically, given the fitness of the seed in the ith species is FSE_i, the probability of this seed to conduct local search is calculated as follows:

$$P_i = \frac{FSE_{max} + |FSE_i| + \eta}{FSE_{max} + |FSE_{min}| + \eta} \tag{2}$$

where FSE_{max} and FSE_{min} are the maximal and minimal fitness values among all seeds, respectively, and η is a very small positive value used to avoid the denominator being zero.

In this way, better seeds have higher probabilities to instigate local search, while those seeds with the same fitness values have the same probabilities to conduct local search. With this mechanism, the local search could be extensively conducted on the top best seeds, and thus many fitness evaluations could be saved, especially when some species fall into local areas.

Step (6) If the termination condition is met, the process stops; otherwise, the process jumps to Step (2) for the next generation.

Within such a framework, any model-based EAs could be embedded to evolve the population to locate multiple optima of an MMOP. In the following, we customize this framework with two probability distribution-based EAs: an explicit one, i.e., estimation of distribution algorithms (EDAs) and an implicit one, i.e., continuous ant colony optimization (ACO).

3.1 Estimation of Distribution Algorithms (EDAs)

EDAs [25] are stochastic optimization techniques that utilize explicit probabilistic models to explore the solution space. This model-based approach equips EDAs with many significant advantages over other types of EAs, such as high population diversity preservation. In particular, EDAs have been extensively studied in the context of locating only one global optimum [16, 67, 79]. A typical EDA works as follows:

Step (1) Sort individuals according to their fitness and select K best individuals from the population;
Step (2) Estimate the probability distribution based on the selected individuals;
Step (3) Sample new individuals based on the estimated distribution;
Step (4) Combine the sampled new individuals and the old population and then select individuals to form a new population for the next generation;
Step (5) If the termination condition is met, this process stops; otherwise, goes to Step (1) to continue.

The core of an EDA is the distribution estimation in Step (2). Different estimation techniques lead to different EDAs, and thus many EDAs have been proposed in recent years. In general, the common estimation techniques could be categorized into four groups: (1) univariate estimation [4], which considers each variable separately and thus estimates the probability distribution of each variable independently; (2) bivariate estimation [47], which considers two variables interacting with each other and thus estimates the joint probability distribution of two variables together; (3) multivariate estimation [16], which takes the interdependency of multiple variables into consideration and thus estimates the joint probability distribution of multiple interacting variables together; and (4) histogram-based estimation [79], which divides the range of each variable into several segments and then calculates the number of individuals locating in each segment.

Although EDAs have been extensively studied and widely utilized to solve many real-world optimization problems, they are all essentially designed for locating only one global optimum. Later, in this section, we will introduce a work [71] that adapts EDAs to multimodal optimization.

3.2 Ant Colony Optimization (ACO)

ACO [17] was inspired by the foraging behavior of ants. Specifically, when an ant finds a food resource, it will deposit pheromone on the ground and the amount of the deposited pheromone depends on the quantity and quality of the food. The amount of the deposited pheromone indicates the degree of attracting other ants to the food resource. By means of the pheromone, ants cooperate with each other indirectly to find the shortest path between their nest and the food resource with high quality and large quantity.

In the beginning, ACO was designed for discrete optimization, and has been widely applied to solve real-world problems [5–7, 42, 66, 74]. It particularly consists of two major steps: (1) solution construction and (2) pheromone update. The former constructs a feasible solution step by step based on the selection probabilities of candidate elements of each variable, while the latter updates the selection probability of each element based on the improvement of solution quality. To adapt it to continuous optimization, ACO$_R$ [54] was developed by Socha and Dorigo by shifting the discrete probability distribution to a continuous one. Specifically, the two major steps of ACO$_R$ are elaborated as follows:

Solution Construction: In ACO$_R$, a new solution is constructed variable by variable. First, before sampling a new value for a variable, each ant randomly selects a solution from the archive containing the already found solutions with probabilities based on the roulette selection. In particular, the selection probability of the jth solution is calculated as

$$P_j = \frac{w_j}{\sum_{i=1}^{NP} w_i} \tag{3}$$

where NP is the archive size and w_j is the weight of the jth solution and is calculated by

$$w_j = \frac{1}{\sigma \times NP \times \sqrt{2 \times \pi}} e^{-\frac{(rank(j)-1)^2}{2 \times \sigma^2 \times NP^2}} \tag{4}$$

where $rank(j)$ is the rank of the jth solution when the population is sorted in the descending order of fitness; and σ is a parameter, which has a significant effect on the weight. Specifically, a small σ indicates that the top-ranked solutions are strongly preferred to be used for offspring generation, while a large σ suggests a uniform probability distribution of solutions [54], which means that all solutions may have nearly equal probabilities to be selected for offspring generation.

Then, given that the selected solution is $\mathbf{x}_j$, an ant samples a new value for each variable (the dth dimension) using the Gaussian distribution as follows:

$$g(x^d, \mu^d, \delta^d) = \frac{1}{\delta^d \times \sqrt{2 \times \pi}} e^{\frac{(x^d - \mu^d)^2}{2 \times (\delta^d)^2}} \tag{5}$$

where d is the dimension index and δ is computed by

$$\delta^d = \xi \sum_{i=1}^{NP} \frac{|x_i^d - x_j^d|}{NP - 1} \tag{6}$$

where ξ is a parameter that has an effect similar to that of the pheromone persistence in the discrete ACO [17]. The higher the value of ξ, the slower ACO$_R$ converges [54]. In ACO$_R$, μ^d is set as the dth dimension of the selected solution. By means of the above process, each ant constructs a new solution randomly based on the Gaussian distribution, which could potentially equip ACO$_R$ with high diversity to escape from local areas.

Pheromone Update: In ACO$_R$, no apparent pheromone representation and updating strategies exist. However, these strategies actually are embedded into the calculation of the weight of each solution in the archive. In Eq. (4), the weight of a solution decreases exponentially with its rank and in Eq. (3), this weight directly determines the probability of the solution chosen by ants. Thus, the weight could be taken as the pheromone. Once *NP* new solutions have been obtained, they are combined with the *NP* solutions in the archive and then *NP* best solutions among the combined $2 \times NP$ solutions are stored in the archive for the next generation. In this way, the search process is biased toward the areas that the best solutions cover. Essentially, the update of the archive plays the role of updating pheromone.

After ACO_R, several variants of continuous ACO [26, 41] have been developed to deal with optimization problems in continuous domains and even mixed ACOs have been designed for mixed-variable optimization problems containing both discrete and continuous variables [40]. However, these ACO variants are still mainly designed for locating a single optimum. Later, in this section, we will introduce a work [72] that adapts ACO to multimodal optimization via taking advantage of the high diversity preservation of ACO, which is very precious for multimodal optimization.

3.3 *Multimodal Estimation of Distribution Algorithms (MEDAs)*

In this subsection, we introduce multimodal EDAs (MEDAs) [71], which are developed by customizing the probability distribution-based optimization framework presented in Fig. 7 utilizing EDAs. Following the major steps listed in the beginning of this section, we elaborate the process of MEDAs as follows.

3.3.1 Dynamic Niching

Within the framework in Fig. 7, any niching strategy could be embedded. To reap the fruit of the latest niching methods, we employ the clustering-based niching method [19], to partition the population into species. Particularly, in [19], the authors developed two clustering strategies for niching: crowding clustering and speciation clustering. Therefore, utilizing these two niching strategies in the framework, we can develop two versions of MEDA, namely, MCEDA incorporating the crowding clustering strategy and MSEDA incorporating the speciation clustering strategy.

In the clustering-based strategies, given the population size is *NP* and the cluster size is M, the population is partitioned into $\lceil NP/M \rceil$ species. One drawback of these two niching strategies is that the cluster size has a significant influence on their performance and also needs to be specified in advance. In particular, for a given population size, if the cluster size is small, a large number of species with a small number of individuals are produced. In this situation, each species covers a narrow range of solution space as shown in black circles in Fig. 8. Therefore, each species could exploit the covered area extensively, but at the same time, this may lead to low diversity of each species, which brings risks to species falling into local areas. Conversely, if the cluster size is large, a small number of species with a large number of individuals are generated. In this situation, each species covers a wide

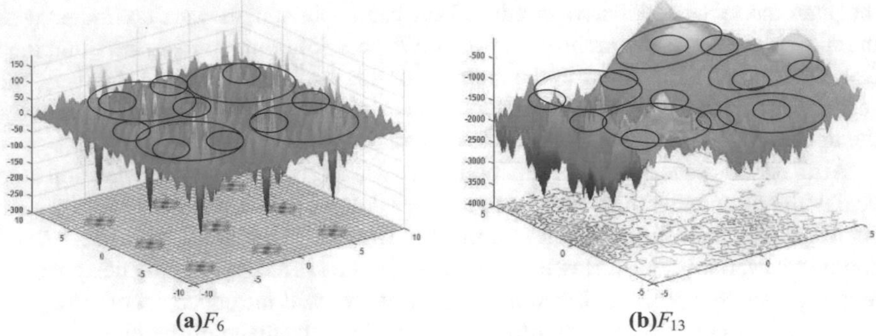

Fig. 8 Landscape of two multimodal functions selected from the CEC'2013 multimodal benchmark set. (**a**) F_6, (**b**) F_{13}

area as displayed in red circles in Fig. 8. Therefore, each species could preserve high diversity during the evolution, but at the same time, the species encounter risks in lacking extensive exploitation to improve the solution quality when mining the promising areas. Therefore, we can see that it is very difficult to set a proper cluster size M in different evolution stages without prior knowledge about the landscapes of optimization problems.

To alleviate the above predicament, a dynamic cluster sizing strategy was developed in [71]. For simplicity, a random-based dynamic cluster sizing is adopted. Concretely, we predefine a cluster size set $\boldsymbol{C}$ containing several different integers representing the candidate cluster sizes. Then, during each generation, before niching, we first randomly select an integer from the predefined set $\boldsymbol{C}$ as the cluster size M. Then, the clustering-based niching is conducted to partition the population into species.

With such a mechanism, MEDAs can potentially balance the exploration and exploitation demands during the evolution. Specifically, when some species converge to local areas, the cluster size M may become larger in the following random selection of M and then the diversity of species could be enhanced via introducing more individuals. In contrast, when more promising areas are discovered, the cluster size M may become small in the following selection of M, and thus species cover narrower ranges to fully exploit the found promising areas. In this way, MEDAs may potentially make a proper balance between exploration and exploitation.

3.3.2 Distribution Estimation

After partitioning the population into species, it comes to the probability distribution estimation of each species. Unlike existing EDAs [4, 16, 79] that estimate the probability distribution of the whole population using a number of selected individuals, MEDAs estimate the probability distribution of each species separately. In this way, multiple distributions covering different areas of the solution space could be obtained, and thus multiple optima could be obtained with the help of different probability distributions. Since a species covers a small area and the number of individuals in each species is generally much smaller than the population size, all members of each species potentially preserve useful information, and thus they all participate in the distribution estimation of that species in MEDAs. In this manner, it is expected that the diversity of each species could be potentially enhanced, which may be beneficial for finding more promising areas.

As described in Sect. 3.1, in the studies of EDAs, many distribution estimation techniques exist, such as univariate estimation [4], multivariate estimation [16], and histogram-based estimation [67, 79]. For simplicity, in MEDAs, we utilize the univariate estimation to estimate the probability distribution of each species, because it preserves a low level of computational complexity [4, 79]. Particularly, the distribution of each species is estimated as follows:

$$\begin{aligned} \mu_i^d &= \tfrac{1}{M}\textstyle\sum_{j=1}^{M} X_{i,j}^d \\ \delta_i^d &= \sqrt{\tfrac{1}{M-1}\textstyle\sum_{j=1}^{M}(X_{i,j}^d - \mu_i^d)^2} \end{aligned} \tag{7}$$

where $\boldsymbol{X}_{i,j} = [X_{i,j}^1, \ldots, X_{i,j}^d, \ldots, X_{i,j}^D]$ is the jth individual in the ith species, $\mu_i = [\mu_i^1, \ldots, \mu_i^d, \ldots, \mu_i^D]$ and $\delta_i = [\delta_i^1, \ldots, \delta_i^d, \ldots, \delta_i^D]$ are, respectively, the mean and standard deviation (std) vectors of the ith species, and D is the dimension size of the multimodal problem.

3.3.3 Offspring Generation

Following the distribution estimation is the offspring generation. In MEDAs, the offspring of each species is also generated separately. In most existing EDAs [4, 16, 25], only the Gaussian distribution is employed to sample points. However, the Gaussian distribution generally has a narrow sampling space, especially when the standard deviation δ is small. Such a way of generating offspring would limit the exploration ability of an EDA. To circumvent this situation, we turn to the Cauchy distribution [73], which has a very similar distribution curve with the Gaussian distribution, but with a long fat tail. It generally produces larger stepsizes than the Gaussian distribution. Armed with this distribution, an EDA may have a better chance to escape from local optima.

Specifically, due to its narrow sampling range, the Gaussian distribution is particularly suitable for the exploitation stage when the population finds more promising areas. As for the Cauchy distribution, taking advantage of its larger sampling stepsizes, it is especially appropriate for the exploration stage when the population falls into local areas and thus needs to jump out. Based on this observation, we consider alternatively utilizing these two distributions to generate offspring for each species. To make it simple, these two distributions are alternatively executed with the same probability for each species. Concretely, the offspring generation of each species is as follows:

$$\begin{cases} \boldsymbol{C}_i = Gaussian(\mu_i, \delta_i) & if\ rand() < 0.5 \\ \boldsymbol{C}_i = Cauchy(\mu_i, \delta_i) & otherwise \end{cases} \tag{8}$$

where $\boldsymbol{C}_i$ is the M offspring of the ith species, and $rand()$ is a uniformly random number generator that generates numbers within [0, 1].

Such alternative usage of these two distributions to generate offspring for each species may afford a potential balance between exploration and exploitation for MEDAs, which is beneficial for seeking multiple optima of an MMOP.

3.3.4 Combination and Selection

After the generation of the offspring for each species, it comes to the selection of individuals for the next generation. Since two clustering-based niching strategies are utilized in MEDAs, leading to MCEDA and MSEDA, respectively, the selection of individuals in these two MEDAs are slightly different.

For MCEDA, the selection procedure in CDE [60] is adopted. Concretely, each offspring is only compared with the nearest individual in the parent population. If it is better, the nearest parent individual to it is replaced. It is worth mentioning that the comparison between an offspring and parent individuals is conducted at the population level, not at the species level.

For MSEDA, different from the selection process in SDE [32], we also adopt the approach from CDE [60] for the consideration of diversity preservation. Nevertheless, different from MCEDA, the comparison between an offspring and the parent individuals is executed at the species level. Specifically, each offspring of one species is only compared with the nearest parent individual in that species. In this way, the replacement is only conducted in the area that one species covers, and thus high diversity can be preserved during the evolution.

3.3.5 Local Search

On account of the sampling strategy, most EDAs have difficulty in improving the accuracy of the obtained solutions. To remedy this issue, many local search strategies [79] have been developed to assist EDAs to find high-quality solutions. In the framework presented in Fig. 7, we have developed an adaptive execution of local search on the seeds of species to improve the accuracy of the best solution of each species. In general, any local search method could be embedded into this framework.

To make it simple and to be consistent with the offspring generation in MEDAs, we adopt the Gaussian distribution with a small local standard deviation to refine the obtained solutions. The Gaussian distribution is employed because of its narrow sampling space, especially when the standard deviation is small, which is beneficial for exploiting the search space in a limited area and thus profitable for refining solutions.

Assume the seeds of all species are stored in a set $\boldsymbol{S}$, containing $s = \lceil NP/M \rceil$ seeds. To guarantee that the solutions could be improved with a high probability, we should sample enough points using the following Gaussian distribution model:

$$\boldsymbol{LC}_i = Gaussian(\boldsymbol{S}_i, \sigma) \tag{9}$$

where $\boldsymbol{LC}_i$ is the sampled individuals of the ith species generated by the local search, and $\boldsymbol{S}_i$ is the seed of the ith species. In MEDAs, $\sigma = 1.0E - 4$ is utilized.

As for the number (denoted as N) of sampled points in the local search method, it should be neither too large nor too small. If N is too large, many fitness evaluations would be wasted if there is no improvement around some seeds. If it is too small, the improvement around the seeds could not be guaranteed because only a few points are sampled. In MEDAs, we experimentally find $N = 5$ is enough to address the above concerns.

Since the sampled individuals $\boldsymbol{LC}_i$ by the local search method are only produced around the seed of the associated species, we might as well compare these sampled points only with the seed. Specifically, the local search is executed as follows. First, it samples a point and compares it with the seed. If the sampled point is better, it replaces the seed. Then, using the updated seed, the local search method samples another point. This process continues until N points are sampled. In this way, we can see that the whole procedure of local search is only related to the seeds of species.

Combining MCEDA and MSEDA with the above local search method and the adaptive execution method presented in the beginning of this section, we can develop LMCEDA and LMSEDA, respectively. With the assistance of the adaptive local search method, the gap between the global optima and the obtained solutions can be narrowed.

3.4 Adaptive Multimodal Ant Colony Optimization (AM-ACO)

In this subsection, we introduce the adaptive multimodal continuous ant colony optimization algorithms (AM-ACO) developed in [72], which incorporates ACO_R as introduced in Sect. 3.2 into the probability distribution-based optimization framework presented in Fig. 7.

Since AM-ACO and MEDAs are customized by the same framework, the dynamic niching, combination and selection, and the adaptive local search method in AM-ACO are all the same as MEDAs introduced in Sect. 3.3. The major difference between AM-ACO and MEDAs lies in the probability estimation and offspring generation, which will be elaborated in the following. It should be mentioned that similar

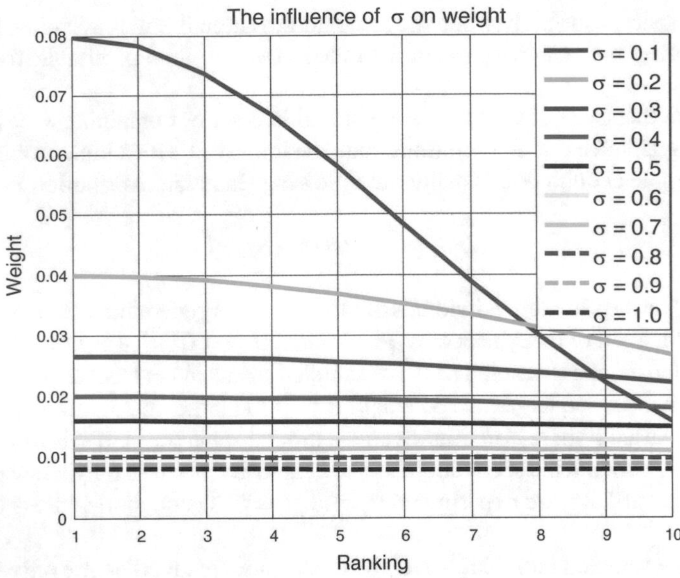

Fig. 9 The influence of σ on the weight of each solution [72]

to MEDAs, the probability estimation and offspring generation in AM-ACO also work at the species level, namely, these two processes are conducted in each species independently.

With respect to the probability estimation, different from MEDAs that estimate a probability distribution for each species, AM-ACO randomly selects the obtained solutions stored in the archive to construct new ones based on their selection probabilities. Therefore, the probability estimation in AM-ACO is actually the calculation of the selection probability of each candidate in the archive.

As elucidated in Sect. 3.2, ACO_R utilizes Eq. (3) and Eq. (4) to compute the probabilities. In particular, in Eq. (4), σ significantly influences the selection probability of each solution in the archive. Figure 9 presents the influence of σ on the weight of each solution with σ varying from 0.1 to 1.0. From this figure, we can see that the smaller the value of σ, the higher the weights of the top ranked solutions. This indicates that in ACO_R, the top-ranked solutions are preferred with a small σ. In accordance with this line, most traditional ACO_R variants for locating only one global optimum adopt a small σ, such as 10^{-4} in [54] and 0.05 in [40].

However, in multimodal optimization, a small σ is not suitable because when seeking multiple global optima simultaneously, it is highly possible that one species may contain several global optima not just one, especially when the number of global optima is larger than that of species. Under such situations, solutions with the same or very similar fitness should have nearly the same possibility to be selected to construct new solutions. Moreover, the solution in different species may preserve different quality, and the number of the promising solutions within each species may be different as well.

Taking the above into consideration, we have proposed an adaptive adjusting strategy for σ in [72] for each species, which is formulated as

$$\sigma_i = 0.1 + 0.3e^{-\frac{FS^i_{max} - FS^i_{min}}{FS_{max} - FS_{min} + \eta}} \tag{10}$$

where σ_i is the σ in Eq. (4) for the ith species; FS^i_{max} and FS^i_{min} are the maximum and minimum fitness values of the ith species, respectively; and FS_{max} and FS_{min} are the maximum and minimum fitness of the whole archive, respectively, and η is a very small positive value used to avoid the zero denominator.

In Eq. (10), we can see that σ_i is different for different species. Particularly, when the solutions in one species differ a lot with respect to fitness, which is reflected by a large value of $FS^i_{max} - FS^i_{min}$, σ_i tends to 0.1. Thus, in this species, the selection is biased to better solutions, which is beneficial for exploitation. On the other hand, when the solutions in one species are very close to each other in terms of fitness, suggested by a small value of $FS^i_{max} - FS^i_{min}$, σ_i has a tendency to 0.4. Therefore, in this species, each solution is nearly unbiased, which is profitable for exploration. In this way, both the difference in solution quality of different species and that of solutions within the same species are taken into consideration for calculating the selection probabilities of solutions for ants when constructing new ones. Therefore, a potential balance between exploration and exploitation can be achieved during the evolution.

In terms of offspring generation, AM-ACO also adopts the similar sampling procedure as in ACO$_R$ presented in Eq. (5). But two changes with respect to the base vector μ are made in AM-ACO: (1) Instead of selecting one solution for each dimension in ACO$_R$, all dimensions of the selected solution are utilized as the base (namely μ in Eq. (5)) to construct a new solution. This change not only reduces the time complexity with respect to solution selection, but also potentially takes the correlation among variables into consideration. In this manner, the potentially useful information in different dimensions could be preserved together. (2) When most solutions in one species are of poor quality or fall into local areas, it is very difficult for the ant colony in this species to escape from local areas. In this situation, utilizing the sampling process in ACO$_R$ may not produce better solutions, but waste precious computation resources. To counteract such a predicament, we introduce a basic DE mutation operator to AM-ACO to shift the base vector (namely μ in Eq. (5)) for an ant to construct new solutions, which is defined as

$$\mu = \boldsymbol{x}_j + F(\boldsymbol{x}_{seed} - \boldsymbol{x}_j) \tag{11}$$

where $\boldsymbol{x}_{seed} = [x^1_{seed}, \ldots, x^d_{seed}, \ldots, x^D_{seed}]$ is the seed of the species that $\boldsymbol{x}_j$ belongs to, and is defined as the best solution in that species, $\boldsymbol{x}_j = [x^1_j, \ldots, x^d_j, \ldots, x^D_j]$ is the selected solution for an ant to construct a new solution based on the selection probability in one species, and F is the scalar factor as in DE. Different from DE operators where the value of F is fixed, here F is randomly generated within (0, 1] [11] to alleviate the sensitivity.

The above update of μ provides a shift for the base vector from $\boldsymbol{x}_j$ to the best solution of the species $\boldsymbol{x}_{seed}$. This is beneficial for the ant to construct a new solution close to the promising area in that species, and thus may improve the chances for the species to escape from local areas and find more promising solutions. However, it should also be noted that such a shift is a little greedy, and may easily drive ants to build solutions close to the best ones in the species. This may result in loss of diversity. Therefore, to guarantee the balance between diversity and convergence, we consider alternatively employing both kinds of μ settings (μ is set as the selected solution or set according to Eq. (11)) with equal probabilities. Thus, the base vector μ is determined by

$$\begin{cases} \mu = \boldsymbol{x}_j & if\ rand() < 0.5 \\ \mu = \boldsymbol{x}_j + F(\boldsymbol{x}_{seed} - \boldsymbol{x}_j) & otherwise \end{cases} \quad (12)$$

In Eq. (5), another factor influencing the sampling of new solutions is δ. Different from ACO_R where δ is computed at the population level as in Eq. (6), the calculation of δ in AM-ACO is at the species level and presented as follows:

$$\delta_{i,j} = \xi \sum_{k=1}^{M} \frac{|\boldsymbol{x}_{i,k} - \boldsymbol{x}_{i,j}|}{M-1} \quad (13)$$

where $\delta_{i,j}$ is the δ associated with the selected solution $\boldsymbol{x}_{i,j}$ in the ith species, M is the cluster size and $\boldsymbol{x}_{i,k}$ is the kth solution in the ith species. As for the parameter ξ, different from ACO_R where a fixed ξ is utilized, in AM-ACO, we set ξ as a uniformly random value generated within (0, 1] for each ant in each species. In this way, ξ may be different for ants within one species or for ants in different species, which is potentially beneficial for balancing exploration and exploitation.

Overall, compared with the original ACO_R [54], AM-ACO operating at the species level is free from the sensitivity to parameters (σ and ξ) by the adaptive adjusting strategy for σ and the random setting for ξ. Since two clustering-based niching strategies (crowding clustering and speciation clustering) are combined with AM-ACO, we can develop two versions, namely, AMC-ACO incorporating the crowding clustering strategy and AMS-ACO incorporating the speciation clustering strategy. In conjunction with the proposed adaptive local search method, LAMC-ACO and LAMS-ACO are, respectively, formed at last.

3.5 Extensive Comparison

In this section, extensive experiments are conducted to verify the effectiveness and efficiency of the above introduced algorithms. Particularly, the CEC'2013 multimodal benchmark function set [35] is utilized to testify the performance of the introduced algorithms. In addition, to make comprehensive comparisons, we select five

Table 2 Common parameter settings for the compared algorithms

Function	Max_Fes	Population Size
$F_1 - F_5$	$5.0E+4$	80
F_6	$2.0E+5$	100
F_7	$2.0E+5$	300
$F_8 - F_9$	$4.0E+5$	300
F_{10}	$2.0E+5$	100
$F_{11} - F_{13}$	$2.0E+5$	200
$F_{14} - F_{20}$	$4.0E+5$	200

recent and popular multimodal algorithms, namely, NCDE [51], Self-CCDE [19], LoICDE [3], PNPCDE [2], and MOMMOP [61], as the compared methods to compare with the introduced algorithms. The first four algorithms are all niching-based methods which utilize different DE variants as the basic optimizers to evolve the population, while the last is a multi-objective based method, which transforms a multimodal optimization problem into a multi-objective optimization problem and then utilizes multi-objective optimization algorithms to seek multiple optima. These five algorithms have shown very promising performance in multimodal optimization. To make fair comparisons, the population size and the allowed maximum number of fitness evaluations are set the same for all algorithms as shown in Table 2. Other key parameters in the compared algorithms are set as recommended in the associated papers.

Figure 10 exhibits the convergence behavior comparison between the introduced algorithms and the five compared algorithms with respect to the number of found global optima at the accuracy level $\varepsilon = 1.0E - 04$. From this figure, we can see that: (1) On $F_1 - F_5$ and F_{10}, all the 9 compared algorithms could find all the global optima of these problems when the maximum number of fitness evaluations is exhausted. However, we find that LAMC-ACO, LAMS-ACO, LMCEDA, and LMSEDA find all the global optima of these problems with fewer fitness evaluations than MOMMOP, LoICDE, and PNPCDE, which demonstrates that the introduced four algorithms could converge faster than these three compared algorithms. (2) On F_6, LAMC-ACO, LAMS-ACO, and LMSEDA outperform the four multimodal algorithms based on DE variants, but are slightly dominated by MOMMOP. On $F_7 - F_9$, the introduced four algorithms are much worse than Self-CCDE and MOMMOP. (3) On the last 10 benchmark functions, namely, $F_{11} - F_{20}$, which are with higher dimensionality and more local optima than the first 10 functions, the introduced four algorithms, especially LAMS-ACO and LMSEDA, obtain significantly better performance than the five compared algorithms. (4) With respect to the comparison among the introduced four algorithms, we find that LAMS-ACO performs slightly better than LMSEDA, and both are better than LAMC-ACO and LMCEDA. In addition, LAMC-ACO performs slightly better than LMCEDA. Particularly, we find that LAMS-ACO and LMSEDA obtain much better performance than LAMC-ACO and LMCEDA respec-

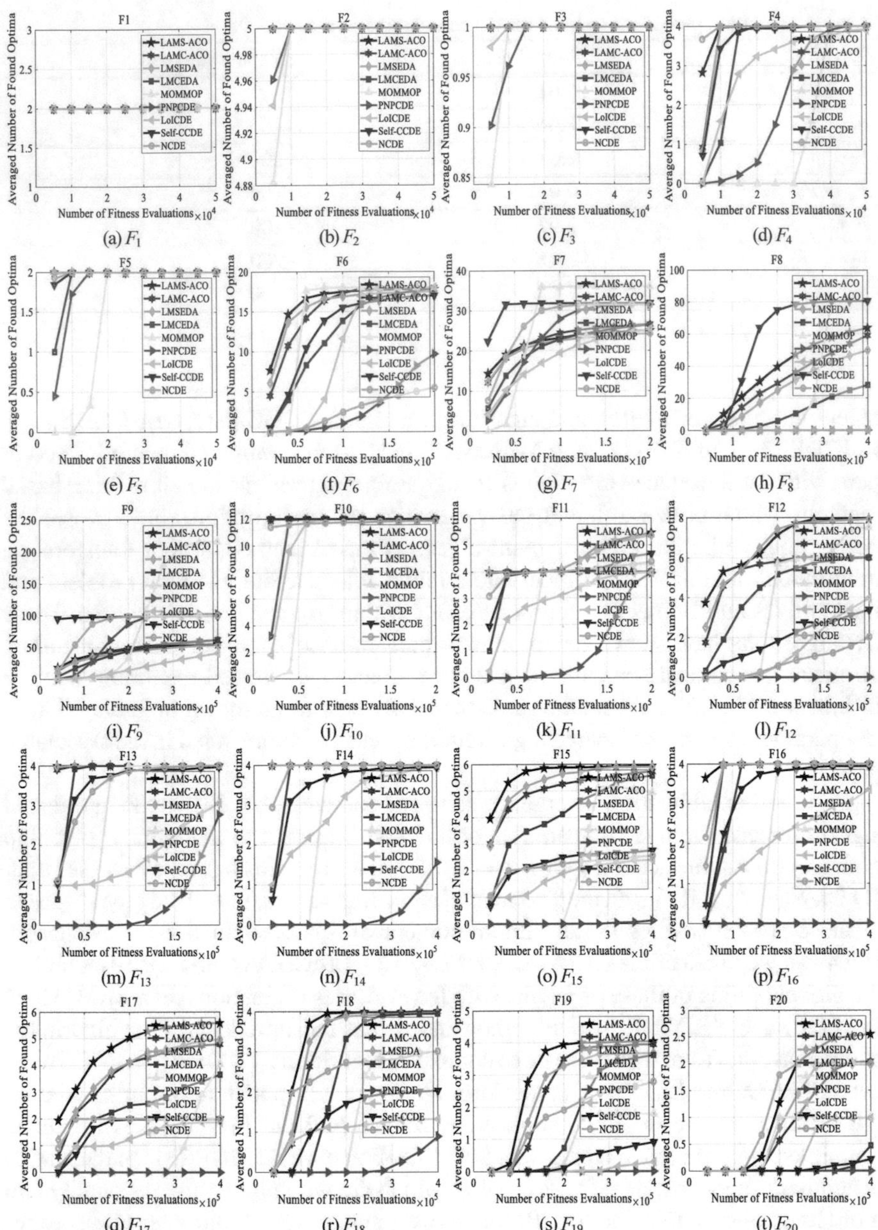

(a) F_1 (b) F_2 (c) F_3 (d) F_4

(e) F_5 (f) F_6 (g) F_7 (h) F_8

(i) F_9 (j) F_{10} (k) F_{11} (l) F_{12}

(m) F_{13} (n) F_{14} (o) F_{15} (p) F_{16}

(q) F_{17} (r) F_{18} (s) F_{19} (t) F_{20}

Fig. 10 The convergence comparison among the compared methods on 20 CEC'2013 benchmark functions at the accuracy level $\varepsilon = 1.0E-04$

tively, especially on higher dimensional multimodal problems, which is demonstrated by Fig. 10k–t on the last 10 benchmark functions.

Additionally, Fig. 11 presents the fitness landscape of the final population of AM-ACOs (LAMC-ACO and LAMS-ACO) and MEDAs (LMCEDA and LMSEDA) when the maximum number of fitness evaluations is exhausted on F_6, F_7, and $F_{10} - F_{13}$ at the accuracy level $\varepsilon = 1.0E - 04$. From this figure, we can see that: (1) As shown in Fig. 11b1–b4 and Fig. 11c1–c4, on functions with no local optima, like F_7 and F_{10}, LAMC-ACO, LAMS-ACO, LMCEDA, and LMSEDA obtain very similar performance. That is the final individuals of these four algorithms are all located around or at the global optima when the maximum number of fitness evaluations is exhausted. (2) When it comes to those multimodal problems with many local optima, like F_6 and $F_{10} - F_{13}$, we find that these four algorithms perform very differently. In particular, the final individuals of LAMS-ACO and LMSEDA gather more extensively around the global optima than those of LAMC-ACO and LMCEDA, respectively. The final population of LAMC-ACO and LMCEDA covers global optima along with more local optima than LAMS-ACO and LMSEDA but with lower accuracy, respectively.

In summary, from the above extensive comparative study, we can conclude that the developed probability distribution-based optimization framework is very promising in tackling multimodal optimization problems. Particularly, we find that this framework is suitable for tackling multimodal optimization problems with many variables and many local optima, which has been demonstrated in the experiments as compared with existing individual-based multimodal EAs.

4 Applications

This section introduces the applications of the above proposed multimodal algorithms. Particularly, we elaborate the application in [59] in detail, which employs the above developed probability distribution-based multimodal optimization framework to deal with multiple pedestrian detection problems.

Pedestrian detection has been extensively researched in computer vision and has been utilized in various domains [21]. However, such a problem is still very challenging, especially for multiple pedestrian detection. In [59], to cope with multiple pedestrian detection efficiently, we transform the multiple pedestrian detection problem into a multimodal optimization problem and then utilize the above customized multimodal estimation of distribution algorithms (MEDA) under the probability distribution-based optimization framework to optimize this problem based on Histograms of Oriented Gradients (HOG) feature and Support Vector Machines (SVM) [10, 46].

Particularly, we embed MEDAs into the detection stage of HOG-SVM [10], which was proposed by Dalal and Triggs and utilizes the HOG feature and the linear SVM to detect pedestrians. Instead of using the exhaustive method in HOG-SVM to search

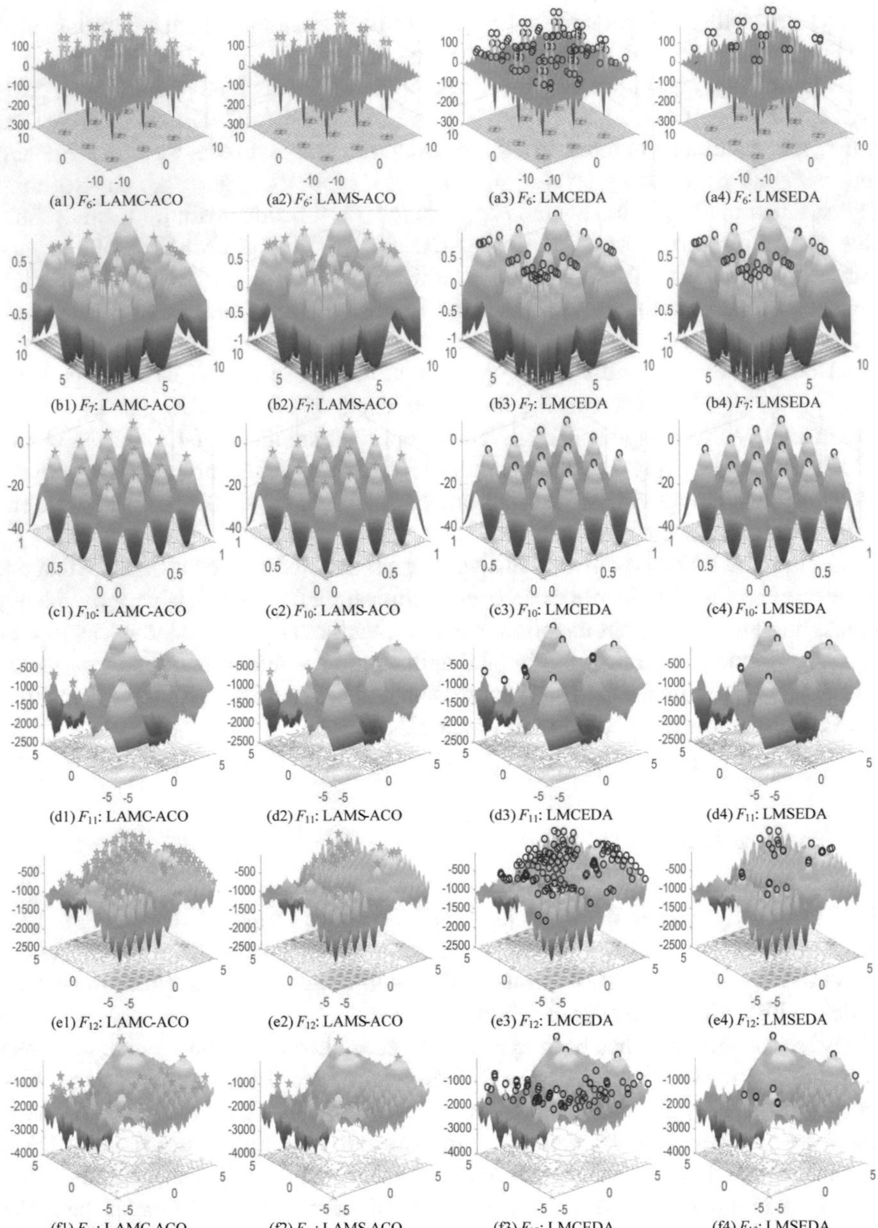

Fig. 11 The landscape comparison between AM-ACOs (LAMC-ACO and LAMS-ACO) and MEDAs (LMCEDA and LMSEDA) on F_6, F_7, and $F_{10} - F_{13}$ at the accuracy level $\varepsilon = 1.0E-04$

for pedestrians, we employ MEDAs to assist HOG-SVM in the detection stage in order to find the rectangular regions containing pedestrians.

Concretely, we adopt a triple vector (rx, ry, rs) to encode an individual in MEDA, where (rx, ry) is the top-left coordinate of a rectangular region in the image and rs is a scalar parameter which transforms the denoted rectangular region into 64×128 pixels. Here, it should be mentioned that the denoted rectangle needs to be transformed into 64×128 pixels, because the detection window remains fixed as 64×128 pixels in the training stage of HOG-SVM. By means of rs, we can obtain the real size of the denoted rectangle in the original image, which is $64 \times rs$ in width and $128 \times rs$ in length.

Given an image with size $w \times h$, we can obtain the range of rs when encoding individuals:

$$rs \in \left(0, min\left(\frac{w}{64}, \frac{h}{128}\right)\right] \tag{14}$$

Accordingly, the ranges of rx and ry are:

$$\begin{aligned} rx &\in [0, w - rs \times 64] \\ ry &\in [0, h - rs \times 128] \end{aligned} \tag{15}$$

To evaluate the individuals, we directly adopt the predicted value calculated by the trained SVM model as the fitness of an individual, which is presented as follows:

$$f(\boldsymbol{x}_i) = \boldsymbol{e} \cdot HOG(\boldsymbol{x}_i) + b \tag{16}$$

where $\boldsymbol{x}_i$ is the ith individual in MEDA and $HOG(\boldsymbol{x}_i)$ is the HOG feature vector of the denoted rectangular region by $\boldsymbol{x}_i$, $\boldsymbol{e}$ and b are model parameters of the trained SVM. Particularly, the larger the fitness of an individual, the higher the probability that the denoted region by this individual contains a pedestrian.

After evolution using MEDAs, the individuals in the population need to be assessed on whether the denoted rectangle regions contain pedestrians, which is the decoding process. In such a process, each individual is evaluated as follows:

$$H(\boldsymbol{x}_i) = \begin{cases} 1 & if \;\; f(\boldsymbol{x}_i) > \theta \\ 0 & otherwise \end{cases} \tag{17}$$

where $\boldsymbol{x}_i$ is the ith individual, $f(\boldsymbol{x}_i)$ is its fitness as calculated in Eq. (16), θ is a threshold parameter and $H(\boldsymbol{x}_i)$ is the binary value after decoding. $H(\boldsymbol{x}_i) = 1$ indicates that the denoted region by individual $\boldsymbol{x}_i$ contains a pedestrian, while $H(\boldsymbol{x}_i)$ means that the denoted region contains no pedestrian.

The above method using MEDAs to cope with multiple pedestrian problems is named as HOG-SVM-MEDA. To evaluate its performance, we conduct experiments on a widely used pedestrian detection dataset called INRIA [10]. This dataset contains 288 high resolution images captured from real-life scenes for testing and 2416 images with 64×128 pixels for training. For comparison, we select the traditional HOG-

Fig. 12 The pedestrian detection result comparison among the three compared methods in different scenarios

SVM and one of its variants in conjunction with NSDE [51], named as HOG-SVM-NSDE [43], to compare with HOG-SVM-MEDA.

In the experiments, the population sizes NP, the niching size NS, the threshold θ in Eq. (17), and the maximum number of iterations $ITER_{max}$ in HOG-SVM-NSDE and HOG-SVM-MEDA are set to be 200, 5, -1, and 700, respectively. Some selected detection results on INRIA are presented in Fig. 12.

From this figure, we can see that: (1) In the scenario where only one pedestrian exists as shown in Fig. 12a, HOG-SVM-MEDA achieves better performance than the other two compared algorithms. In particular, HOG-SVM-MEDA could not only find the only one pedestrian, but also locate the pedestrian with a more accurate rectangle. (2) When it comes to the scenario where two pedestrians exist as displayed in Fig. 12b, we find that HOG-SVM-MEDA could locate the two pedestrians more accurately than the other two methods. Specifically, in the last three images in Fig. 12b, even though HOG-SVM-MEDA finds the two pedestrians with three rectangles (one redundant rectangle), it is still better than the other two methods, because the other two methods only find one or even no pedestrian. (3) With respect to the more complicated scenario where multiple pedestrians exist as presented in Fig. 12c, we can see that HOG-SVM-MEDA still outperforms the other two algorithms significantly.

Overall, we can see that the developed HOG-SVM-MEDA could locate pedestrians accurately no matter in scenarios with only one pedestrian or in those with multiple pedestrians, and thus it is very promising for multiple pedestrian detection.

In addition, it is also noteworthy that the proposed AM-ACOs have also been employed to solve real-world optimization problems. For instance, in [65], the authors utilized AM-ACOs to deal with multisensor remote sensing image registration problems and the experimental results substantiated that AM-ACOs are much more promising than the existing methods.

5 Discussion and Future Work

From our experimental studies, we can see the potential of probabilistic multimodal optimization techniques in finding multiple optima of MMOP. First, probabilistic multimodal optimization techniques could be employed in the niching stage to promote the efficiency of niching, so that the niching speed could be potentially accelerated and at the same time, the diversity of partitioned species could also be potentially enhanced. Second, they can also be employed in the optimization stage to improve the optimization efficiency and effectiveness. The most advantageous superiority of probabilistic multimodal optimization techniques is that they can offer high diversity to the population, so that the search space of multimodal optimization problems can be fully explored and falling into local areas can be avoided simultaneously.

Though we have introduced several probabilistic multimodal optimization techniques, the development of probabilistic multimodal optimization is far from complete. Much work needs to be done in the future. Specifically, as far as we are concerned, the following directions deserve particular attention:

(1) Novel probability-based niching methods. In this chapter, we have introduced the LSH-based fast niching method, which only utilizes a simple probability model for niching. Other advanced probability models could be utilized to accelerate niching. In addition to the niching speed and the niching accuracy, the diversity of the partitioned species should also deserve attention because such diversity is beneficial for the species to escape from local areas. Therefore, special care

needs to be devoted to achieving a good balance among these three aspects when designing probability distribution-based niching approaches.

(2) Parallel probabilistic multimodal optimization approaches. Recently, parallel or distributed EAs [20, 29, 55, 70] have attracted plenty of attention from researchers to promote the optimization efficiency of EAs. In multimodal optimization, since each species is separately evolved after niching, it is also appealing to evolve these species in parallel to accelerate the execution of multimodal algorithms. Therefore, developing parallel probabilistic multimodal optimization approaches is necessary, especially in the situation where one multimodal optimization problem is of high time complexity. Along this line, some researchers have made some early attempts to develop distributed multimodal optimization methods [9, 63]. However, this direction is still in its infancy and needs great attention for boosting. Particularly, how to design the parallel models and how to let species communicate with each other efficiently deserve special attention.

(3) Applications with probabilistic multimodal methods. In many real-world applications, decision makers need multiple optimal schemes to make a comprehensive choice, such as protein structure prediction [64], holographic design [49], tourist itinerary planning [27, 28], and electromagnetic design [15]. In addition, many machine learning and data mining problems are actually multimodal problems, such as clustering [44, 52] and neural network optimization [24]. Therefore, using probabilistic multimodal optimization methods to solve real-world application problems like the aforementioned ones also deserves extensive attention in the future.

6 Conclusion

This book chapter introduces probabilistic multimodal optimization methods. In particular, we have presented a probability-based fast niching framework using LSH and a probability distribution-based optimization framework for multimodal optimization. By means of projecting individuals into different buckets probabilistically, the former can not only accelerate the niching speed for existing multimodal algorithms via avoiding the calculation of pairwise Euclidean distance between individuals in the whole population, but also offer diversity to each species to some extent. Taking advantage of probability distributions, the latter could offer high search diversity to the population, and thus can fully explore the multimodal space. By means of the customized methods under the above two frameworks, we have demonstrated their great efficiency and effectiveness in seeking multiple global optima. In addition, we have also elaborated the application of the developed probability distribution-based optimization framework on multiple pedestrian detection problems. At last, we have discussed some potential directions that deserve special attention in the future. We hope this chapter will attract more researchers to boost the development of probabilistic multimodal optimization.

Acknowledgements This work was supported in part by the Key Project of Science and Technology Innovation 2030 supported by the Ministry of Science and Technology of China (Grant No. 2018AAA0101300), in part by the National Natural Science Foundation of China under Grant 61976093, 62006124, and 61873097, in part by the Science and Technology Plan Project of Guangdong Province 2018B050502006, in part by Guangdong Natural Science Foundation Research Team 2018B030312003, in part by the Natural Science Foundation of Jiangsu Province under Project BK20200811, in part by the Natural Science Foundation of the Jiangsu Higher Education Institutions of China under Grant 20KJB520006 and in part by the Startup Foundation for Introducing Talent of NUIST.

References

1. Ali, M.Z., Awad, N.H., Suganthan, P.N., Reynolds, R.G.: An adaptive multipopulation differential evolution with dynamic population reduction. IEEE Trans. Cybern. **47**(9), 2768–2779 (2017)
2. Biswas, S., Kundu, S., Das, S.: An improved parent-centric mutation with normalized neighborhoods for inducing niching behavior in differential evolution. IEEE Trans. Cybern. **44**(10), 1726–1737 (2014)
3. Biswas, S., Kundu, S., Das, S.: Inducing niching behavior in differential evolution through local information sharing. IEEE Trans. Evol. Comput. **19**(2), 246–263 (2015)
4. Bosman, P.A., Thierens, D.: Expanding from discrete to continuous estimation of distribution algorithms: the idea. In: International Conference on Parallel Problem Solving from Nature. Springer, pp. 767–776 (2000)
5. Chen, W.N., Zhang, J.: An ant colony optimization approach to a grid workflow scheduling problem with various qos requirements. IEEE Trans. Syst. Man Cybern. Part C (Appl. Rev.) **39**(1), 29–43 (2009)
6. Chen, W.N., Zhang, J.: Ant colony optimization for software project scheduling and staffing with an event-based scheduler. IEEE Trans. Softw. Eng. **39**(1), 1–17 (2013)
7. Chen, W.N., Zhang, J., Chung, H.S.H., Huang, R.Z., Liu, O.: Optimizing discounted cash flows in project scheduling–an ant colony optimization approach. IEEE Trans. Syst. Man Cybern. Part C (Appl. Rev.) **40**(1), 64–77 (2010)
8. Chen, W.N., Zhang, J., Lin, Y., Chen, N., Zhan, Z.H., Chung, H.S.H., Li, Y., Shi, Y.H.: Particle swarm optimization with an aging leader and challengers. IEEE Trans. Evol. Comput. **17**(2), 241–258 (2013)
9. Chen, Z.G., Zhan, Z.H., Wang, H., Zhang, J.: Distributed individuals for multiple peaks: a novel differential evolution for multimodal optimization problems. IEEE Trans. Evolut. Comput. **24**(4), 708–719 (2020)
10. Dalal, N., Triggs, B.: In: Histograms of oriented gradients for human detection, vol. 1, pp. 886–893. IEEE (2005)
11. Das, S., Konar, A., Chakraborty, U.K.: Two improved differential evolution schemes for faster global search. In: The Annual Conference on Genetic and Evolutionary Computation. ACM, pp. 991–998 (2005)
12. Das, S., Suganthan, P.N.: Differential evolution: a survey of the state-of-the-art. IEEE Trans. Evol. Comput. **15**(1), 4–31 (2011)
13. Datar, M., Immorlica, N., Indyk, P., Mirrokni, V.S.: Locality-sensitive hashing scheme based on p-stable distributions. In: The Annual Symposium on Computational Geometry. ACM, pp. 253–262 (2004)
14. De Jong, K.A.: Analysis of the Behavior of a Class of Genetic Adaptive Systems. University of Michigan (1975)
15. Dilettoso, E., Salerno, N.: A self-adaptive niching genetic algorithm for multimodal optimization of electromagnetic devices. IEEE Trans. Mag. **42**(4), 1203–1206 (2006)

16. Dong, W., Chen, T., Tino, P., Yao, X.: Scaling up estimation of distribution algorithms for continuous optimization. IEEE Trans. Evol. Comput. **17**(6), 797–822 (2013)
17. Dorigo, M., Stützle, T.: Ant colony optimization: overview and recent advances. In: Handbook of Metaheuristics, pp. 227–263. Springer (2010)
18. François, O.: An evolutionary strategy for global minimization and its markov chain analysis. IEEE Trans. Evol. Comput. **2**(3), 77–90 (1998)
19. Gao, W., Yen, G.G., Liu, S.: A cluster-based differential evolution with self-adaptive strategy for multimodal optimization. IEEE Trans. Cybern. **44**(8), 1314–1327 (2014)
20. Ge, Y.F., Yu, W.J., Lin, Y., Gong, Y.J., Zhan, Z.H., Chen, W.N., Zhang, J.: Distributed differential evolution based on adaptive mergence and split for large-scale optimization. IEEE Trans. Cybern. **48**(7), 2166–2180 (2018)
21. Geronimo, D., Lopez, A.M., Sappa, A.D., Graf, T.: Survey of pedestrian detection for advanced driver assistance systems. IEEE Trans. Pattern Anal. Mach. Intell. **32**(7), 1239–1258 (2010)
22. Gionis, A., Indyk, P., Motwani, R., et al.: Similarity search in high dimensions via hashing. VLDB **99**, 518–529 (1999)
23. Goldberg, D.E., Richardson, J., et al.: Genetic algorithms with sharing for multimodal function optimization. In: International Conference on Genetic Algorithms and Their Applications, pp. 41–49. Lawrence Erlbaum, Hillsdale, NJ (1987)
24. Gong, Y.J., Zhang, J., Zhou, Y.: Learning multimodal parameters: a bare-bones niching differential evolution approach. IEEE Trans. Neural Netw. Learn. Syst. **29**(7), 2944–2959 (2018)
25. Hauschild, M., Pelikan, M.: An introduction and survey of estimation of distribution algorithms. Swarm Evol. Comput. **1**(3), 111–128 (2011)
26. Hu, X.M., Zhang, J., Chung, H.S.H., Li, Y., Liu, O.: Samaco: variable sampling ant colony optimization algorithm for continuous optimization. IEEE Trans. Syst. Man Cybern. Part B (Cybern.) **40**(6), 1555–1566 (2010)
27. Huang, T., Gong, Y.J., Kwong, S., Wang, H., Zhang, J.: A niching memetic algorithm for multi-solution traveling salesman problem. IEEE Trans. Evolut. Comput.**24**(3), 508–522 (2020)
28. Huang, T., Gong, Y.J., Zhang, Y.H., Zhan, Z.H., Zhang, J.: Automatic planning of multiple itineraries: A niching genetic evolution approach. IEEE Trans. Intell. Transp. Syst. **21**(10), 4225–4240 (2020)
29. Jia, Y.H., Chen, W.N., Gu, T.L., Zhang, H.X., Yuan, H.Q., Kwong, S., Zhang, J.: Distributed cooperative co-evolution with adaptive computing resource allocation for large scale optimization. IEEE Trans. Evol. Comput. **23**(2), 188–202 (2019)
30. Li, C., Yang, S.: A general framework of multipopulation methods with clustering in undetectable dynamic environments. IEEE Trans. Evol. Comput. **16**(4), 556–577 (2012)
31. Li, J.P., Balazs, M.E., Parks, G.T., Clarkson, P.J.: A species conserving genetic algorithm for multimodal function optimization. Evol. Comput. **10**(3), 207–234 (2002)
32. Li, X.: Efficient differential evolution using speciation for multimodal function optimization. In: The Annual Conference on Genetic and Evolutionary Computation. ACM, pp. 873–880 (2005)
33. Li, X.: A multimodal particle swarm optimizer based on fitness euclidean-distance ratio. In: the Annual Conference on Genetic and Evolutionary Computation. ACM, pp. 78–85 (2007)
34. Li, X.: Niching without niching parameters: particle swarm optimization using a ring topology. IEEE Trans. Evol. Comput. **14**(1), 150–169 (2010)
35. Li, X., Engelbrecht, A., Epitropakis, M.G.: Benchmark functions for cec'2013 special session and competition on niching methods for multimodal function optimization. RMIT University, Evolutionary Computation and Machine Learning Group, Australia, Technical Report (2013)
36. Li, X., Epitropakis, M.G., Deb, K., Engelbrecht, A.: Seeking multiple solutions: an updated survey on niching methods and their applications. IEEE Trans. Evol. Comput. **21**(4), 518–538 (2017)
37. Li, Y.K., Chen, Y.L., Zhong, J.H., Huang, Z.X.: Niching particle swarm optimization with equilibrium factor for multi-modal optimization. Inf. Sci. **494**, 233–246 (2019)
38. Li, Y.L., Zhan, Z.H., Gong, Y.J., Chen, W.N., Zhang, J., Li, Y.: Differential evolution with an evolution path: A deep evolutionary algorithm. IEEE Trans. Cybern. **45**(9), 1798–1810 (2015)

39. Liang, J.J., Suganthan, P.N.: Dynamic multi-swarm particle swarm optimizer with local search. In: IEEE Congress on Evolutionary Computation, vol. 1, pp. 522–528. IEEE (2005)
40. Liao, T., Socha, K., de Oca, M.A.M., Stützle, T., Dorigo, M.: Ant colony optimization for mixed-variable optimization problems. IEEE Trans. Evol. Comput. **18**(4), 503–518 (2014)
41. Liao, T., Stützle, T., de Oca, M.A.M., Dorigo, M.: A unified ant colony optimization algorithm for continuous optimization. Eur. J. Oper. Res. **234**(3), 597–609 (2014)
42. Lin, Y., Zhang, J., Chung, H.S.H., Ip, W.H., Li, Y., Shi, Y.H.: An ant colony optimization approach for maximizing the lifetime of heterogeneous wireless sensor networks. IEEE Trans. Syst. Man Cybern. Part C (Appl. Rev.) **42**(3), 408–420 (2012)
43. Lin, Z.J., Chen, W.N., Zhang, J., Li, J.J.: In: Fast multiple human detection with neighborhood-based speciation differential evolution. In: International Conference on Information Science and Technology, pp. 200–207. IEEE (2017)
44. Ling, H.L., Wu, J.S., Zhou, Y., Zheng, W.S.: How many clusters? a robust pso-based local density model. Neurocomputing **207**, 264–275 (2016)
45. Mahfoud, S.W.: Niching methods for genetic algorithms. Urbana **51**(95001), 62–94 (1995)
46. Maji, S., Berg, A.C., Malik, J.: In: Classification using intersection kernel support vector machines is efficient. IN: IEEE Conference on Computer Vision and Pattern Recognition, pp. 1–8. IEEE (2008)
47. Pelikan, M., Mühlenbein, H.: The bivariate marginal distribution algorithm. In: Advances in Soft Computing, pp. 521–535. Springer (1999)
48. Pétrowski, A.: A clearing procedure as a niching method for genetic algorithms. In: IEEE International Conference on Evolutionary Computation, pp. 798–803. IEEE (1996)
49. Qing, L., Gang, W., Qiuping, W.: Restricted evolution based multimodal function optimization in holographic grating design. In: IEEE Congress on Evolutionary Computation, vol. 1, pp. 789–794. IEEE (2005)
50. Qu, B.Y., Suganthan, P.N., Das, S.: A distance-based locally informed particle swarm model for multimodal optimization. IEEE Trans. Evol. Comput. **17**(3), 387–402 (2013)
51. Qu, B.Y., Suganthan, P.N., Liang, J.J.: Differential evolution with neighborhood mutation for multimodal optimization. IEEE Trans. Evol. Comput. **16**(5), 601–614 (2012)
52. Sheng, W., Swift, S., Zhang, L., Liu, X.: A weighted sum validity function for clustering with a hybrid niching genetic algorithm. IEEE Trans. Syst. Man Cybern. Part B (Cybern.) **35**(6), 1156–1167 (2005)
53. Shi, Y., Eberhart, R.: A modified particle swarm optimizer. In: IEEE Congress on Evolutionary Computation, pp. 69–73. IEEE (1998)
54. Socha, K., Dorigo, M.: Ant colony optimization for continuous domains. Eur. J. Oper. Res. **185**(3), 1155–1173 (2008)
55. Song, A., Chen, W.N., Gu, T.L., Yuan, H.Q., Kwong, S., Zhang, J.: Distributed virtual network embedding system with historical archives and set-based particle swarm optimization. IEEE Trans. Syst. Man Cybern.: Syst. **51**(2), 927–942 (2021)
56. Srinivas, M., Patnaik, L.M.: Genetic algorithms: a survey. Computer **27**(6), 17–26 (1994)
57. Stoean, C., Preuss, M., Stoean, R., Dumitrescu, D.: Multimodal optimization by means of a topological species conservation algorithm. IEEE Trans. Evol. Comput. **14**(6), 842–864 (2010)
58. Suganthan, P.N.: Particle swarm optimiser with neighbourhood operator. In: IEEE Congress on Evolutionary Computation, vol. 3, pp. 1958–1962. IEEE (1999)
59. Tan, D.Z., Chen, W.N., Zhang, J., Yu, W.J.: Fast pedestrian detection using multimodal estimation of distribution algorithms. In: The Genetic and Evolutionary Computation Conference, pp. 1248–1255. ACM (2017)
60. Thomsen, R.: Multimodal optimization using crowding-based differential evolution. In: IEEE Congress on Evolutionary Computation, vol. 2, pp. 1382–1389. IEEE (2004)
61. Wang, Y., Li, H.X., Yen, G.G., Song, W.: Mommop: multiobjective optimization for locating multiple optimal solutions of multimodal optimization problems. IEEE Trans. Cybern. **45**(4), 830–843 (2015)
62. Wang, Z.J., Zhan, Z.H., Lin, Y., Yu, W.J., Wang, H., Kwong, S., Zhang, J.: Automatic niching differential evolution with contour prediction approach for multimodal optimization problems. IEEE Trans. Evol. Comput. **24**(1), 114–128 (2020)

63. Wang, Z.J., Zhan, Z.H., Zhang, J.: Distributed minimum spanning tree differential evolution for multimodal optimization problems. Soft. Comput. **23**(24), 13339–13349 (2019)
64. Wong, K.C., Leung, K.S., Wong, M.H.: Protein structure prediction on a lattice model via multimodal optimization techniques. In: The Annual Conference on Genetic and Evolutionary Computation, pp. 155–162. ACM (2010)
65. Wu, Y., Ma, W., Miao, Q., Wang, S.: Multimodal continuous ant colony optimization for multisensor remote sensing image registration with local search. Swarm Evol. Comput. **47**, 89–95 (2019)
66. Xing, L.N., Rohlfshagen, P., Chen, Y.W., Yao, X.: A hybrid ant colony optimization algorithm for the extended capacitated arc routing problem. IEEE Trans. Syst. Man Cybern. Part B (Cybern.) **41**(4), 1110–1123 (2011)
67. Yang, P., Tang, K., Lu, X.: Improving estimation of distribution algorithm on multimodal problems by detecting promising areas. IEEE Trans. Cybern. **45**(8), 1438–1449 (2015)
68. Yang, Q., Chen, W.N., Da Deng, J., Li, Y., Gu, T., Zhang, J.: A level-based learning swarm optimizer for large scale optimization. IEEE Trans. Evolut. Comput. (2017)
69. Yang, Q., Chen, W.N., Gu, T., Zhang, H., Deng, J.D., Li, Y., Zhang, J.: Segment-based predominant learning swarm optimizer for large-scale optimization. IEEE Trans. Cybern. **47**(9), 2896–2910 (2017)
70. Yang, Q., Chen, W.N., Gu, T.L., Zhang, H.X., Yuan, H.Q., Kwong, S., Zhang, J.: A distributed swarm optimizer with adaptive communication for large-scale optimization. IEEE Trans. Cybern. **50**(7), 3393–3408 (2020)
71. Yang, Q., Chen, W.N., Li, Y., Chen, C.P., Xu, X.M., Zhang, J.: Multimodal estimation of distribution algorithms. IEEE Trans. Cybern. **47**(3), 636–650 (2017)
72. Yang, Q., Chen, W.N., Yu, Z., Gu, T., Li, Y., Zhang, H., Zhang, J.: Adaptive multimodal continuous ant colony optimization. IEEE Trans. Evol. Comput. **21**(2), 191–205 (2017)
73. Yao, X., Liu, Y., Lin, G.: Evolutionary programming made faster. IEEE Trans. Evol. Comput. **3**(2), 82–102 (1999)
74. Yu, X., Chen, W.N., Gu, T.L., Yuan, H.Q., Zhang, H.X., Zhang, J.: Aco-a*: Ant colony optimization plus a* for 3-d traveling in environments with dense obstacles. IEEE Trans. Evol. Comput. **23**(4), 617–631 (2019)
75. Zhang, Y.H., Gong, Y.J., Gao, Y., Wang, H., Zhang, J.: Parameter-free voronoi neighborhood for evolutionary multimodal optimization. IEEE Trans. Evolut. Comput. **24**(2), 335–349 (2020)
76. Zhang, Y.H., Gong, Y.J., Yuan, H.Q., Zhang, J.: A tree-structured random walking swarm optimizer for multimodal optimization. Appl. Soft Comput. **78**, 94–108 (2019)
77. Zhang, Y.H., Gong, Y.J., Zhang, H.X., Gu, T.L., Zhang, J.: Toward fast niching evolutionary algorithms: a locality sensitive hashing-based approach. IEEE Trans. Evol. Comput. **21**(3), 347–362 (2017)
78. Zhao, H., Zhan, Z.H., Lin, Y., Chen, X.F., Luo, X.N., Zhang, J., Kwong, S., Zhang, J.: Local binary pattern-based adaptive differential evolution for multimodal optimization problems. IEEE Trans. Cybern. **50**(7), 3343–3357 (2020)
79. Zhou, A., Sun, J., Zhang, Q.: An estimation of distribution algorithm with cheap and expensive local search methods. IEEE Trans. Evol. Comput. **19**(6), 807–822 (2015)

Reduced Models of Gene Regulatory Networks: Visualising Multi-modal Landscapes

Khulood Alyahya, Kevin Doherty, Ozgur E. Akman, and Jonathan E. Fieldsend

Abstract We present a prototypical multimodal optimisation problem from the systems biology domain—tuning the kinetic parameters of a reduced order gene regulatory network (GRN) model to obtain optimal fits to gene expression timeseries. After introducing the problem, the chapter then illustrates different fitness landscapes of the GRN parameter fitting problem using various statistical plots of landscape features, along with local optima networks (LONs)—graphs representing local optima (modes), their basin sizes and connectivity across the landscape. In a typical multimodal optimisation process, the problem owners get presented with a putative list of modal solutions from which to verify and select a design. We argue in this chapter that it is often useful to present a characterisation of the search landscape itself along with the list of modal solutions. The characterisation of the search landscape can provide insight into the domain, and may guide, for instance, problem reformulations, or the final mode selection based on broader features than simply mode performance, e.g. basin size if robustness of modes is a concern.

K. Alyahya (✉) · K. Doherty · O. E. Akman · J. E. Fieldsend
Computer Science, University of Exeter, North Park Road, EX4 4QF Exeter, UK
e-mail: K.Alyahya@exeter.ac.uk; kalyahya1@ksu.edu.sa

K. Doherty
e-mail: K.Doherty@exeter.ac.uk

O. E. Akman
e-mail: O.E.Akman@exeter.ac.uk

J. E. Fieldsend
e-mail: J.E.Fieldsend@exeter.ac.uk

K. Alyahya
King Saud University, Riyadh 11451, Saudi Arabia

M. Preuss et al. (eds.), *Metaheuristics for Finding Multiple Solutions*,
Natural Computing Series,
https://doi.org/10.1007/978-3-030-79553-5_10

1 Introduction

Heuristic optimisation approaches, such as those within the field of evolutionary computation, attempt to find the optimal value of an objective function through a sampling of points in the design space where the points chosen for sampling are informed by the fitness of previously sampled points. The goal of any heuristic optimisation algorithm is to search over this fitness landscape in order to find the point in design space that *maximises* the fitness. This can be thought of as querying the heights of points in a physical landscape and exploring the landscape through repeated querying, with the aim of finding the highest point. The idea of a fitness landscape originally comes from evolutionary biology and the work of Sewall Wright [34], in which the space of gene combinations result in a fitness landscape that can be traversed through selection, mutation and recombination—an obvious parallel for evolutionary computation.

A recognised issue with heuristic search algorithms is the need for a decision on the values of algorithm parameters, the so-called *hyperparameters*. However, knowledge of fitness landscape features can guide the problem owner to an appropriate choice. For example, the most recent winner of the Black-Box Optimisation Competition (BBComp) for single-objective functions, uses a per-instance algorithm configuration of the popular CMA-ES algorithm [10], in which a fraction of the evaluation budget is reserved for sampling the fitness landscape, in order to select the hyperparameters [20]. Various methods exist for extracting landscape features. In [10], a model is trained to choose parameters based on the results of sampling. However, various straightforward calculations can provide useful information. For example, Alyahya and Rowe [7] show that Simple Random Sampling provides an unbiased estimate of the number of local optima, which can have an important effect on the performance of optimisation algorithms.

One way to summarise features of the landscape is to construct a *Local Optima Network* (LON) [26]. This describes the fitness landscape as a graph, where the vertices are local optima and the edges define the probability of moving from one optimum to another. This same idea is commonly used to characterise the chemical potential energy landscapes of molecule configurations [16, 29], where vertices are local potential energy minima and edges correspond to the transition states between minima. Inferring a LON from the fitness landscape of a given optimisation problem allows us to compress the landscape information into a more concise mathematical structure that minimises information loss, as opposed to calculating summary statistical measures, such as autocorrelation or fitness distance correlation [28]. This can inform us as to how difficult a landscape may be for a particular optimiser. A LON is usually constructed by sampling the space and using greedy local search in order to locate optima. From this, we also get an estimate of the basin size associated with each optimum.

In this chapter, we apply the LON analysis to a data-driven application problem: fitting a gene regulatory network (GRN) model to data. GRNs describe the complex network of interactions between a set of genes carried out by intermediates such as mRNA and proteins. Mathematical models of GRNs can be used to infer qualitative information regarding the system, through identifying parameter values. They can also be used to make predictions and inform future experiments. The values of the parameters in these models are typically found by fitting to data [1–5, 15, 17, 30]. Finding the best fit between a model and data defines an optimisation problem that can be multimodal [15]. Due to the complexity of the cost function, which is typically the result of a numerical simulation, it is usually considered as a black-box.

A LON is not only useful for informing a search algorithm, but can also provide insight into the real-world system under consideration, beyond what can be achieved by the optimisation algorithm alone. In the case of a GRN model, the LON can tell us if there is a contiguous region of parameter space—a plateau—for which the model can describe the data equally well. This provides information on the identifiability of certain model parameters. If we can move in a path along the fitness landscape where the model provides comparably good fits to the data, then we have determined some model parameters whose values cannot be uniquely determined. Likewise, the LON may suggest that there are separate modes in parameter space (disjoint sets of parameter values) that provide equally good fits to the data. Moreover, if the fitness of distinct modes is similar but the basin sizes are significantly different, the LON will provide information regarding the robustness of different modes. Model solutions situated at modes with larger basin sizes are more robust to variation in parameters. In the case of GRN models, the desired solution may be the more robust one, even if it is not a global optimum, as these modes are more likely to viable within an inherently noisy biological environment.

We analyse a reduced model of a GRN, that of the circadian clock in the fungus *Neurospora crassa*, initially developed in [5]. This model is constructed using Boolean Delay Equations (BDEs) [5, 14, 18, 27], in which the time-dependent activity of three variables—an mRNA, a protein and a light input to the model—are logical-valued. The model contains three real-valued *delay parameters*, which describe the time taken for a change in the activity of one variable to result in a change in the activity of another. Also, associated with the model are two *threshold parameters*, which describe the concentration above which the mRNA or protein is considered to be active. These are used to discretise the data before fitting. We also define two logical-valued *gate parameters*, the values of which determine whether a particular reaction in the model is activating or inhibiting. These parameters determine the regulatory logic of the model itself, thereby giving rise to four different model architectures in this case. Using LONs, we can analyse the landscapes defined by the different model architectures, allowing us to gain an understanding of the relative difficulty of the resulting global optimisation problems.

The fitness landscape obtained for the *Neurospora* model with a noise-free dataset has previously been shown to be multimodal [15]. In this chapter, we use LONs to compare the fitness landscapes induced by optimising to data with and without noise, using two different neighbourhood operators: the Von Neumann neighbourhood and the Moore neighbourhood. We also compare LONs where the threshold parameters are fixed at the values that result in the maximum entropy of thresholded data, and where the fitness of a point in delay space is averaged over a set of threshold parameters. We demonstrate that whilst fixing the maximum entropy thresholds may seem like a reasonable and simple choice, it results in a fitness landscape with a high degree of neutrality , which can be difficult for optimisers to navigate.

We also demonstrate that, optima with a larger basin size tend to have a higher fitness. This is a desirable feature for an optimisation algorithm as it results in a landscape that is easier to search. It is also desirable from the point of view of modelling, as it means that the global optimum is also one of the most robust to parameter perturbations.

2 Data-Driven Application: Gene Regulatory Network Models

2.1 Introduction to Gene Regulatory Networks and Circadian Rhythms

The central dogma of molecular biology describes the flow of information in gene expression: a gene produces messenger RNA (mRNA), through *transcription* and mRNA produces protein through *translation*. Proteins carry out myriad tasks within cells, often regulating the expression of other genes, or even the gene from which it was produced. Through *regulators*, such as proteins and mRNAs, genes are involved in a complex network of interactions, termed gene regulatory networks (GRNs). Mathematical models of GRNs provide valuable insight into the behaviour of these systems. They can provide a useful qualitative description of the network from a systems-level point of view, or can be used for quantitative analysis, to test hypotheses in a relatively quick and cheap manner (compared with wet-lab implementations) and inform future experiments. A variety of types of mathematical models are used to represent GRNs [13], notably ordinary differential equation (ODE) models, logic models and stochastic models. However, modelling presents a number of challenges. Typically, models contain a large number of parameters. These parameters quantify biological processes represented in the model (e.g. rates of transcription/translation/degradation for ODEs), but it is often exceedingly difficult or costly to measure these parameters in practice. Hence, the majority of the model parameters are usually unknown and hence must be inferred in some way [1].

Estimates of some parameter values can sometimes be obtained from measurements taken in similar systems—or in slightly different experimental conditions—or from parameter values used in a similar or reduced model. However, it is usual to search for reasonable estimates by fitting the output of the model directly to data, for example, by minimising the least-squares fit of model output to experimental expression profiles [9]. The process of fitting to data presents its own difficulties, and often, a "good enough" fit, determined by inspection of the resulting expression profile, is all that is sought. But it is recognised that many established systems biology models are "sloppy" [19], meaning that large variations in some directions in parameter space can result in negligible change in the behaviour of the model. This insensitivity to changes in some parameter directions creates large neutral areas in the search space [12]. Hence, a global characterisation of the fitness landscape is of great interest to practitioners of mathematical modelling of GRNs. This can help to develop algorithms designed or tuned specifically for certain classes of models (e.g. if we can say that the parameter landscape of these models are usually rugged or multimodal) and can also inform the practitioner's understanding of the models themselves (e.g. if there are distinct regions of parameter space—i.e. distinct model behaviours—that can describe the data equally well).

In this chapter, we use local optima network (LON) analysis to characterise the landscape defined by the problem of fitting GRN models to data and, as a result, provide insight into the GRN models themselves. We focus here on a prototypical example from biology: *circadian rhythms*. Ubiquitous across organisms is an internal timekeeping mechanism, the circadian clock, which is regulated by the earth's natural light-dark cycle and tunes many of an organism's behaviours and internal processes to a roughly 24 h period. The complexity of the GRNs underlying circadian clocks varies considerably among organisms, but some features are common. First, the system is *entrained* by an external light stimulus. This forces the system into a daily periodic cycle. Once this forcing is removed, the system drifts into a free-running (autonomous) oscillation, but will maintain a period close to 24 h for some time. Secondly, the system contains *negative feedback*. As an example of this, consider again the central dogma. A protein may inhibit the transcription of the mRNA from which it was translated, reducing mRNA levels which, after some time delay, results in a reduction in protein levels (as protein is being constantly degraded). This, in turn, allows the rate of mRNA transcription to increase, thus enhancing again the translation of protein. The negative feedback between the mRNA and protein results in a periodic cycle of mRNA and protein activity. This negative feedback need not only be between regulators produced by one gene, as in this example, but may be between different genes in the GRN.

In the remainder of this section, we begin by describing an exemplar circadian clock based on Boolean logic, before analysing the LONs for the problem of fitting the output of the models to data. The Boolean model contains a combination of real-valued and logical-valued parameters—hence the search space can be considered a mixed-integer one.

2.2 Boolean Delay Equations

Boolean Delay Equations (BDEs) describe the logical-valued (0/1 or OFF/ON) states of a set of time-dependent variables [2, 5, 14, 15, 18, 27]. The state of a variable x_i at time t is determined by the states of variables x_k at times $t - \tau_{ik}$ in the past, for $i, k = 1, \ldots, n_V$, where τ_{ik} denote some positive real-valued *delay parameters* and n_V is the number of model variables. The corresponding set of BDEs can be written as

$$x_i(t) = f_i(x_1(t - \tau_{i1}), x_2(t - \tau_{i2}), \ldots, x_n(t - \tau_{in_V})); \quad t \geq t_0,$$

where $x_i(t) \in \{0, 1\}$ denotes the state of the ith variable at time $t \in \mathbb{R}$ and $\tau_{ik} \in \mathbb{R}$ is the signalling delay that defines the time it takes for x_k to affect x_i. In order to generate a model prediction beginning from time t_0, the states of the model variables over the range $[t_0 - \max(\tau_{ik}), t_0]$, termed the *history*, must be provided, and this can be taken from experimental data, where available.

BDE models reduce complexity, compared with common approaches for modelling time-dependent processes, such as ordinary differential equations or stochastic simulations. In particular, when concerned with the problem of optimising parameter values to data, BDEs offer a number of advantages [5]:

1 They contain significantly fewer parameters: the exemplar model we analyse here contains only three delay parameters, allowing us to view the distribution of local optima in three dimensions and exhaustively evaluate the cost function over a discretised mesh, covering the entire delay space. (The mesh width being the limiting factor on this enumeration.)
2 There is a natural cost function for fitting to data: the Hamming distance. This is not always the case in other modelling approaches where there are a number of possible cost functions, each of which defines a different optimisation problem.
3 It is easier to decide on reasonable parameter bounds: in our circadian clock problem, the model output should be oscillatory with period equal to approximately 24 h. Hence, it is natural to bound delay values in a 0–24 h range. Similarly, by definition, discretisation thresholds lie between 0% and 100%.
4 BDEs allow for the enumeration of alternative model architectures: the structure of the model itself can easily be parameterised by logical parameters. The values of these parameters can define whether the effect of a reaction is activating or inhibiting, for example. Or in the case where the activity of one variable is affected by more than one variable, the particular combination of input variable activities required to elicit a response (e.g. in the case of two input variables, the logical AND gate represents the case where both inputs need to be on). This makes BDE models useful tools for determining suitable model architectures that are consistent with data. This also allows for the simultaneous optimisation of parameter values and model architectures, leading to a novel mixed-integer optimisation problem.

For these reasons, the optimisation of parameters in these models makes a good data-driven test case for LONs. In what follows, we describe the exemplar BDE model of circadian rhythms and how the cost function is constructed.

2.3 An Exemplar Computational Model of Circadian Rhythms Based on BDEs

In previous work, Akman et al. introduced a general scheme for modelling biochemical oscillators using the BDE formalism, constructing Boolean versions of several established ODE clock models [5]. This included an ODE model of the clock in the filamentous fungus *Neurospora crassa*. This is based on a single negative feedback loop in which the gene *FREQUENCY* (*FRQ*) is repressed by its protein product. *FRQ* transcription is upregulated by light, thereby providing a mechanism for light entrainment [23]. The model comprises three differential equations describing the dynamics of *FRQ* mRNA and the cytoplasmic and nuclear forms of FRQ protein

$$\begin{aligned}\frac{dM}{dt} &= (v_s + L(t))\frac{k_I^N}{k_I^N + P_n^N} - v_m\frac{M}{k_m + M},\\ \frac{dP_c}{dt} &= k_s M - v_d\frac{P_c}{k_d + P_c} - k_1 P_c + k_2 P_n,\\ \frac{dP_n}{dt} &= k_1 P_c - k_2 P_n.\end{aligned} \tag{1}$$

A circuit diagram of the ODE model can be seen in Fig. 1a. Collectively, the reactions are parameterised by 10 kinetic constants: v_s, k_I and N (transcription); k_s (translation); v_m, k_m and v_d (degradation); k_1 and k_2 (nuclear translocation). In this study, we use the parameter values reported in the original modelling study [23]. In the M equation of system (1), the forcing term $L(t)$ models the effect of light, which perturbs the clock by upregulating *FRQ* transcription. L switches between 0 and a maximum value 1 at lights-on (t_{DAWN}), and then switches back to 0 at lights-off (t_{DUSK}):

$$L(t) = \begin{cases} 1 & \text{if } t_{DAWN} \le \text{mod}(t, 24) \le t_{DUSK},\\ 0 & \text{otherwise.}\end{cases} \tag{2}$$

Akman et al. showed that the *Neurospora* ODE model (1) can be expressed in general BDE form as follows:

$$\begin{aligned} x_M(t) &= G(x_P(t-\tau_2), g_2) \text{ OR } L(t-\tau_3),\\ x_P(t) &= G(x_M(t-\tau_1), g_1).\end{aligned} \tag{3}$$

The circuit diagram for this model is shown in Fig. 1b. In (3), x_M is *FRQ* mRNA, x_P is lumped FRQ protein (combining the cytoplasmic and nuclear forms) and $\boldsymbol{\tau} = (\tau_1, \tau_2, \tau_3)$ are the signalling delays. The light input $L(t)$ is given by Eq. (2).

As mentioned above, a key advantages of BDE modelling is that it allows the enumeration of different model architectures. This is achieved by defining Boolean functions (*logic gates*), which determine the state of a model variable given one or more inputs. The logic gates can take either one or two inputs, as well as a binary *gate parameter*. The model architecture is, therefore, defined by a bitstring of gate parameters, which are termed the *logic gate configuration (LGC)* of the model. A logic gate configuration can also be represented in its decimal form for convenience and clarity [5]. As an example of this framework, the logic gate $G(x, g)$ in (3) implements the identity or NOT gate, modelling activation and repression by x, respectively, depending on the value of the bit g: $G(x, 0) = x$, $G(x, 1) =$

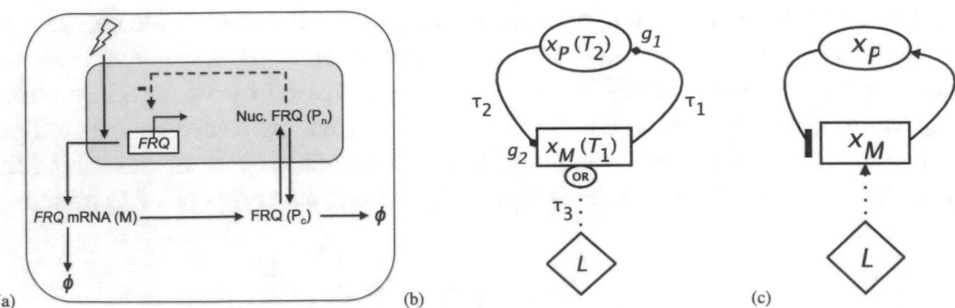

Fig. 1 **a** Circuit diagram for the ODE model (1) of the *N. crassa* clock. *FRQ* mRNA (M) is translated into protein (P_c) in the cytoplasm and then transported into the nucleus (P_n) where it represses the transcription of *FRQ*. Light (flash symbol) entrains the model by increasing the transcription rate. **b** Circuit diagram of the general Boolean formulation (3) of the circuit in **a**. x_M denotes *FRQ* mRNA and x_P denotes bulk FRQ protein (cytoplasmic and nuclear forms combined). L indicates the light input, $\{\tau_1, \tau_2, \tau_3\}$ represent the signalling delays and $\{T_1, T_2\}$ are the discretisation thresholds used to fit the discrete model to continuous data. g_1 and g_2 index logic gates $g_i \in \{0, 1\}$ that can be varied to generate different regulatory structures. The logic configuration consistent with the pattern of activation and inhibition of the ODE model used to generate fitting data (the DE LC) is $g_1 = 0, g_2 = 1$: the circuit diagram of the DE LC is plotted in **c**

NOT x. The bitstring $\mathbf{g} = g_1 g_2$ thus specifies the logic configuration. There are, therefore, four possible LCs, consistent with the underlying directed graph of the ODE model (i.e. the graph with connections $M \to P_c \to P_n \to M$; cf. Fig. 1a). Of these LCs, only one–namely, $\mathbf{g} = 01$—is consistent with the pattern of activation and inhibition of the ODE model (activation of x_P by x_M and repression of x_M by x_P; cf. Fig. 1c). This configuration is, therefore, referred to as the *DE LC*. The other three configurations represent alternative regulatory structures consistent with the ODE model graph: for example, $\mathbf{g} = 00$ corresponds to a circuit in which *frq mRNA* activates FRQ protein, but FRQ protein activates *FRQ* mRNA (a double positive feedback loop).

2.4 Parameter Optimisation of the BDE Model

In this section, we describe the steps necessary for optimising the parameters of the BDE model (3). An illustrative example is shown in Fig. 2.

2.4.1 Light Conditions

Often, a desired outcome of a mathematical model is to be able to predict the behaviour of a biological system under environmental or genetic perturbations. It is common that when fitting to data, we must fit to data collected in a number of different experimental conditions. In the case of circadian clock models, we wish to

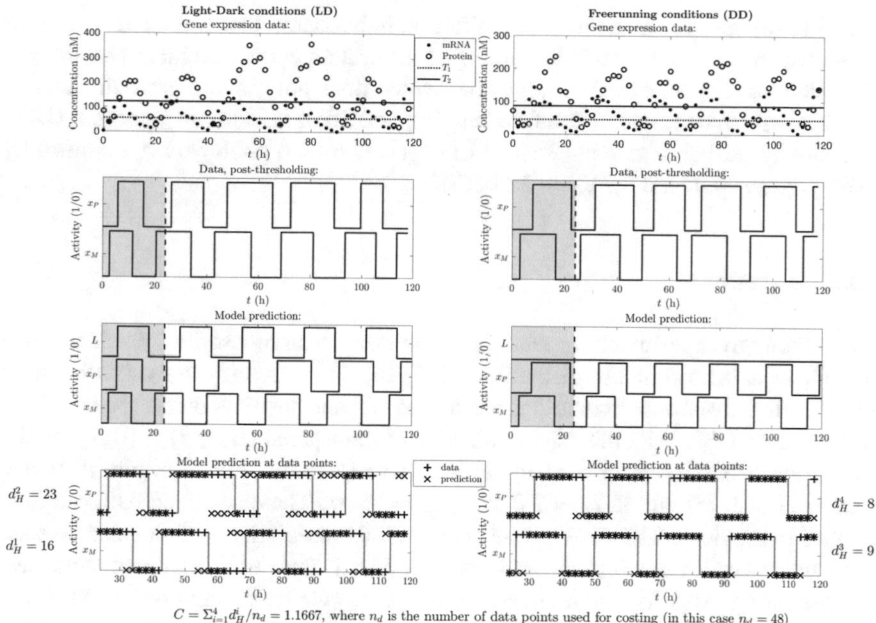

Fig. 2 An illustrative example of calculating a cost function value for the *Neurospora* model for a given set of delay and threshold parameters. Here, $\tau_1 = 4.46$ h, $\tau_2 = 6.28$ h, $\tau_3 = 10.4$ h, $T_1 = 0.3$ and $T_2 = 0.3$, and we use the data-generating gate configuration $g = 01$. The data provided here are the mRNA and protein concentrations in two light conditions: LD (left column) and DD (right column). The data is the output of a run of the Gillespie algorithm and is sampled at 2 h intervals for clarity. Note that this dataset is used for illustrative purposes and is not the dataset used to generate the LONs, which is sampled at 0.5 h intervals. In the first step (row 1 to row 2), the data is thresholded. Note that we assume a different threshold value for each model variable, but these are shared among experimental conditions. Where two consecutive data points cross a threshold, we calculate the switch point as the time at which the linear interpolant between the two data points crosses the threshold. We take the first 24 h of the thresholded data (row 2) as model history, indicated here as the shaded region. In the next step (row 3), we generate a model prediction, using the history from row 2, again indicated by the shaded region. In row 3, we include the light variable $L(t)$, which is set to 0 in (3) to simulate DD, and is specified by Eq. (2) with $t_{DAWN} = 6$ and $t_{DUSK} = 18$ to simulate LD. In the final step (row 4), we evaluate the model at each of the time points given by the data. Note that in the plots in row 4, we do not include the history (as it is not used for costing). The Hamming distance values for each variable in each light condition, d_H^i for $i = 1, 2, 3, 4$, are shown next to the plots. The cost function value, shown at the bottom, is equal to the sum of the four Hamming distance values, normalised by the number of data points used for costing, n_d

fit them to expression data under two different light conditions: 1. A light-dark (LD) cycle, which defines the periodic forcing of the system under normal environmental conditions; and 2. A free-running cycle, which does not include periodic forcing. For the *Neurospora* model, the free-running condition is constant darkness (DD), obtained by setting L to 0 in (3). The LD cycle is a 12-h photoperiod, obtained by setting $t_{DAWN} = 6$ and $t_{DUSK} = 18$ in (2).

2.4.2 Thresholds

When fitting to experimentally recorded gene expression timeseries, which are real-valued, the data must first be discretised [2, 5, 15]. This can be done by thresholding: any data that is above a threshold is defined as ON and any data below the threshold is defined as OFF. We define real-valued threshold parameters, $T_i \in [0, 1]$, as the proportion of the range of data for x_i that is ON or OFF. For example, if $T_i = 0$ then $x_i(t) = 1,\ \forall t$ and if $T_i = 1$ then $x_i(t) = 0,\ \forall t$. The discretised data can then be used to provide a history for generating a model prediction and can also be used to calculate a cost value for this prediction [2, 5, 15]. The threshold parameters have the interesting property that their values affect the cost function in two ways. First, they determine the model prediction, through the calculation of the history from the data and, second, they determine the structure of the Boolean data that is used in the calculation of the Hamming distance.

Any choice of fixed threshold values is an arbitrary one but we would like to ensure that a solution found through optimisation should be robust to the choice of threshold parameters. We, therefore, explore two different approaches:

1. We choose the threshold parameters that maximise the joint entropy of the discretised data [15]. A bitstring has low entropy if it contains little information, meaning it has a high proportion of either ones or zeros. Hence, our motivation for this approach is that a desirable solution should contain significant information, otherwise it will not have the desired periodic behaviour that is typical of a circadian clock. We denote the proportion of data points for variable i that exceed a given threshold value in light condition j, by p_{ij}. The joint entropy, $\mathscr{E}$, over variables $i = 1, \ldots, n_V$ and light conditions $j = 1, \ldots, n_L$ is then given by

$$\mathscr{E} = \sum_{i=1}^{n_V} \sum_{j=1}^{n_L} \left(-p_{ij} \log_2(p_{ij}) - (1 - p_{ij}) \log_2(1 - p_{ij}) \right).$$

2. The cost value for a set of delays is averaged over the individual costs for a set of threshold values.

This approach is described further in Sect. 2.4.4.

2.4.3 Datasets

We present here two synthetic datasets [5], to which we fit our BDE model. Each dataset contains expression data in two light conditions, DD and LD as described in Sect. 2.4.1. The datasets are proxies for experimental timeseries that have been recorded with an experimentally viable sampling rate in the presence of noise, thereby enabling us to systematically assess the extent to which the landscape analysis captures core features of the optimisation problem, and the extent to which this is affected by experimental variation. They are generated as described below:

D1 Deterministic dataset: this data is the output of the corresponding ODE model (1) sampled at regular intervals of 0.5 h. Hence, these timeseries contain no noise and provide a somewhat idealised description of the system. Fitting to this data is a useful test as it allows us to analyse the optimisation problem in the absence of noise, and can inform us as to whether the optimisation can recover the same model architecture as that which generated the data.

D2 Stochastic datasets: this data is the output of the Gillespie algorithm for the same reaction system as D1 [5], with the same sampling interval of 0.5 h. Comparing LONs for D1 and D2 will inform us as to how noise influences the search landscape.

2.4.4 Optimisation

Following [15], we define the cost function, C for a given set of delays, $\boldsymbol{\tau}$, thresholds, $\mathbf{T} = (T_1, T_2)$, and logic gate configuration, $\mathbf{g} = g_1 g_2$, as

$$C(\boldsymbol{\tau}, \mathbf{g}, \mathbf{T}) = \frac{1}{n_d} \sum_{j=1}^{n_L} \sum_{i=1}^{n_V} d_H(\mathbf{F}_{ij}(\boldsymbol{\tau}, \mathbf{g}, \mathbf{D}_{ij}(\mathbf{T})), \mathbf{D}_{ij}(\mathbf{T})), \tag{4}$$

where $d_H(\cdot, \cdot)$ denotes the Hamming distance between two bitstrings, $\mathbf{D}_{ij}$ are the discretised data values for variable i in light condition j, $\mathbf{F}_{ij}$ are the model prediction values at the corresponding time points in $\mathbf{D}_{ij}$, and n_d are the number of time points over which the Hamming distance is calculated.

In this chapter, the landscape of the following two cost functions are studied:

- $C_{ME}(\boldsymbol{\tau}, \mathbf{g}) = C(\boldsymbol{\tau}, \mathbf{g}, \mathbf{T}^{ME})$: The threshold values are fixed at the values that maximise the joint entropy of the data (as described in Sect. 2.4.2) over each model variable in each light condition.
- $C_A(\boldsymbol{\tau}, \mathbf{g}) = \frac{1}{n_T} \sum_{l=1}^{n_T} C(\boldsymbol{\tau}, \mathbf{g}, \mathbf{T}^l)$: The cost is averaged over a set of threshold values, as described in the following section.

For each delay parameter τ_k, we assume that $0 < \tau_k \leq 24$. We also assume that the sum of the delays in the feedback loop does not exceed 24 h

$$\tau_1 + \tau_2 \leq 24. \tag{5}$$

Note that the cost functions C_{ME} and C_A induce mixed-integer landscapes, due to the discrete nature of the gate configuration parameters, g_j. A fundamentally important question when analysing the landscape results in this case is whether we can efficiently search across different gate configurations.

3 Landscape Analysis

Formally, a landscape is a triple $(\mathscr{X}, N, f)$, where $\mathscr{X}$ is the set of candidate solutions, N is a neighbourhood operator specifying the connectivity between candidate solutions and f is the objective function. We consider two search spaces in this chapter. The first one is the continuous space of the delays $\boldsymbol{\tau} \in (0, 24]^3$. We approximate this continuous space by discretising it into uniformly spaced grids with step sizes of 0.5. The choice of this step size in the $\boldsymbol{\tau}$ space was guided by the sampling rate of the data. This discretising allows us to enumerate the whole search space at the given grid resolution and perform exact landscape analysis. We note, however, that as the number of dimensions d increases, exact landscape analysis of continuous spaces becomes infeasible, even after discretisation at relatively coarse grid sizes, meaning that sampling is necessary. The second search space is a mixed-integer (MI) space of the gates $\boldsymbol{g}$ and delays $\boldsymbol{\tau}$.

For the third element of the landscape triple, namely, the objective function f, we examine two functions $C_A(\boldsymbol{\tau}, \mathbf{g})$ and $C_{ME}(\boldsymbol{\tau}, \mathbf{g})$, the calculation of both involves a continuous threshold space $\mathbf{T} \in [0, 1]^2$ which we discretise into uniformly spaced grids with step sizes of 0.05. $C_A(\boldsymbol{\tau}, \mathbf{g})$ is then calculated by averaging over this threshold space, and $C_{ME}(\boldsymbol{\tau}, \mathbf{g})$ by selecting the value that maximises the joint entropy of the data.

To highlight the important effect the neighbourhood function has on the structure of the resulting landscape, we perform a comparative analysis of two neighbourhood functions. The neighbourhood functions we consider are natural choices in gridded spaces, namely the Moore neighbourhood (MN) and the von Neumann neighbourhood (VN) [33]. We define the VN for radius r as $\text{VN}(\mathbf{x}, r) = \{\mathbf{y} \in \mathscr{X} \mid 0 < \|\mathbf{x} - \mathbf{y}\|_1 \leq r\}$ and the MN as $\text{MN}(\mathbf{x}, r) = \{\mathbf{y} \in \mathscr{X} \mid 0 < \|\mathbf{x} - \mathbf{y}\|_\infty \leq r\}$. The size of $\text{VN}(\mathbf{x}, r)$ is the Delannoy number $D(d, r) = \sum_{k=0}^{\min(d,r)} \binom{d+r-k}{d}\binom{d}{k} - 1$ [11]. In a 2D grid, the size is the centred square number $2r(r+1)$. The size of $\text{MN}(\mathbf{x}, r)$ is $(2r+1)^d - 1$. Note that the size of the MN grows exponentially with d. We study landscapes induced by both neighbourhoods for $r = 1, 2$ and consider greedy (best improving move) local search. The landscape properties studied in this chapter consider only the feasible region ($0 < \tau_k \leq 24$ and $\tau_1 + \tau_2 < 24$). In the mixed-integer (MI) space of the gates ($\boldsymbol{g}$) and delays ($\boldsymbol{\tau}$), we still examine the MN and VN neighbourhoods in the $\boldsymbol{\tau}$ subspace, but with the addition of the Hamming neighbourhood (H) in the gate subspace. Formally, we define the mixed-integer neighbourhood of $\mathbf{x}$ with radius r as $\text{MI-VN}(\mathbf{x}, r) = \{\mathbf{y} \in \mathscr{X} \mid 0 < \|\mathbf{x}_\tau - \mathbf{y}_\tau\|_1 + d_{\text{H}}(\mathbf{x}_g, \mathbf{y}_g) \leq r\}$ for the von Neumann neighbourhood, and $\text{MI-MN}(\mathbf{x}, r) = \{\mathbf{y} \in$

$\mathscr{X} \mid 0 < \|\mathbf{x}_\tau - \mathbf{y}_\tau\|_\infty + d_{\mathrm{H}}(\mathbf{x}_g, \mathbf{y}_g) \leq r\}$ the Moore neighbourhood, where d_{H} donates the Hamming distance in both cases.

There are many interesting properties of a landscape structure that can yield useful insights into the studied problem itself, and also into the performance of search algorithms. One such intrinsic property is the distribution of configuration types with respect to search. For a given point $\mathbf{x} \in \mathscr{X}$ in the landscape, according to the topology and fitness values of its direct neighbourhood, it can belong to one of seven different types of search position [21]. These types are:

- Strict local minimum (SLMIN): $\forall \mathbf{y} \in N(\mathbf{x}),\ f(\mathbf{y}) > f(\mathbf{x})$.
- Non-strict local minimum (NSLMIN): $\forall \mathbf{y} \in N(\mathbf{x}),\ f(\mathbf{y}) \geq f(\mathbf{x})$, and $\exists\, \mathbf{u}, \mathbf{z} \in N(\mathbf{x})$, such that $f(\mathbf{u}) = f(\mathbf{x})$, and $f(\mathbf{z}) > f(\mathbf{x})$.
- Interior plateau (IPLAT): $\forall\ \mathbf{y} \in N(\mathbf{x}),\ f(\mathbf{y}) = f(\mathbf{x})$.
- Ledge (LEDGE): $\exists\ \mathbf{u}, \mathbf{y}, \mathbf{z} \in N(\mathbf{x})$, such that $f(\mathbf{u}) = f(\mathbf{x}), f(\mathbf{y}) > f(\mathbf{x})$, and $f(\mathbf{z}) < f(\mathbf{x})$.
- Slope (SLOPE): $\forall \mathbf{y} \in N(\mathbf{x}),\ f(\mathbf{y}) \neq f(\mathbf{x})$, and $\exists\, \mathbf{u}, \mathbf{z} \in N(\mathbf{x})$, such that $f(\mathbf{u}) < f(\mathbf{x})$, and $f(\mathbf{z}) > f(\mathbf{x})$.
- Non-strict local maximum (NSLMAX): $\forall \mathbf{y} \in N(\mathbf{x}),\ f(\mathbf{y}) \leq f(\mathbf{x})$, and $\exists\, \mathbf{u}, \mathbf{z} \in N(\mathbf{x})$, such that $f(\mathbf{u}) = f(\mathbf{x})$, and $f(\mathbf{z}) < f(\mathbf{x})$.
- Strict local maximum (SLMAX): $\forall \mathbf{y} \in N(\mathbf{x}),\ f(\mathbf{y}) < f(\mathbf{x})$.

Across all the different landscapes studied in this chapter—namely, across the deterministic and stochastic datasets D1 and D2, different cost functions and different neighbourhood operators—the search space was found to mainly be composed of the SLOPE configurations (see Fig. 3). No IPLAT configurations were found, apart from for the LGC $\mathbf{g} = 00$ with the maximum entropy cost function $C_{ME}(\boldsymbol{\tau}, \mathbf{g})$. This could be due to the coarser grid taken over $\boldsymbol{\tau}$-space—usually as the step size of the grid decreases, the IPLAT configurations increases. The number of LEDGE configurations is significantly larger for the maximum entropy cost function, $C_{ME}(\boldsymbol{\tau}, \mathbf{g})$. In fact, for the LGC $\mathbf{g} = 00$, there are more LEDGE configurations than SLOPE.

Another important feature of a given problem's landscape is neutrality, due in part to its potentially disruptive effect on a search algorithm's performance. Here, we study neutral regions and distinguish between open and closed ones, following [6, 8]. In particular, a plateau is a set of connected non-strict local optima, with or without interior plateau points. An exit is a neighbour of one or more configurations in the plateau, which shares the same fitness value of the plateau, but has an improving neighbour. When a plateau has at least one exit we call it *open*, otherwise we call it *closed*. Algorithm 1 shows how a plateau is explored exhaustively starting from a non-strict local optimum or an interior plateau configuration, where U is the set of unvisited non-strict local optima, interior plateau configurations and exit configurations, P is the set of visited non-strict local optima and interior plateau configurations, and E is the set of exits found. After exploring the entire plateau, the algorithm then returns the sets P and E.

When considering the threshold-averaged cost function $C_A(\boldsymbol{\tau}, \mathbf{g})$, plateaus were found only in the landscapes for LGCs $\mathbf{g} = 00$ and $\mathbf{g} = 11$ (see Figs. 6, 7, 8 and 9). Both open and closed plateaus were found with the largest plateau size being 30.

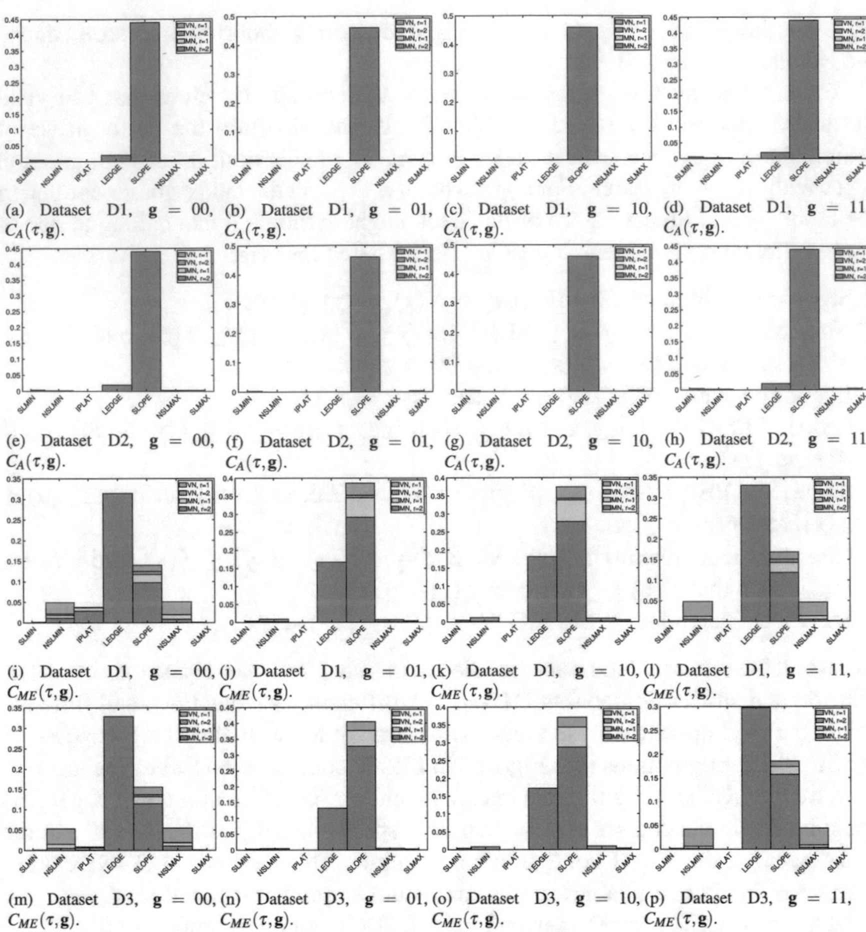

(a) Dataset D1, $\mathbf{g} = 00$, $C_A(\tau, \mathbf{g})$. (b) Dataset D1, $\mathbf{g} = 01$, $C_A(\tau, \mathbf{g})$. (c) Dataset D1, $\mathbf{g} = 10$, $C_A(\tau, \mathbf{g})$. (d) Dataset D1, $\mathbf{g} = 11$, $C_A(\tau, \mathbf{g})$.

(e) Dataset D2, $\mathbf{g} = 00$, $C_A(\tau, \mathbf{g})$. (f) Dataset D2, $\mathbf{g} = 01$, $C_A(\tau, \mathbf{g})$. (g) Dataset D2, $\mathbf{g} = 10$, $C_A(\tau, \mathbf{g})$. (h) Dataset D2, $\mathbf{g} = 11$, $C_A(\tau, \mathbf{g})$.

(i) Dataset D1, $\mathbf{g} = 00$, $C_{ME}(\tau, \mathbf{g})$. (j) Dataset D1, $\mathbf{g} = 01$, $C_{ME}(\tau, \mathbf{g})$. (k) Dataset D1, $\mathbf{g} = 10$, $C_{ME}(\tau, \mathbf{g})$. (l) Dataset D1, $\mathbf{g} = 11$, $C_{ME}(\tau, \mathbf{g})$.

(m) Dataset D3, $\mathbf{g} = 00$, $C_{ME}(\tau, \mathbf{g})$. (n) Dataset D3, $\mathbf{g} = 01$, $C_{ME}(\tau, \mathbf{g})$. (o) Dataset D3, $\mathbf{g} = 10$, $C_{ME}(\tau, \mathbf{g})$. (p) Dataset D3, $\mathbf{g} = 11$, $C_{ME}(\tau, \mathbf{g})$.

Fig. 3 Search position types found in the landscapes obtained with different cost functions (threshold-averaged—C_A; maximum entropy—C_{ME}), neighbourhood types (VN—von Neumann neighbourhood; MN—Moore neighbourhood) and neighbourhood size r ($r = 1, 2$)

More plateaus of larger size were found in the landscapes of the maximum entropy cost function $C_{ME}(\boldsymbol{\tau}, \mathbf{g})$. This is true across all LGCs, with the largest plateau being of size ~1000. As expected, using larger neighbourhood functions results in fewer optima (see Fig. 4). However, apart from a few cases (e.g. $\mathbf{g} = 10$), the distribution of optima quality was found to be similar. In all landscapes, there is a trend of weak to moderate negative correlation between the basin size, the volume of $\mathscr{X}$ for which applying greedy local search under the neighbourhood function leads to the corresponding optimum, and the optimum cost. This suggests that these landscapes are not deceptive, and that fitter optima do, in general, have larger basin sizes and are, therefore, potentially easier to locate.

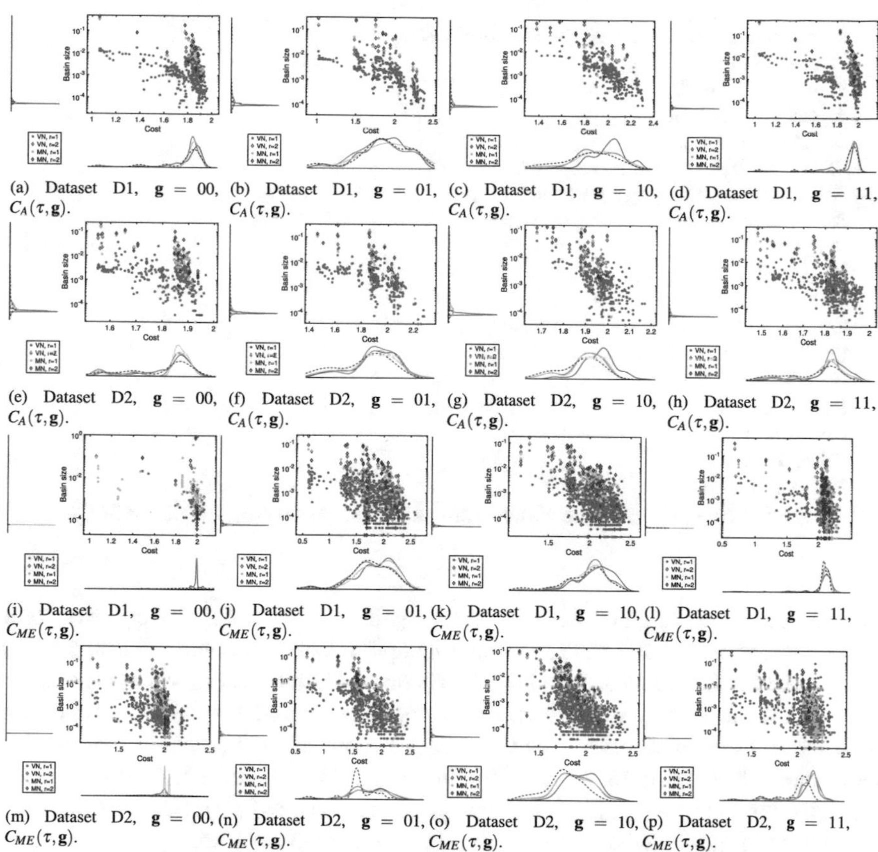

(a) Dataset D1, $\mathbf{g} = 00$, $C_A(\tau, \mathbf{g})$. (b) Dataset D1, $\mathbf{g} = 01$, $C_A(\tau, \mathbf{g})$. (c) Dataset D1, $\mathbf{g} = 10$, $C_A(\tau, \mathbf{g})$. (d) Dataset D1, $\mathbf{g} = 11$, $C_A(\tau, \mathbf{g})$.

(e) Dataset D2, $\mathbf{g} = 00$, $C_A(\tau, \mathbf{g})$. (f) Dataset D2, $\mathbf{g} = 01$, $C_A(\tau, \mathbf{g})$. (g) Dataset D2, $\mathbf{g} = 10$, $C_A(\tau, \mathbf{g})$. (h) Dataset D2, $\mathbf{g} = 11$, $C_A(\tau, \mathbf{g})$.

(i) Dataset D1, $\mathbf{g} = 00$, $C_{ME}(\tau, \mathbf{g})$. (j) Dataset D1, $\mathbf{g} = 01$, $C_{ME}(\tau, \mathbf{g})$. (k) Dataset D1, $\mathbf{g} = 10$, $C_{ME}(\tau, \mathbf{g})$. (l) Dataset D1, $\mathbf{g} = 11$, $C_{ME}(\tau, \mathbf{g})$.

(m) Dataset D2, $\mathbf{g} = 00$, $C_{ME}(\tau, \mathbf{g})$. (n) Dataset D2, $\mathbf{g} = 01$, $C_{ME}(\tau, \mathbf{g})$. (o) Dataset D2, $\mathbf{g} = 10$, $C_{ME}(\tau, \mathbf{g})$. (p) Dataset D2, $\mathbf{g} = 11$, $C_{ME}(\tau, \mathbf{g})$.

Fig. 4 Basin size (log scale), as a proportion of the feasible search space plotted against mode cost for the landscapes obtained with different cost functions (threshold-averaged—C_A; maximum entropy—C_{ME}), neighbourhood types (VN—von Neumann neighbourhood; MN—Moore neighbourhood) and neighbourhood size r ($r = 1, 2$). Each data point represents either a SLMIN or a plateau. The correlation coefficient between basin size and cost is calculated using Kendall's τ

4 Local Optima Networks

Beyond 3 dimensions, the fitness landscape is difficult to visualise; hence, compact visualisations of landscapes are helpful to both optimisation researchers (e.g. to help understand the problem in the context of optimiser behaviour) and to problem owners. In recent years, the local optima network (LON) has been developed as an intuitive and informative graph-based visualisation of the fitness landscape that is independent of its dimensionality [24]. In a LON, the graph vertices correspond to local modes—the more vertices in the graph the greater the number of local attractors in the landscape. Often the vertex size indicates the corresponding basin size of a mode, and the vertices are coloured according to the mode quality. The occurrence

```
start with x , where x is a NSLMIN or IPLAT
c ← f(x)
P ← ∅, U ← {x}, E ← ∅
while |U| > 0 do
  Choose y ∈ U
  U ← U/{y}
  if c > arg min_{z∈N(y)} f(z) then
    E ← E ∪ {y}
  else
    P ← P ∪ {y}
    for all z ∈ N(y) do
      if z ∉ P and f(z) = c then
        U ← U ∪ {z}
      end if
    end for
  end if
end while
return (P, E)
```

Algorithm 1: Exhaustive Plateau Exploration (assuming minimisation).

of a directed and weighted edge between a pair of vertices i and j conveys some measure of how *easy* it is to move from one optima/basin i to optima/basin j.

Formally, a LON is a graph $G = (V, E)$ in which the vertices V represent strict local optima (those solutions which are no worse than any of its neighbours) and their respective basins of attraction[1] and the edges E represent the probability of moving from one optima/basin to another. The two most widely used edge definitions are basin transitions (see e.g. [24]) and escape edges (see e.g. [31]). Here, we consider the latter definition, where the weight of an edge from vertex V_i to another V_j is the proportion of the solutions from $\mathscr{X}$ who lie in the jth basin *and* whose distance to the ith mode location is $\leq m$. Typically, m is chosen, so that it is equal or greater than the distance of neighbours r under the neighbourhood operator $N(\mathbf{x})$. For instance, where the design space alphabet is {0, 1}, the operator $N(\mathbf{x})$ used in greedy moves may return all single bit flip (Hamming distance 1) neighbours of $\mathbf{x}$, whereas the escape edge calculation may be for $m = 2$, and therefore, include both single- and two-bit flip neighbours ($r = 1$ and $r = 2$, respectively)—representing more exploratory moves, but limiting the start location to be local optima locations. Figure 5 shows two LONs generated for a random instance of an NK model [22] with $N = 14$ and $K = 3$ for different values of r and m. The NK-model for landscapes is a class of tunably rugged fitness landscapes that was proposed by Kauffman [22]. The landscape can be configured using two parameters N and K, where parameter N denotes the number of variables (string length) and parameter K denotes the number of variables that influence a particular variable (i.e. the degree of epistatic interactions). NK-landscapes can be tuned to be smooth or rugged by changing the value of parameter K from 0 (very smooth) to $N - 1$ (very rugged).

[1] This definition of vertex has been extended to account for neutrality in the search space [32].

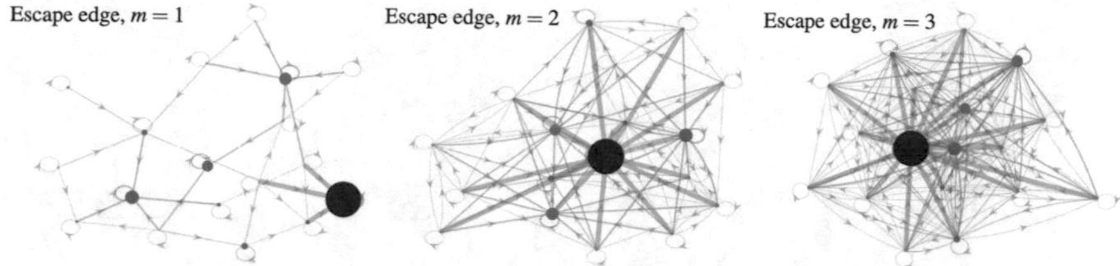

Fig. 5 Exact LON of a random NK model instance with $N = 14$ and $K = 3$ using escape edges with $m = 1$ (left), $m = 2$ (middle), and $m = 3$ (right). In this instance, there are 18 local optima, with a range of basin sizes. The *connectivity* of the LON graph varies depending on the edge definition used. The vertex size is scaled proportionally to the corresponding mode's basin size and the thickness of a directed edge is in proportion to the number of possible paths between one basin and the other

In this chapter, we account for neutrality in the landscape when calculating the LONs. Every node in the LONs is either a SLMIN or a plateau. In and out edges from a plateau node are calculated and weighted over all configurations in that plateau. Figures 6, 7, 8, 9, 10, 11, 12 and 13 show contour plots of the 3-d $\boldsymbol{\tau}$-space along with the optima and plateaus that are found using the VN and MN neighbourhood operators.[2] The figures also show the corresponding LONs where the escape edges are calculated using the same neighbourhood r that generated the vertices, but where the range of perturbation m can vary. Note that the optima, plateaus and attraction basins shown are all in the feasible region of the search space. In the LON figures, we mark improving edges in grey, worsening edges in red and equal edges (i.e. those leading to an optimum with the same cost) in blue. Indicating the different kinds of edges in this manner highlights monotonic and compressed monotonic LONs, which can be helpful in identifying funnel structures [25]. Monotonic LONs are LONs with only non-deteriorating transitions between local optima. Compressed Monotonic Local Optima Networks (CMLON) are those where the nodes with blue edges are compressed into a single node. As expected, increasing m increases the connectivity of the graph, since increasing the range of perturbation increases the likelihood of falling into other optima or plateau basins. All LONs are strictly-monotonic when $m \leq r$, apart from some LONs extracted from the landscape with the cost function $C_{ME}(\boldsymbol{\tau}, \mathbf{g})$, which are non-strictly-monotonic. See for example: (e), (f), (h), (t) in Fig. 11 and (e), (h), (q) and (r) in Fig. 13, where $r = 2, m = 1$. The LONs in these figures have a few equal escape edges (i.e. non-strictly-monotonic LONs). This is perhaps due to the higher number of local optima and plateaus in that landscape. Apart from losing the spatial information, the LONs seem to give a good reduction of the landscape, generating a compact summary of the structure of the optima and plateaus. For example, in Fig. 10, the LONs of the VN landscape for $\mathbf{g} = 00$ capture the extremely large plateau and its connectivity to the other optima, which have comparatively much smaller basins. However, the underlying spatial information—specifically the fact that τ_3 has no effect on the solutions in this particular region of $\boldsymbol{\tau}$-space—is lost.

[2] All figures are available in high quality under http://pop-project.ex.ac.uk/grn_lons.html.

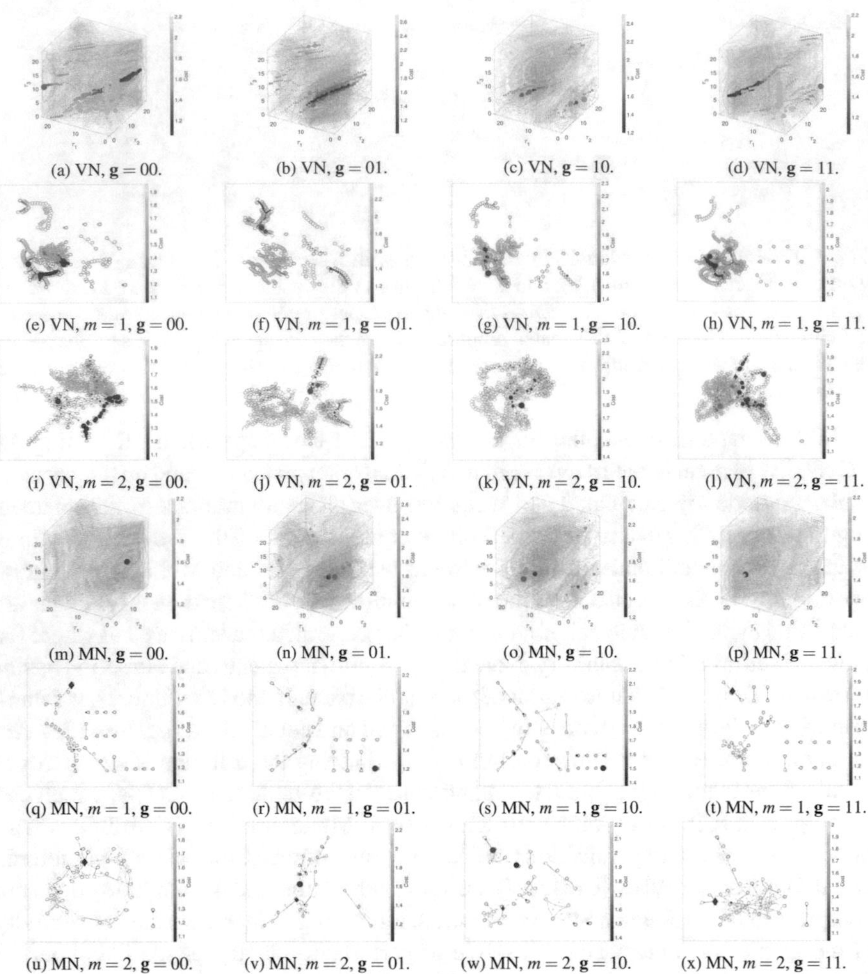

(a) VN, $\mathbf{g} = 00$. (b) VN, $\mathbf{g} = 01$. (c) VN, $\mathbf{g} = 10$. (d) VN, $\mathbf{g} = 11$.

(e) VN, $m = 1$, $\mathbf{g} = 00$. (f) VN, $m = 1$, $\mathbf{g} = 01$. (g) VN, $m = 1$, $\mathbf{g} = 10$. (h) VN, $m = 1$, $\mathbf{g} = 11$.

(i) VN, $m = 2$, $\mathbf{g} = 00$. (j) VN, $m = 2$, $\mathbf{g} = 01$. (k) VN, $m = 2$, $\mathbf{g} = 10$. (l) VN, $m = 2$, $\mathbf{g} = 11$.

(m) MN, $\mathbf{g} = 00$. (n) MN, $\mathbf{g} = 01$. (o) MN, $\mathbf{g} = 10$. (p) MN, $\mathbf{g} = 11$.

(q) MN, $m = 1$, $\mathbf{g} = 00$. (r) MN, $m = 1$, $\mathbf{g} = 01$. (s) MN, $m = 1$, $\mathbf{g} = 10$. (t) MN, $m = 1$, $\mathbf{g} = 11$.

(u) MN, $m = 2$, $\mathbf{g} = 00$. (v) MN, $m = 2$, $\mathbf{g} = 01$. (w) MN, $m = 2$, $\mathbf{g} = 10$. (x) MN, $m = 2$, $\mathbf{g} = 11$.

Fig. 6 **a–d** and **m–p**. Contour plots in $\boldsymbol{\tau}$-space for dataset D1 with the threshold-averaged cost function $C_A(\boldsymbol{\tau}, \mathbf{g})$. Spheres indicate optima and plateaus are shown as lines with widths corresponding to the basin size of the entire plateau. **e–l** and **q–x**. The corresponding LONs obtained with $m = 1$ and $m = 2$. Squares denote closed plateaus, diamonds denote open plateaus and the globally optimal solutions are marked with a red arrow labelled with the optimal cost value. The width of the edges reflects the probability of transition and their lengths are scaled proportionally to the Euclidean distance between the nodes. Self-loops are removed for clarity. Improving edges are coloured in grey, worsening edges are coloured in red and edges leading to an optimum with the same quality are coloured in blue. The range of both VN and MN is $r = 1$ in each plot. Each optimum and plateau shown is scaled proportionally to its basin size

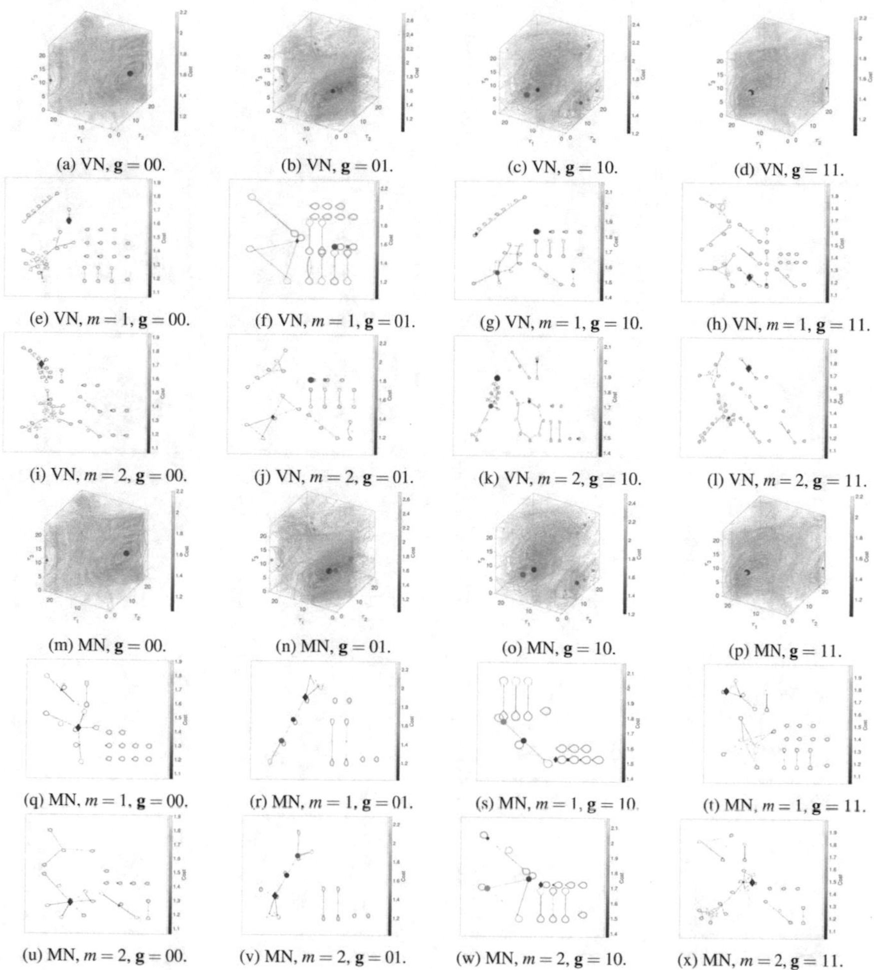

(a) VN, $\mathbf{g} = 00$. (b) VN, $\mathbf{g} = 01$. (c) VN, $\mathbf{g} = 10$. (d) VN, $\mathbf{g} = 11$.

(e) VN, $m = 1$, $\mathbf{g} = 00$. (f) VN, $m = 1$, $\mathbf{g} = 01$. (g) VN, $m = 1$, $\mathbf{g} = 10$. (h) VN, $m = 1$, $\mathbf{g} = 11$.

(i) VN, $m = 2$, $\mathbf{g} = 00$. (j) VN, $m = 2$, $\mathbf{g} = 01$. (k) VN, $m = 2$, $\mathbf{g} = 10$. (l) VN, $m = 2$, $\mathbf{g} = 11$.

(m) MN, $\mathbf{g} = 00$. (n) MN, $\mathbf{g} = 01$. (o) MN, $\mathbf{g} = 10$. (p) MN, $\mathbf{g} = 11$.

(q) MN, $m = 1$, $\mathbf{g} = 00$. (r) MN, $m = 1$, $\mathbf{g} = 01$. (s) MN, $m = 1$, $\mathbf{g} = 10$. (t) MN, $m = 1$, $\mathbf{g} = 11$.

(u) MN, $m = 2$, $\mathbf{g} = 00$. (v) MN, $m = 2$, $\mathbf{g} = 01$. (w) MN, $m = 2$, $\mathbf{g} = 10$. (x) MN, $m = 2$, $\mathbf{g} = 11$.

Fig. 7 **a–d** and **m–p**. Contour plots in $\boldsymbol{\tau}$-space for dataset D1 with the threshold-averaged cost function $C_A(\boldsymbol{\tau}, \mathbf{g})$. Details as in the caption of Fig. 6, except the range of both VN and MN is $r = 2$ in each plot

The LONs in Figs. 6, 7, 8, 9, 10, 11, 12 and 13 allow us to draw comparisons between different choices of neighbourhood operator, r, datasets and cost functions. We summarise the changes as follows:

- C_A versus C_{ME}: Compare Figs. 6, 7, 8 and 9 with Figs. 10, 11, 12 and 13. Comparing the contour plots in these figures shows that C_{ME} (Figs. 10, 11, 12 and 13) results in much larger regions of neutrality compared with C_A (Figs. 6, 7, 8 and 9). Comparing the LONs, we see that C_{ME} also results in a noticeably different network topology, with a much larger number of highly connected local optima.

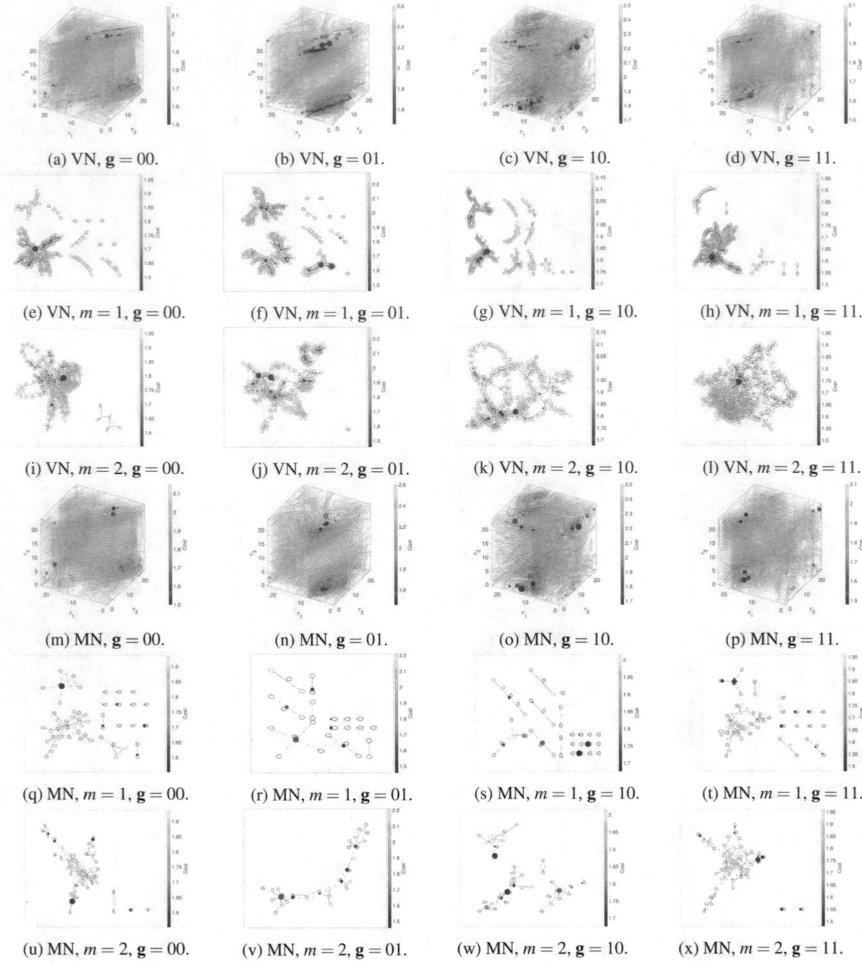

(a) VN, $\mathbf{g} = 00$. (b) VN, $\mathbf{g} = 01$. (c) VN, $\mathbf{g} = 10$. (d) VN, $\mathbf{g} = 11$.

(e) VN, $m = 1$, $\mathbf{g} = 00$. (f) VN, $m = 1$, $\mathbf{g} = 01$. (g) VN, $m = 1$, $\mathbf{g} = 10$. (h) VN, $m = 1$, $\mathbf{g} = 11$.

(i) VN, $m = 2$, $\mathbf{g} = 00$. (j) VN, $m = 2$, $\mathbf{g} = 01$. (k) VN, $m = 2$, $\mathbf{g} = 10$. (l) VN, $m = 2$, $\mathbf{g} = 11$.

(m) MN, $\mathbf{g} = 00$. (n) MN, $\mathbf{g} = 01$. (o) MN, $\mathbf{g} = 10$. (p) MN, $\mathbf{g} = 11$.

(q) MN, $m = 1$, $\mathbf{g} = 00$. (r) MN, $m = 1$, $\mathbf{g} = 01$. (s) MN, $m = 1$, $\mathbf{g} = 10$. (t) MN, $m = 1$, $\mathbf{g} = 11$.

(u) MN, $m = 2$, $\mathbf{g} = 00$. (v) MN, $m = 2$, $\mathbf{g} = 01$. (w) MN, $m = 2$, $\mathbf{g} = 10$. (x) MN, $m = 2$, $\mathbf{g} = 11$.

Fig. 8 **a–d** and **m–p**. Contour plots in $\boldsymbol{\tau}$-space for dataset D2 with the threshold-averaged cost function $C_A(\boldsymbol{\tau}, \mathbf{g})$. Details as in the caption of Fig. 6

- VN versus MN: Compare (a)–(l) with (m)–(x) in Figs. 6, 7, 8, 9, 10 and 11. It can be seen that MN results in fewer local optima. This is as expected, since MN is a larger neighbourhood and so will generate a more connected space.
- $r = 1$ versus $r = 2$: Compare Fig. 6 with Fig. 7, Fig. 8 with Fig. 9, Fig. 10 with Fig. 11 and Fig. 12 with Fig. 13. Increasing the radius of the neighbourhood results in fewer local optima, as expected.
- D1 versus D2: Compare Figs. 6 and 7 with Figs. 8 and 9 and compare Figs. 10 and 11 with Figs. 12 and 13. The values on the colourbars indicate that noise in the data in general degrades the quality of solutions. The contour plots indicate that the locations of the local optima with largest basin sizes can be quite different.

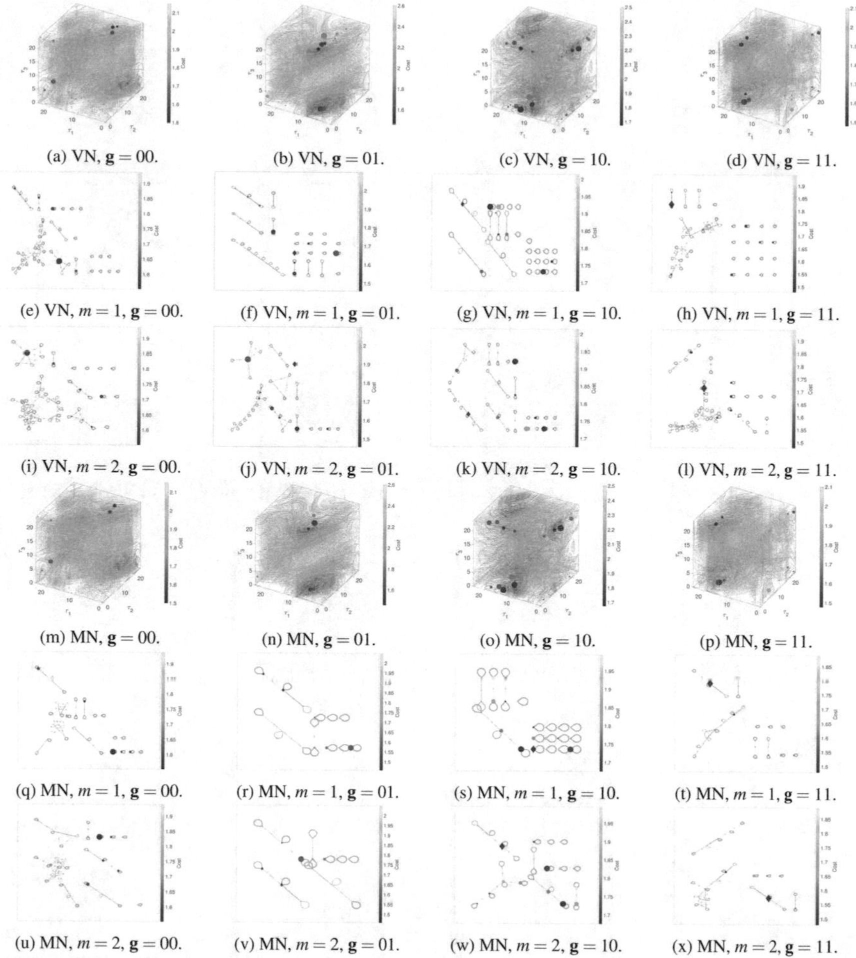

(a) VN, $\mathbf{g} = 00$. (b) VN, $\mathbf{g} = 01$. (c) VN, $\mathbf{g} = 10$. (d) VN, $\mathbf{g} = 11$.

(e) VN, $m = 1$, $\mathbf{g} = 00$. (f) VN, $m = 1$, $\mathbf{g} = 01$. (g) VN, $m = 1$, $\mathbf{g} = 10$. (h) VN, $m = 1$, $\mathbf{g} = 11$.

(i) VN, $m = 2$, $\mathbf{g} = 00$. (j) VN, $m = 2$, $\mathbf{g} = 01$. (k) VN, $m = 2$, $\mathbf{g} = 10$. (l) VN, $m = 2$, $\mathbf{g} = 11$.

(m) MN, $\mathbf{g} = 00$. (n) MN, $\mathbf{g} = 01$. (o) MN, $\mathbf{g} = 10$. (p) MN, $\mathbf{g} = 11$.

(q) MN, $m = 1$, $\mathbf{g} = 00$. (r) MN, $m = 1$, $\mathbf{g} = 01$. (s) MN, $m = 1$, $\mathbf{g} = 10$. (t) MN, $m = 1$, $\mathbf{g} = 11$.

(u) MN, $m = 2$, $\mathbf{g} = 00$. (v) MN, $m = 2$, $\mathbf{g} = 01$. (w) MN, $m = 2$, $\mathbf{g} = 10$. (x) MN, $m = 2$, $\mathbf{g} = 11$.

Fig. 9 **a–d** and **m–p**. Contour plots in $\boldsymbol{\tau}$-space for dataset D2 with the threshold-averaged cost function $C_A(\boldsymbol{\tau}, \mathbf{g})$. Details as in the caption of Fig. 6, except the range of both VN and MN is $r = 2$ in each plot

However, the LONs in many cases appear to have comparable numbers of local optima and connectivity.

Figure 14 shows LONs extracted from the mixed-integer space of the gates $\boldsymbol{g}$ and delays $\boldsymbol{\tau}$. The figures shows that searching over the gates can make the problem more difficult, given that the basin size of the optimal solution/s that lie in the data-generating subspace ($\mathbf{g} = 01$) arc smallcr than the modes and plateaus in the suboptimal subspaces. However, as in the single gate landscapes, the correlation between basin size and cost is still weak to moderate, as shown in Fig. 15. Local optima in C_{ME} have a wider range of cost values (both higher and lower) than in

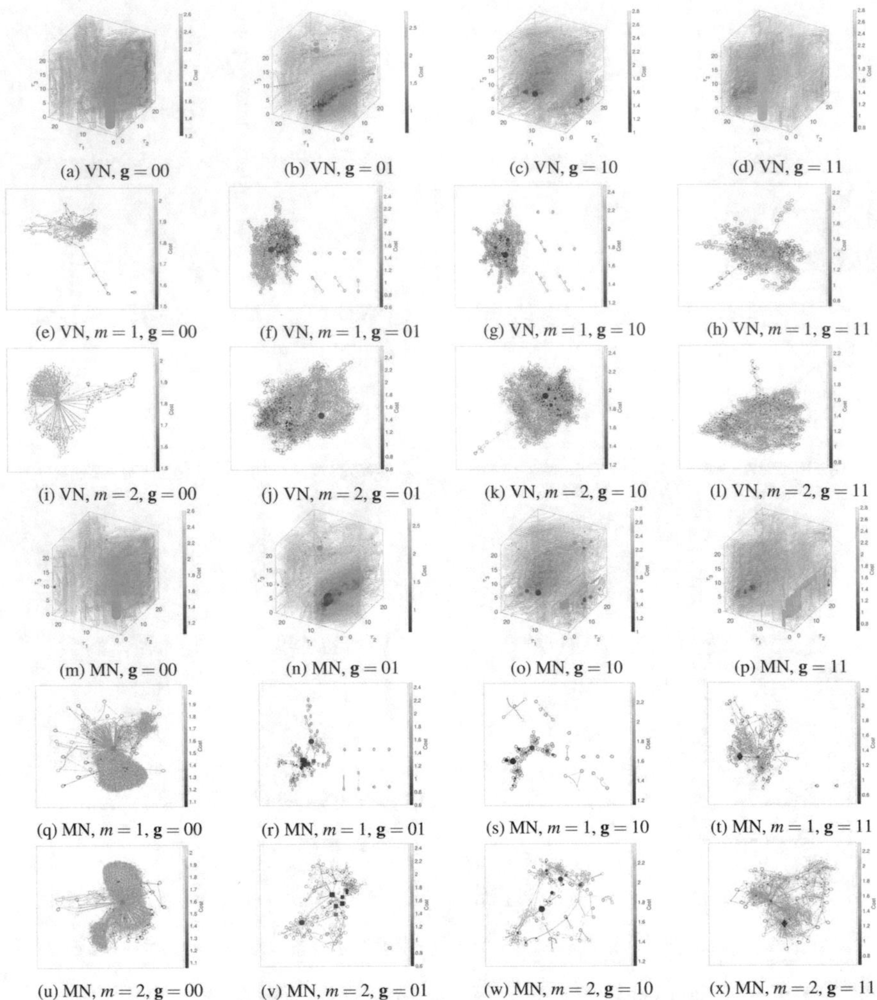

(a) VN, $\mathbf{g} = 00$ (b) VN, $\mathbf{g} = 01$ (c) VN, $\mathbf{g} = 10$ (d) VN, $\mathbf{g} = 11$

(e) VN, $m = 1$, $\mathbf{g} = 00$ (f) VN, $m = 1$, $\mathbf{g} = 01$ (g) VN, $m = 1$, $\mathbf{g} = 10$ (h) VN, $m = 1$, $\mathbf{g} = 11$

(i) VN, $m = 2$, $\mathbf{g} = 00$ (j) VN, $m = 2$, $\mathbf{g} = 01$ (k) VN, $m = 2$, $\mathbf{g} = 10$ (l) VN, $m = 2$, $\mathbf{g} = 11$

(m) MN, $\mathbf{g} = 00$ (n) MN, $\mathbf{g} = 01$ (o) MN, $\mathbf{g} = 10$ (p) MN, $\mathbf{g} = 11$

(q) MN, $m = 1$, $\mathbf{g} = 00$ (r) MN, $m = 1$, $\mathbf{g} = 01$ (s) MN, $m = 1$, $\mathbf{g} = 10$ (t) MN, $m = 1$, $\mathbf{g} = 11$

(u) MN, $m = 2$, $\mathbf{g} = 00$ (v) MN, $m = 2$, $\mathbf{g} = 01$ (w) MN, $m = 2$, $\mathbf{g} = 10$ (x) MN, $m = 2$, $\mathbf{g} = 11$

Fig. 10 **a–d** and **m–p**. Contour plots in $\boldsymbol{\tau}$-space for dataset D1 with the threshold-averaged cost function $C_{ME}(\boldsymbol{\tau}, \mathbf{g})$. Other details as in the caption of Fig. 6

C_A. Figure 14 also explains our previous finding that a GA searching this MI space does not converge to any of the optima in the $\mathbf{g} = 10$ subspace [15]—this is due to the poor quality of these optima and their small basin sizes.

Finally, Fig. 16 presents a comparison between the MI landscape and the landscape of the optimal (data-generating) gate, $\mathbf{g} = 01$. The first two properties are the global basin(s) size and the number of local optima. We also show how other properties which can be extracted from such landscapes vary. These features describe the global structure of the landscape, and are are extracted from the CMLONs of the escape LONs. A funnel [25] consists of a funnel sink (that is a node with a zero out-degree)

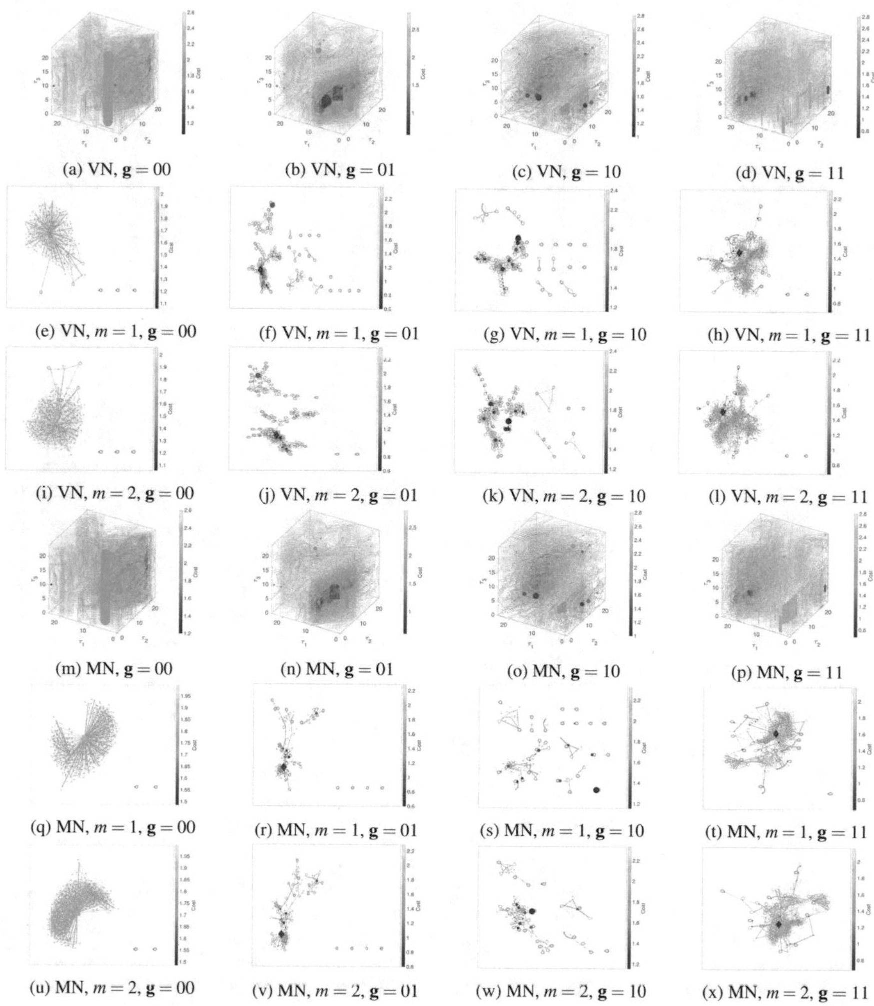

Fig. 11 **a–d** and **m–p**. Contour plots in $\boldsymbol{\tau}$-space for dataset D1 with the threshold-averaged cost function $C_{ME}(\boldsymbol{\tau}, \mathbf{g})$. Details as in the caption of Fig. 6, except the range of both VN and MN is $r = 2$ in each plot

and all the nodes that have a monotonic sequence leading to it. A primary funnel is the part of the funnel where all optima lead to the sink of that funnel. A multi-path optimum is an optimum that belongs to more than one funnel. The size of the funnel basins in the figure is calculated by summing the attraction basin sizes of all nodes in the funnel. Again, we see that using the maximum entropy cost function, C_{ME} consistently induces landscapes with a larger number of local optima across all neighbourhood functions in both the MI and the optimal gate landscapes, compared to the threshold-averaged cost function C_A. In general, there are also more multi-

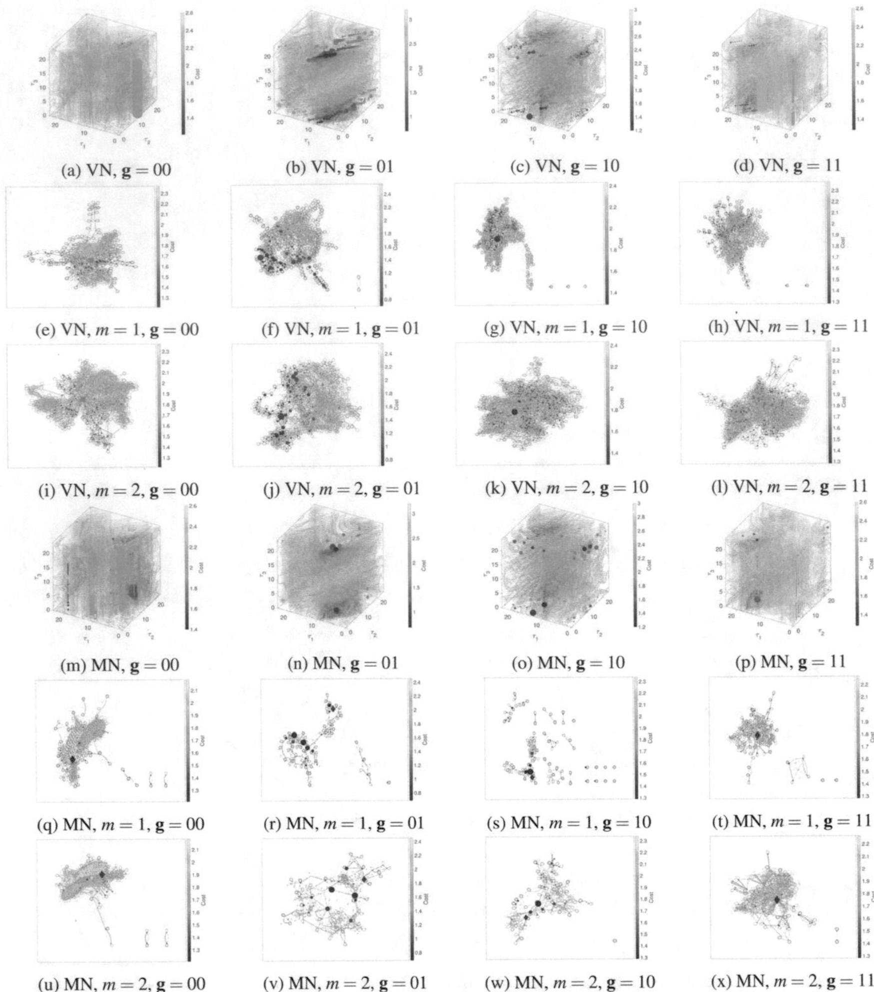

(a) VN, $\mathbf{g} = 00$ (b) VN, $\mathbf{g} = 01$ (c) VN, $\mathbf{g} = 10$ (d) VN, $\mathbf{g} = 11$

(e) VN, $m = 1$, $\mathbf{g} = 00$ (f) VN, $m = 1$, $\mathbf{g} = 01$ (g) VN, $m = 1$, $\mathbf{g} = 10$ (h) VN, $m = 1$, $\mathbf{g} = 11$

(i) VN, $m = 2$, $\mathbf{g} = 00$ (j) VN, $m = 2$, $\mathbf{g} = 01$ (k) VN, $m = 2$, $\mathbf{g} = 10$ (l) VN, $m = 2$, $\mathbf{g} = 11$

(m) MN, $\mathbf{g} = 00$ (n) MN, $\mathbf{g} = 01$ (o) MN, $\mathbf{g} = 10$ (p) MN, $\mathbf{g} = 11$

(q) MN, $m = 1$, $\mathbf{g} = 00$ (r) MN, $m = 1$, $\mathbf{g} = 01$ (s) MN, $m = 1$, $\mathbf{g} = 10$ (t) MN, $m = 1$, $\mathbf{g} = 11$

(u) MN, $m = 2$, $\mathbf{g} = 00$ (v) MN, $m = 2$, $\mathbf{g} = 01$ (w) MN, $m = 2$, $\mathbf{g} = 10$ (x) MN, $m = 2$, $\mathbf{g} = 11$

Fig. 12 **a–d** and **m–p**. Contour plots in $\boldsymbol{\tau}$-space for dataset D2 with the threshold-averaged cost function $C_{ME}(\boldsymbol{\tau}, \mathbf{g})$. Details as in the caption of Fig. 6

path local optima in the C_{ME} landscape. The figure clearly indicates that it is much easier to find the optimal solutions in the single gate space than it is in the MI space, as quantified by the corresponding basin and funnel sizes. In the single gate space, the global basin size in D1 using the largest neighbourhood represents 30% of the feasible search space, whilst the funnel size is around 80% of the feasible space. This finding suggests that perturbation can be a good mechanism to escape local optima and might have a higher success in locating the global optimum. Once more, the landscape analysis provides some insight into our previous optimisation results [15], where we found that it was more difficult to locate the optimal gate using a GA with no measures to preserve diversity (e.g. fitness sharing).

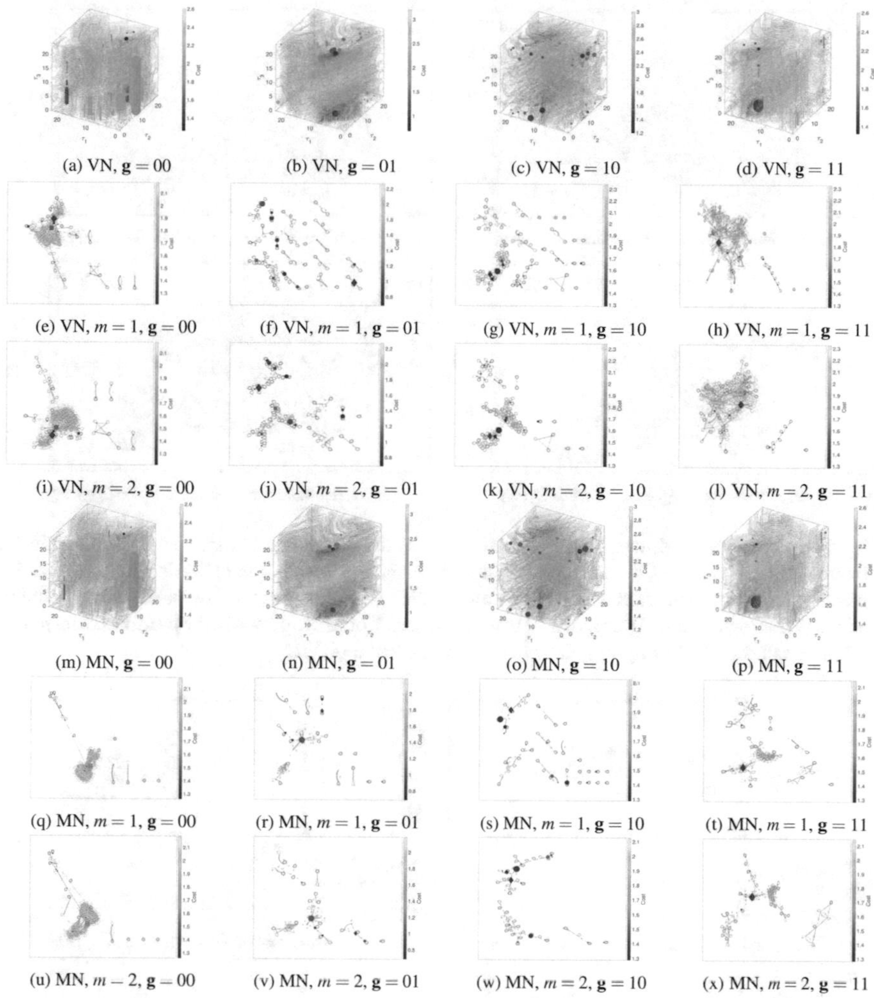

(a) VN, $\mathbf{g} = 00$ (b) VN, $\mathbf{g} = 01$ (c) VN, $\mathbf{g} = 10$ (d) VN, $\mathbf{g} = 11$

(e) VN, $m = 1, \mathbf{g} = 00$ (f) VN, $m = 1, \mathbf{g} = 01$ (g) VN, $m = 1, \mathbf{g} = 10$ (h) VN, $m = 1, \mathbf{g} = 11$

(i) VN, $m = 2, \mathbf{g} = 00$ (j) VN, $m = 2, \mathbf{g} = 01$ (k) VN, $m = 2, \mathbf{g} = 10$ (l) VN, $m = 2, \mathbf{g} = 11$

(m) MN, $\mathbf{g} = 00$ (n) MN, $\mathbf{g} = 01$ (o) MN, $\mathbf{g} = 10$ (p) MN, $\mathbf{g} = 11$

(q) MN, $m = 1, \mathbf{g} = 00$ (r) MN, $m = 1, \mathbf{g} = 01$ (s) MN, $m = 1, \mathbf{g} = 10$ (t) MN, $m = 1, \mathbf{g} = 11$

(u) MN, $m = 2, \mathbf{g} = 00$ (v) MN, $m = 2, \mathbf{g} = 01$ (w) MN, $m = 2, \mathbf{g} = 10$ (x) MN, $m = 2, \mathbf{g} = 11$

Fig. 13 **a–d** and **m–p**. Contour plots in $\boldsymbol{\tau}$-space for dataset D2 with the threshold-averaged cost function $C_{ME}(\boldsymbol{\tau}, \mathbf{g})$. Details as in the caption of Fig. 6, except the range of both VN and MN is $r = 2$ in each plot

5 Discussion

In this chapter, we used landscape analysis tools, including local optima networks, to study a data-driven optimisation problem from the computational systems biology field—the optimisation of an exemplar circadian clock model to synthetic experimental gene/protein expression profiles. We carried out a comparative study of different landscapes induced by different cost functions and neighbourhood operators.

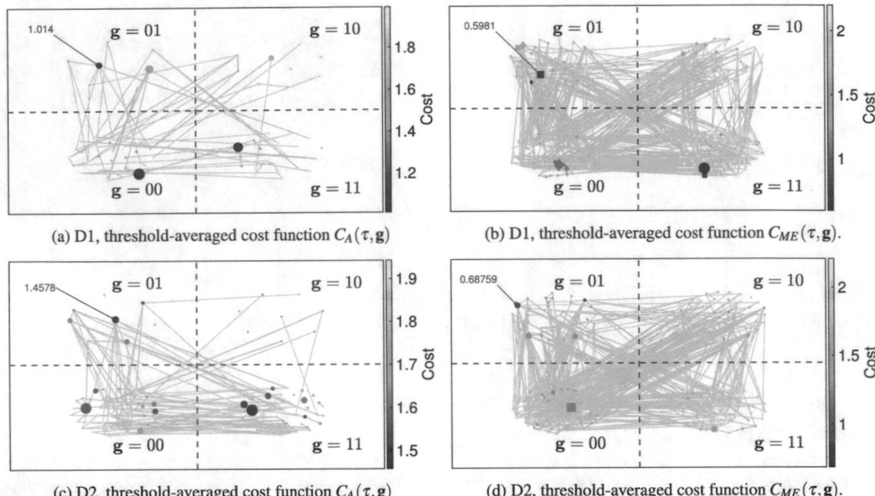

(a) D1, threshold-averaged cost function $C_A(\tau, \mathbf{g})$

(b) D1, threshold-averaged cost function $C_{ME}(\tau, \mathbf{g})$.

(c) D2, threshold-averaged cost function $C_A(\tau, \mathbf{g})$

(d) D2, threshold-averaged cost function $C_{ME}(\tau, \mathbf{g})$.

Fig. 14 LONs obtained with MN, $r = 1$, $m = 1$, for the mixed-integer space of $\boldsymbol{\tau}$ and $\mathbf{g}$. Circles donate a SLMIN, squares denote closed plateaus, diamonds denote open plateaus and the globally optimal solutions are marked with a red arrow labelled with the optimal cost value. Each optimum and plateau shown is scaled proportional to its basin size. Nodes are drawn in the box that corresponds to their $\mathbf{g}$ values, or majority of gate values in the case of a plateau

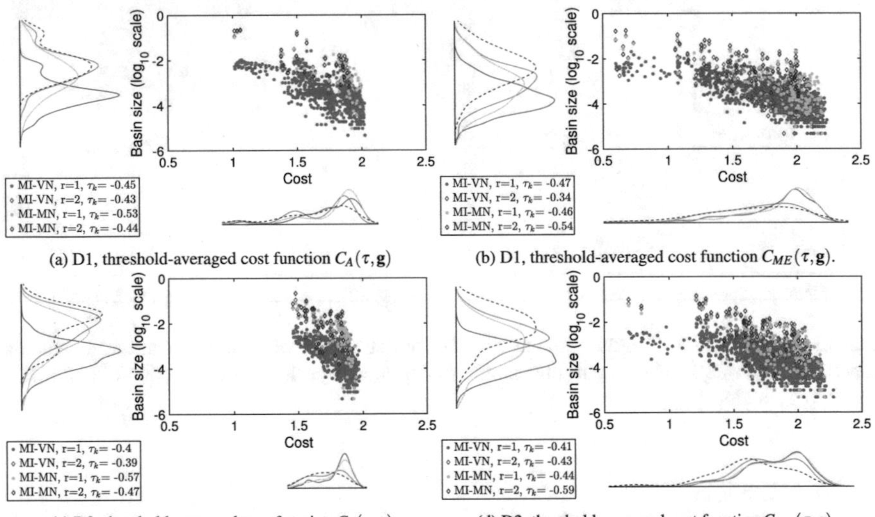

(a) D1, threshold-averaged cost function $C_A(\tau, \mathbf{g})$

(b) D1, threshold-averaged cost function $C_{ME}(\tau, \mathbf{g})$.

(c) D2, threshold-averaged cost function $C_A(\tau, \mathbf{g})$

(d) D2, threshold-averaged cost function $C_{ME}(\tau, \mathbf{g})$.

Fig. 15 Basin size (log scale), as a proportion of the feasible search space plotted against mode cost for the mixed-integer landscapes of $\boldsymbol{\tau}$ and $\mathbf{g}$ obtained with different cost functions (threshold-averaged—C_A; maximum entropy—C_{ME}), neighbourhood types (MI-VN—mixed integer Von Neumann neighbourhood; MI-MN—mixed integer Moore neighbourhood) and neighbourhood size r ($r = 1, 2$). Each data point represents either a SLMIN or a plateau. The correlation coefficient between basin size and cost is calculated using Kendall's τ

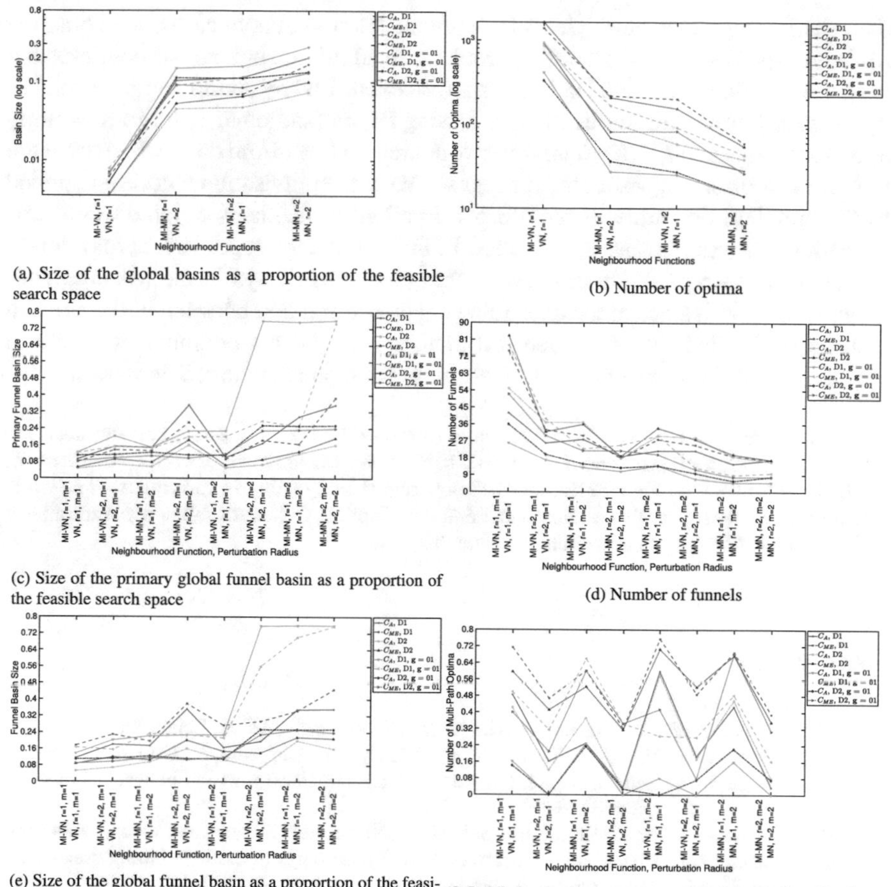

(a) Size of the global basins as a proportion of the feasible search space

(b) Number of optima

(c) Size of the primary global funnel basin as a proportion of the feasible search space

(d) Number of funnels

(e) Size of the global funnel basin as a proportion of the feasible search space

(f) Saddle local optima as a proportion of the number of optima

Fig. 16 Global basin and funnel size, and the number of local optima and funnels in the mixed-integer landscapes of $\boldsymbol{\tau}$ and $\mathbf{g}$ and the landscape of the data-generating gate $\mathbf{g} = 01$ for all cost functions and datasets. The x-axis ticks in **a–b** are sorted according to the size of the neighbourhood, and in **c–f** are sorted according to the perturbation size used to generate the corresponding escape LONs

Our results demonstrate that the landscape analysis and LONs can provide novel insights into the intrinsic properties of the studied problem, and furthermore could be used to guide the selection of both the search operators and the objective function in future optimisation studies. For example, the properties of the landscape generated by the two different cost functions studied here, i.e. averaging across threshold combinations versus selecting the threshold combination that maximises the entropy of the data, indicates that the former approach is preferable, since it induces a less deceptive landscape with fewer optima and plateaus. Interestingly, comparing the mixed-integer and single gate landscapes confirmed that whilst, as expected, the data-generating architecture yields the optimal solution in both cases, searching the MI landscape is a more challenging problem. In addition, our analysis confirms that both the MI and single gate landscapes are characterised by significant multimodality. This has important implications for using BDEs (and other reduced modelling approaches—e.g. [17, 30]) to infer network architectures from data, when these are not known a priori. In particular, it suggests that any optimisation algorithm applied to this problem domain will require perturbation mechanisms capable of causing transitions between basins of attraction in both landscape types (i.e. across delays in the single gate problem and across both gates and delays in the MI problem). Moreover, the existence of a reasonable negative correlation between basin size and solution quality indicates that such perturbation mechanisms, combined with a local search method, have the potential to return useful modes to the problem owner.

Acknowledgements This work was financially supported by the Engineering and Physical Sciences Research Council [grant number EP/N017846/1]. We would like to acknowledge the use of the University of Exeter High-Performance Computing (HPC) facility in carrying out this work. The authors would like to thank Sébastien Vérel and Gabriela Ochoa for providing inspirational invited talks on LONs at their institution during this grant.

References

1. Adams, R., Clark, A., Yamaguchi, A., Hanlon, N., Tsorman, N., Ali, S., Lebedeva, G., Goltsov, A., Sorokin, A.A., Akman, O.E., Troein, C., Millar, A.J., Goryanin, I., Gilmore, S.: SBSI: an extensible distributed software infrastructure for parameter estimation in systems biology. Bioinformatics **29**(5), 664–665 (2013)
2. Akman, O.E., Fieldsend, J.E.: Multi-objective optimisation of gene regulatory networks: insights from a boolean circadian clock model. In: Proceedings of the 12th International Conference on Bioinformatics and Computational Biology, vol. 70, pp. 149–162 (2020)
3. Akman, O.E., Locke, J.C.W., Tang, S., Carré, I., Millar, A.J., Rand, D.A.: Isoform switching facilitates period control in the Neurospora crassa circadian clock. Mol. Syst. Biol. **4**, 64 (2008)
4. Akman, O.E., Rand, D.A., Brown, P.E., Millar, A.J.: Robustness from flexibility in the fungal circadian clock. BMC Syst. Biol. **4**, 88 (2010)
5. Akman, O.E., Watterson, S., Parton, A., Binns, N., Millar, A.J., Ghazal, P.: Digital clocks: simple Boolean models can quantitatively describe circadian systems. J. Roy. Soc. Interface **9**(74), 2365–2382 (2012)
6. Alyahya, K.: Fitness landscape analysis of a class of NP-hard problems. Ph.D. thesis, University of Birmingham, July (2016)
7. Alyahya, K., Rowe, J.E.: Simple random sampling estimation of the number of local optima. In: Handl, J., Hart, E., Lewis, P.R., López-Ibáñez, M., Ochoa, G., Paechter, B. (eds.) Parallel Problem Solving from Nature - PPSN XIV, pp. 932–941. Springer International Publishing, Cham (2016)
8. Alyahya, K., Rowe, J.E.: Landscape analysis of a class of NP-hard binary packing problems. Evol. Comput. **27**(1), 47–73 (2019)
9. Banga, J.R.: Optimization in computational systems biology. BMC Syst. Biol. **2**, 47 (2008)
10. Belkhir, N., Dréo, J., Savéant, P., Schoenauer, M.: Per instance algorithm configuration of cma-es with limited budget. In: Proceedings of the Genetic and Evolutionary Computation Conference, GECCO '17, pp. 681–688. ACM, New York, NY, USA (2017)
11. Breukelaar, R., Bäck, T.: Using a genetic algorithm to evolve behavior in multi dimensional cellular automata: emergence of behavior. In: Proceedings of the 7th Annual Conference on

Genetic and Evolutionary Computation, GECCO '05, pp. 107–114. ACM, New York, NY, USA (2005)
12. Daniels, B.C., Chen, Y.-J., Sethna, J.P., Gutenkunst, R.N., Myers, C.R.: Sloppiness, robustness, and evolvability in systems biology. Curr. Opin. Biotechnol. **19**(4), 389–395 (2008). Protein technologies/Systems biology
13. de Jong, H.: Modeling and simulation of genetic regulatory systems: a literature review. J. Comput. Biol. **9**(1), 67–103 (2002). PMID: 11911796
14. Dee, D.P., Ghil, M.: Boolean difference equations, i: formulation and dynamic behavior. SIAM J. Appl. Math. **44**(1), 111–126 (1984)
15. Doherty, K., Alyahya, K., Akman, O.E., Fieldsend, J.E.: Optimisation and landscape analysis of computational biology models: a case study. In: Proceedings of the Genetic and Evolutionary Computation Conference Companion, GECCO '17, pp. 1644–1651. ACM, New York, NY, USA (2017)
16. Doye, J.P.K.: Network topology of a potential energy landscape: a static scale-free network. Phys. Rev. Lett. **88**, 23:238701 (2002)
17. Foo, M., Bates, D.G., Akman, O.E.: A simplified modelling framework facilitates more complex representations of plant circadian clocks. PLoS Comput. Biol. **16**(3), e1007671 (2020)
18. Ghil, M., Zaliapin, I., Coluzzi, B.: Boolean delay equations: a simple way of looking at complex systems. Phys. D: Nonlinear Phenom. **237**(23), 2967–2986 (2008)
19. Gutenkunst, R.N., Waterfall, J.J., Casey, F.P., Brown, K.S., Myers, C.R., Sethna, J.P.: Universally sloppy parameter sensitivities in systems biology models. PLoS Comput. Biol. **3**(10), e189 (2007)
20. Hansen, N., Ostermeier, A.: Completely derandomized self-adaptation in evolution strategies. Evol. Comput. **9**(2), 159–195 (2001)
21. Hoos, H.H., Stützle, T.: Stochastic Local Search: Foundations and Applications. Elsevier (2004)
22. Kauffman, S., Weinberger, E.: The NK model of rugged fitness landscapes and its application to the maturation of the immune response. J. Theor. Biol. **141**(2), 211–245 (1989)
23. Leloup, J.C., Gonze, D., Goldbeter, A.: Limit cycle models for circadian rhythms based on transcriptional regulation in Drosophila and Neurospora. J. Biol. Rhythms **14**(6), 433–448 (1999)
24. Ochoa, G., Tomassini, M., Vérel, S., Darabos, C.: A study of NK landscapes' basins and local optima networks. In: Proceedings of the 10th Annual Conference on Genetic and Evolutionary Computation, pp. 555–562. ACM (2008)
25. Ochoa, G., Veerapen, N., Daolio, F., Tomassini, M.: Understanding phase transitions with local optima networks: number partitioning as a case study. In: Hu, B., López-Ibáñez, M. (eds.) Evolutionary Computation in Combinatorial Optimization, pp. 233–248. Springer International Publishing, Cham (2017)
26. Ochoa, G., Verel, S., Daolio, F., Tomassini, M.: Local optima networks: a new model of combinatorial fitness landscapes. In: Richter, H., Engelbrecht, A. (eds.) Recent Advances in the Theory and Application of Fitness Landscapes, pp. 233–262. Springer, Berlin Heidelberg, Berlin, Heidelberg (2014)
27. Öktem, H., Pearson, R., Egiazarian, K.: An adjustable aperiodic model class of genomic interactions using continuous time Boolean networks (Boolean delay equations). Chaos **13**(4), 1167–1174 (2003)
28. Pitzer, E., Affenzeller, M.: A comprehensive survey on fitness landscape analysis. In: Fodor, J., Klempous, R., Araujo, S., Paz, C. (eds.) Recent Advances in Intelligent Engineering Systems, vol. 378, pp. 161–191. Springer, Berlin (2012)
29. Stillinger, F.H., Weber, T.A.: Packing structures and transitions in liquids and solids. Science **225**(4666), 983–989 (1984)
30. Tokuda, I.T., Akman, O.E., Locke, J.C.: Reducing the complexity of mathematical models for the plant circadian clock by distributed delays. J. Theor. Biol. **463**, 155–166 (2019)
31. Vérel, S., Daolio, F., Ochoa, G., Tomassini, M.: Local optima networks with escape edges. In: Hao, J.-K., Legrand, P., Collet, P., Monmarché, N., Lutton, E., Schoenauer, M. (eds.) Artificial Evolution, pp. 49–60. Springer, Berlin Heidelberg (2012)

32. Verel, S., Ochoa, G., Tomassini, M.: Local optima networks of NK landscapes with neutrality. IEEE Trans. Evol. Comput. **15**(6), 783–797 (2011)
33. Wolfram, S.: Statistical mechanics of cellular automata. Rev. Mod. Phys. **55**, 601–644 (1983)
34. Wright, S.: The roles of mutation, inbreeding, crossbreeding, and selection in evolution. In: Proceedings of 6th International Congress on Genetics, vol. 1, pp. 356–366 (1932)

Grammar-Based Multi-objective Genetic Programming with Token Competition and Its Applications in Financial Fraud Detection

Haibing Li and Man-Leung Wong

Abstract In this study, we propose a new approach based on Grammar-based Genetic Programming (GBGP), token competition, multi-objective optimization, and ensemble learning for solving Financial Fraud Detection (FFD) problems. Token competition is a niching technique to maintain diversity among individuals. It can be used to adjust the objective values of each individual, and the individuals with similar objective values but different meanings are separated. Financial fraud is a serious problem that often produces destructive results in the world and it is exacerbating swiftly in many countries. It refers to many activities including credit card fraud, money laundering, insurance fraud, corporate fraud, etc. The major consequences of financial fraud are loss of billions of dollars each year, investor confidence, and corporate reputation. Therefore, a research area called FFD is obligatory, in order to prevent the destructive results caused by financial fraud. We comprehensively compare the proposed approach with Logistic Regression, Neural Networks, Support Vector Machine, Bayesian Networks, Decision Trees, AdaBoost, Bagging, and LogitBoost on four FFD datasets including two real-life datasets. The experimental results showed the effectiveness of the new approach. It outperforms existing data mining methods in different aspects.

1 Introduction

Genetic Programming (GP) is a population-based optimization method that extends traditional genetic algorithms [10, 12] to automatically induce computer programs [17, 19]. Unlike Genetic Algorithms (GA), GP uses a tree structure to represent an individual in a population.

Comparing Grammar-Based Genetic Programming (GBGP) with traditional GP, the concept of grammar is employed [24], which is used to control the structures

H. Li · M.-L. Wong (✉)
Department of Computing and Decision Sciences, Lingnan University, Tuen Mun, Hong Kong
e-mail: mlwong@ln.edu.hk

H. Li
e-mail: haibingli@ln.edu.hk

M. Preuss et al. (eds.), *Metaheuristics for Finding Multiple Solutions*, Natural Computing Series,
https://doi.org/10.1007/978-3-030-79553-5_11

evolved during the evolutionary process. GBGP supports logic grammars, context-free grammars (CFGs), and context-sensitive grammars [13] to generate tree-based programs. A suitable grammar is designed for solving a particular problem.

GBGP can learn programs in various programming languages and induce knowledge in different representations such as fuzzy Petri nets and first-order logical relations [23]. The system is also powerful enough to represent context-sensitive information and domain-dependent knowledge. This knowledge can be used to accelerate the learning speed and/or improve the quality of the induced programs and knowledge [25].

As GBGP uses only a single fitness/objective function, it cannot handle problems with more than one objective. Moreover, a single individual with high fitness value may dominate the whole population and prevent other individuals with similar fitness values to survive.

In order to generate better individuals for problems with more than one objective, multi-objective optimization methods [4, 7, 8, 15, 20] and token competition [25] are integrated with GBGP to produce a set of non-dominated individuals on all objectives. A novel ensemble learning method is then used to select rules to form an ensemble of solutions. The proposed new approach is evaluated on the Financial Fraud Detection (FFD) problem and it is observed that the new approach outperforms existing data mining methods and the original GBGP in different aspects.

The rest of this paper is organized as follows. In Sect. 2, the background and literature review of this work are discussed. In Sect. 3, the proposed approach is described in detail. The motivations of different techniques and the framework of the proposed approach are discussed. In Sect. 4, a number of experiments are conducted to compare the performance of the proposed approach with other data mining methods. The experiment results are presented and discussed comprehensively in this section. The contributions and the future research direction of this study are discussed in the last section.

2 Background

2.1 Multi-objective Optimization Problems

Many real-world problems can be regarded as multi-objective optimization problems and it is usually difficult to obtain a single solution for these problems. For example, it is not difficult for a stock buyer to choose a stock with the highest expected return. However, if the buyer is also concerned about the risk of the selected stock, the problem becomes more complicated because there is a relationship between expected return and risk. The stock selected by the buyer may also have very high risk, thus the buyer may consider other alternatives with smaller expected return and risk. In general, researchers search for a single solution by assigning different weights for each objective. The weight is used to describe the importance of the objective. For example, if users want to optimize Eq. (1)

$$y = Maximize(w_1 * objective_1 + w_2 * objective_2) \tag{1}$$

where w_1 is the weight for $objective_1$, w_2 is the weight for $objective_2$, and $w_1 + w_2 = 1$, the first issue is to assign the value for each weight. If the users think $objective_1$ is more important than the other, then the weight of w_1 should be higher than w_2. However, it is not easy to determine the right values for these weights and the solutions will be significantly deteriorated if inappropriate weight values are used. Moreover, if there are more objectives, the problem of assigning weight values becomes even more challenging. Therefore, methods to solve multiobjectives optimization problems are required.

The main purpose of these methods is to find a number of trade-off solutions (i.e., Pareto solutions or non-dominated solutions), which can meet all objectives, and then the users can make a final decision based on the optimal solutions obtained [4]. The optimal solutions form a curve, called Pareto front, among all objectives, which is shown in Fig. 1.

Without loss of generality, we define multi-objective maximization problems as follows.

$$maximize\ f(x) = [f_1(x), f_x(x)...f_k(x)] \tag{2}$$

subject to

$$g_i(x) \leq 0\ i = 1, 2, ..., m \tag{3}$$

$$h_j(x) = 0\ j = 1, 2, ..., p \tag{4}$$

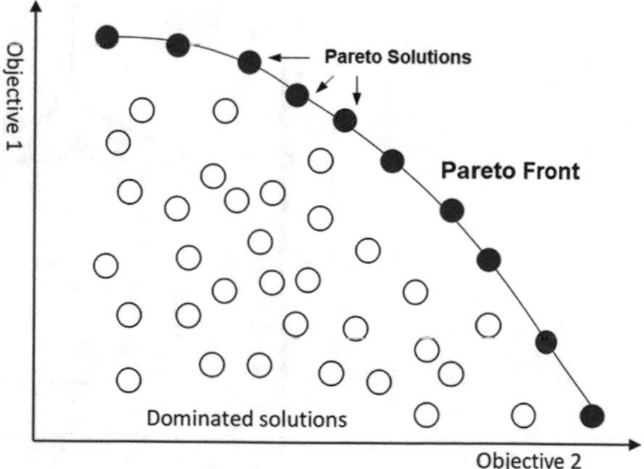

Fig. 1 Example of Pareto Front

where $x = [x_1, x_2, ..., x_n]$ is a vector of n decision variables, $f(x) = [f_1(x), f_x(x)... f_k(x)]$ is called objective functions, and g_i, h_j, i = 1, 2, ..., m, and j = 1, 2, ..., p are the constraint functions of the problem. The number of p (i.e., equality constraints) must be smaller than the number of n (i.e., objectives); otherwise the problem is over-constrained (i.e., no solutions) [4]. A Pareto or non-dominated solution x satisfies all constraint functions and there exists no other feasible solution x' which would increase some objective values without causing a simultaneous decrease in at least one of the other objectives. These Pareto or non-dominated solutions are good tradeoffs for the multi-objective optimization problem.

2.2 *Genetic Programming (GP)*

The overall evolutionary process of GP is depicted in Fig. 2.

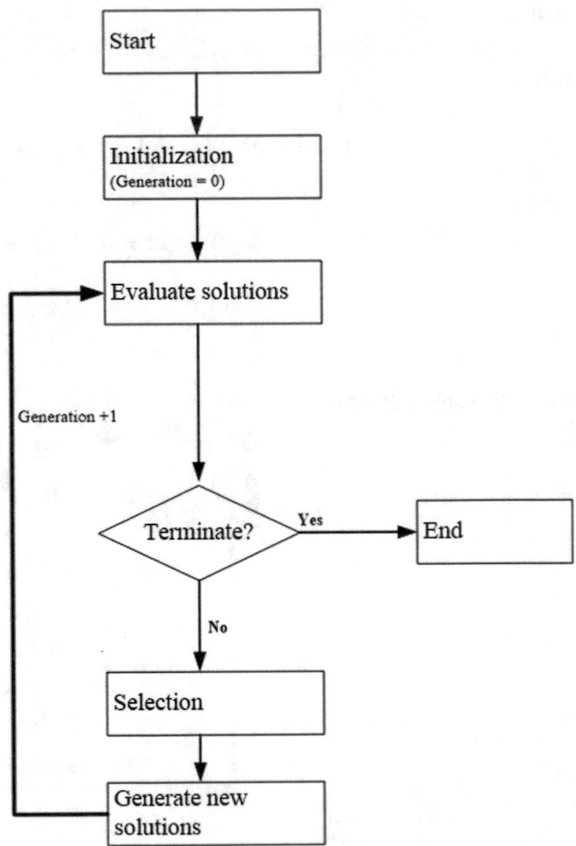

Fig. 2 General process of Genetic Programming (GP)

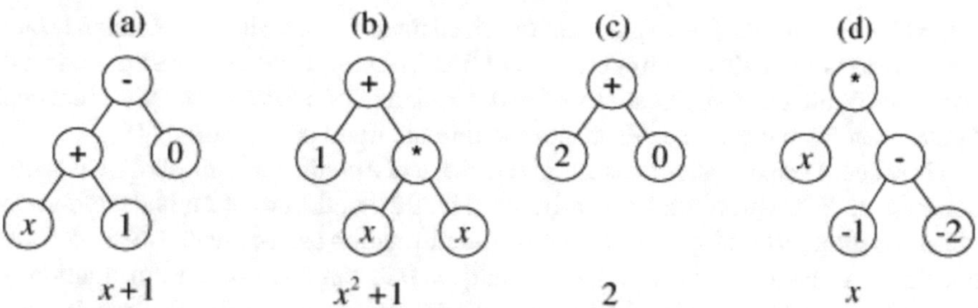

Fig. 3 Four individuals in GP

Firstly, GP randomly creates a number of computer programs (i.e., individuals), which are composed of functions and terminals to form an initial population in generation 0 [14]. Some possible computer programs with a function set and terminal set are shown in Fig. 3. For these programs, the terminal set is {0,1,2,-1,-2,x}, and the value of each terminal node is selected from the terminal set, which is located in the leaves of individual program trees. For example, in Fig. 3a, x,1 and 0 are the terminal nodes. The function set is {+,-,*}, and the value of a function node is selected from the function set, which is the connector of some other nodes, such as − and + in Fig. 3a.

Afterward, it iteratively selects some individuals based on their fitness values and breeds them into a new generation of individuals by using different genetic operators. Fitness value is the score to measure the quality of the individual, which means that a good computer program has a high fitness value. The selection is based on fitness values; therefore poor computer programs have lower probability of being selected. When one or two individuals are selected, genetic operators will be applied to produce new individuals. Genetic operators include crossover, mutation, and reproduction.

The new individuals (i.e., offspring) will replace some individuals (i.e., parent) according to the altering scheme at each generation. In general, poor individuals have high probability of being replaced by new individuals. The evolution is repeated until the termination criterion, such as the number of iterations executed, is satisfied. Finally, the evolved population contains a number of good individuals to solve the given problem [17]. In order to apply GP for a problem, the user needs to specify a set of primitive functions **F**, a set of terminals **T**, a fitness function, a set of related parameters for evolution (e.g., crossover rate, mutation rate, and selection rate) and the termination criteria [25].

2.3 *Financial Fraud Detection*

Financial fraud is a serious problem that often produces destructive results in the world and it is exacerbating swiftly in many countries, such as China. It is a criminal act, which violates the law to gain unauthorized financial benefit [18]. Financial fraud refers to many activities, such as credit card fraud, money laundering, insurance fraud, corporate fraud, etc. Credit card fraud and corporate fraud have attracted a great deal of attention from the year 1998 and are still in the trend of escalation [18].

Credit card fraud is about unauthorized usage of a credit card, unusual transaction behavior or transactions on an inactive card [3, 22]. In the era of rapid development of information technology, a vast volume of information can be created every second, but there can be a lack of powerful techniques that can analyze the information. It is costly to detect the potential fraudulent transactions manually. The results may be destructive if one chooses to ignore them or detect them incorrectly. At the same time, credit cards are the most popular transaction method with increasing users, but the credit card fraud rate is also increasing.

Corporate securities fraud is related to corporate fraud in listed firms. For example, it may be perpetrated to increase the stock prices of fraudulent firms, to obtain more loans from banks or repay lesser dividends to shareholders [21]. In the U.S., financial analysts have been confirmed to contribute to corporate fraud detection. Effective external monitoring can increase the confidence of shareholders or investors, which is crucial to the functioning of any capital market [6]. It is also important for China's securities market, as corporate fraud can impede China's economic development since it has serious consequences for shareholders, employees, and society [6]. No matter what type of fraud is involved, it results in losses of billions of dollars every year [16]. Since the amount of fraud is increasing rapidly, the workload of auditors is also increasing. They have become overburdened with the task of detection of fraud. Various efficient financial fraud detection techniques are required to detect which ones will commit a fraud.

Financial Fraud Detection (FFD) is vital to prevent the destructive consequences of financial fraud. It can distinguish fraudulent information from data, thereby discovering fraudulent activities or behavior and enabling decision-makers to develop appropriate policies and strategies to decrease the influences of fraud [18].

3 Approach

The general framework of the proposed method for solving financial fraud problems is shown in Table 1. Three major components are included in this framework. The first consists of Grammar-based Multi-objective Genetic Programming (GBMGP) with Token Competition, which is described in Sect. 3.1. The second consists of ensemble learning, which is described in Sect. 3.2. The third consists of prediction in model testing.

Table 1 General framework of using GBMGP with token competition

The input to the system:
• Datasets: Training and testing datasets.
• Objectives: A number of objectives, and maximization or minimization of each objective.
• Pre-defined grammar for the specific problem.
• Parameters for evolution: Number of generations and number of individuals.
• Ensemble learning technique.
1. Grammar-based Multi-objective Genetic Programming (GBMGP) with Token Competition:
• Applying genetic programming to learn classification rules from the training dataset.
• Training is guided by the pre-defined grammar.
• Applying Token Competition to enhance diversity.
• Output the population with evolved classification rules.
2. Generating ensemble:
• Applying ensemble approach for the population.
3. Testing:
• Applying the final ensemble on the testing dataset.

3.1 Grammar-Based Multi-objective Genetic Programming (GBMGP) with Token Competition

This subsection describes the key components of Grammar-Based Multi-objective Genetic Programming (GBMGP) with Token Competition and the corresponding motivations, designs, and implementations in detail.

Figure 4 shows the general process of GBMGP with Token Competition. Compared with traditional Genetic Programming [17], the method has three more components, which are Step 4, Step 5, and the Grammar, to handle multi-objective problems, maintain the diversity of classification rules, and guide the evolutionary process, respectively. The well-known multi-objective learning algorithm called Non-dominated Sorting Genetic Algorithm II (NSGA-II) is applied in Step 4 [7]. A diversity maintenance scheme called Token Competition is applied in Step 5. Section 3.1.1 shows how Grammar-Based Genetic Programming (GBGP) [25] is used in the entire evolutionary process. Section 3.1.2 elaborates the multi-objective approach. Section 3.1.3 elaborates Token Competitions among classification rules (i.e., Step 5).

3.1.1 Grammar-Based Genetic Programming (GBGP)

GBGP [24] employs the concept of grammar. Table 2 shows an example of a simple grammar. The genetic operations (e.g., crossover and mutation) will be executed based on the grammar, so that the new offspring generated must be valid according to the grammar.

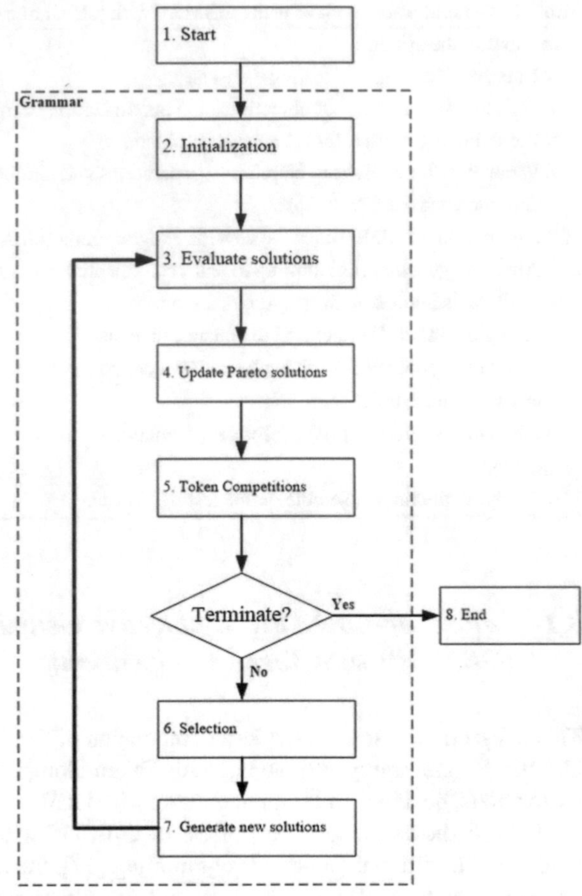

Fig. 4 General process of GBMGP with Token Competition

Table 2 Example of a grammar to control the evolutionary process

If-Exp → *Boolean-Exp Then Else*
Boolean-Exp → *Operator Term Value*
Boolean-Exp → true | false
Term → meeting | board
Value → [0, 100]
Operator → = | >= | <= | > | <
Then → yes
Else → no

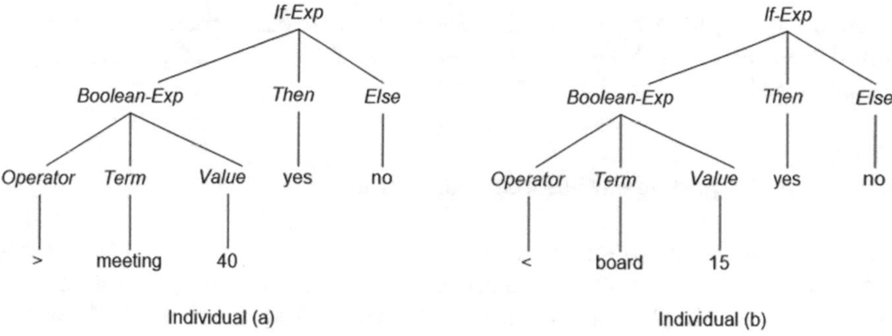

Fig. 5 Example of two individuals

For example, consider two individuals shown in Fig. 5, which are generated based on the grammar in Table 2. Individual (a) indicates that if the number of board meetings is greater than 40, then the *If-Exp* expression returns "yes" (i.e., the firm is fraudulent); otherwise it returns "no" (i.e., the firm is not fraudulent). Individual (b) indicates that if the number of board members is smaller than 15, then the *If-Exp* expression returns "yes"; otherwise it returns "no".

The original GBGP has been applied in classification rule learning [25]. A classification rule is a statement in the format of "If *antecedents* then *consequent*", which is commonly used by human to represent knowledge. Classification rule learning tries to learn rules from a dataset. In GBGP, a tree structure is used to represent a rule, and grammars for rules have been developed to create the appropriate tree structures. A fitness function has been used to evaluate the evolved rules. The fitness function applies the support-confident framework proposed by [1]. In this framework, two objectives, *support* and *confidence*, are considered at the same time.

- *Support* is used to evaluate the coverage of a rule. A good rule should have an appropriate support value so that it covers a number of cases of a dataset. Support is a ratio of the number of cases matching both the *antecedents* and the *consequent* to the total number of cases.
- *Confidence* measures accuracy. It is a ratio of the number of cases matching both the *antecedents* and the *consequent* to the number of cases fulfilling the *antecedents* only.

GBGP evaluates the support and confidence values of each rule and calculates the *normalized confidence* which is $confidence * log(\frac{confidence}{prob})$ where *prob* is the ratio of the number of cases matching the *consequent* to the number of cases. Since there are two different objective values but GBGP has only one fitness function, GBGP combines the two objective values into a single fitness value by using the following equation,

$$fitness = w_1 * support + w_2 * normalized\ confidence \quad (5)$$

where w_1 and w_2 are weights to control the balance between support and normalized confidence. GBGP has been applied to learn rules from medical datasets and good performance has been obtained [25].

3.1.2 Multi-objective Evolutionary Algorithms

As we have discussed in the previous subsection, each rule is evaluated by using two objectives. However the two objectives are conflicting and a good rule may be eliminated if we simply combine the two objective values of a rule to be its fitness value, because the values of w_1 and w_2 may be inappropriate. However, it is not easy to determine the right values of w_1 and w_2 so that good rules can be maintained. Multi-objective optimization methods can thus be used to keep these good rules. The main idea of multi-objective optimization methods is to obtain a number of non-dominated solutions, which are also called Pareto solutions. As shown in Fig. 6, the non-dominated solutions (i.e., black points) will form a curve called a Pareto front. These non-dominated solutions should be good classification rules. A number of Multi-objective Evolutionary Algorithms (MOEAs) can be applied and we use Non-dominated Sorting Genetic Algorithm II (NSGA-II) [7].

There are two main features in NSGA-II [7], as shown in Fig. 6. The first feature is to *sort* the individuals into different level of fronts (i.e., ranks). The individuals in the first front are the non-dominated solutions (i.e., Pareto solutions). The individuals in subsequent ranks are poorer than the individuals in previous ranks. The second feature is to measure the *crowding distance* between an individual and its neighbors.

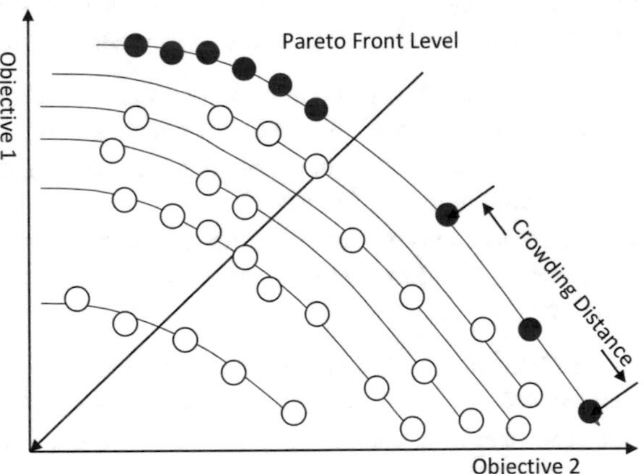

Fig. 6 Example of a population sorted by NSGA-II

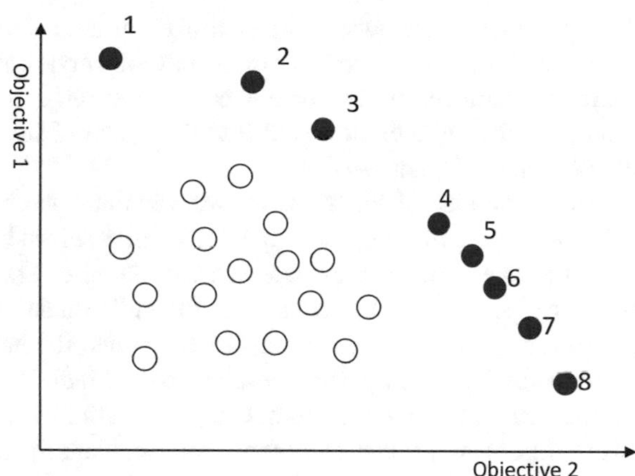

Fig. 7 Example of a population

3.1.3 Token Competition

Although all individuals are expected to be optimal on both objective functions, some potential good individuals may be ignored during the learning process. Figure 7 shows an example of a population.

In Fig. 7, the eight black points are the Pareto solutions with index number. Points 1, 2, and 3 can be regarded as the less crowded region, and points 4, 5, 6, 7, and 8 are located in the crowded region. Points 1 and 8 can be selected because they are boundary points (i.e., only have one neighbor) and therefore they have the largest crowding distances. The situation of point 4 is better than points 5, 6, and 7, since it is relatively far away from point 3, so its crowding distance is also higher than the other points in crowded regions (i.e., points with small crowding distances). Points 5,6, and 7 may not be selected because they have very small crowding distances. Although the points 5,6, and 7 are on the Pareto front (i.e., non-dominated solutions), they may not be selected during the learning process. Each point is a classification rule. Even though rules 5, 6, and 7 have similar support and confidence values, they may have different meanings. For example, there are three rules that indicate points 5,6, and 7 in the following:

1. if Location is Shanghai, then class = yes
2. if Industry is manufacturing, then class = no
3. if Number of employees is between 500 and 1000, then class = no

These three classification rules may have similar values for confidence and support, but they have totally different meanings. Therefore, we want to apply a mechanism to further evaluate this kind of individual by adjusting their locations. Token competition can be used to achieve this purpose.

The token competition technique is employed in our approach to increase the diversity, so that good individuals in different niches are maintained in the population.

The concept is as follows: In the natural environment, once an individual has found a good place for living, it will try to exploit this niche and prevent other newcomers to share the resources, unless the newcomer is stronger than it is. The other individuals are hence forced to explore and find their own niches. In this way, the diversity of the population is increased.

The basic idea of token competition is that each training case has a token. If a rule can cover a training case, then its token is seized by the rule. Other rules cannot take the token even if they can classify the case. The priority of obtaining tokens is determined by the performance of the individuals. In other words, all individuals are sorted from high to low ranks. In each rank, the individuals are also sorted from large to small crowding distances. The sorted individuals will be evaluated for each training case at each generation. Only the first individual that can cover the training case will obtain a token and others cannot. Consequently, a strong individual with high rank and a large crowding distance can exploit the niche by seizing as many tokens as it can. The other rules entering the same niche will have their strength decreased because they cannot compete with the stronger rule.

After evaluating all individuals in all training cases, each individual will know the number of tokens it obtained. For each individual, its support and confidence values are then modified according to the formula shown in Eq. (6).

$$adjusted\_objective\_value_m = raw\_objective\_value_m \times \frac{count}{ideal\_total} \tag{6}$$

where $raw\_objective\_value_m$ is the original value of objective m (e.g., confidence or support), $count$ is the number of tokens that the rule obtained and $ideal\_total$ is the total number of tokens that it can obtain ideally.

Effectively, the location of each individual in the objective value space is changed. Figure 8 shows an example of three individuals with adjusted objective values.

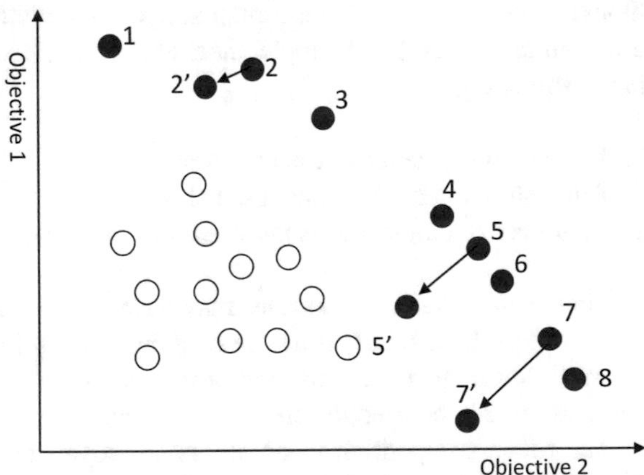

Fig. 8 Example of a population with tokens

In Fig. 8, the locations of the three individuals 2, 5, and 7 are changed based on the number of obtained tokens, respectively. Then the density of the region with individuals 4, 5, 6, 7, and 8 is decreased; consequently the remaining individuals 4, 6, and 8 can have higher probabilities of being selected during the learning process. For example, the original individual 5 is a non-dominated solution and its crowding distance is relatively small because it is close to individuals 4 and 6. Therefore, individuals 5 and 6 have low probability of being selected, because they are in the same front and similar (i.e., close) to each other in the objective value space. After adjusting by token competition, the location of individual 5 is changed to 5' if it obtains few tokens. Therefore, individual 5 is dominated by other individuals. The crowding distance of individual 6 is increased because its neighbors are changed to individuals 4 and 8. Compared to the original Pareto front, the probability of selecting individual 6 is increased.

Token competition is a greedy operation. It favors strong rules as their chance of survival is maintained, while their close competitors are weakened as they cannot get the tokens in the niche. From another point of view, each rule contributes to the population by covering training cases. If a training case has already been covered by one rule, then another rule covering the same training cases will make no contribution to the population. Thus the fitness of the latter rule should be discounted. Token competition is a simple method to force the increase of the diversity of the population. It has an advantage that it does not require a distance function. In fitness sharing [9], it is required to define a similarity or a distance function, so as to measure the similarity or dissimilarity between two individuals. However, it may be difficult to define how one individual is similar to another individual, especially in Genetic Programming. Genetic Algorithms use a fixed length binary string as the chromosome. Thus the genotypic difference (i.e., difference in the bits) can be used as a general similarity measurement. However this is not valid in the tree structure of Genetic Programming. Moreover, the similarity in genotype may not truly reflect the similarity of the individuals. Token competition simplifies the problem by simply regarding two individuals to be similar if they cover similar sets of training cases.

The execution of token competition is faster than that of fitness sharing [9]. To calculate the fitness score of one individual in fitness sharing, the similarity scores of all other individuals with respect to this individual have to be calculated. If a similarity score can be computed in time $O(t)$, and the population size is p, each individual needs a time $O(p * t)$ to calculate the similarity score, and the time needed to complete fitness sharing in each generation is $O(p^2 * t)$. On the other hand, calculations of similarity are not needed in token competition. The required information of token counting is the list of training cases that each individual covered. This information is already stored during the evaluation process. If an individual covers m training cases, a time of $O(m)$ is needed to seize the tokens, and token competition in each generation can be completed in $O(M * p)$, where M is the average value of m. This computation is straight forward and can be faster than fitness sharing if $O(M) < O(p * t)$.

As a result of token competition, there are rules that cannot seize any token. These rules are redundant as all of its training cases are already covered by the stronger rules. They can be replaced by new individuals. Introducing these new individuals can inject a larger degree of diversity into the population and provide extra chances for generating good rules.

3.2 Statistical Selection Learning

In this study, a number of classification rules are evolved by GBMGP with Token Competition. It is difficult to determine which of them should be selected eventually. If all of them are selected as members of an ensemble, the classification performance of the ensemble may be deteriorated, because some poor individuals may produce incorrect results. In general, the performance could be improved if an appropriate ensemble method is used. Therefore, we adopt the ideas from ensemble learning and develop a novel ensemble technique for solving the FFD problems.

The new method selects and combines a number of evolved rules into an ensemble for achieving two different goals. The first one is to improve the classification performance of the final ensemble. The second is to accomplish diversity maintenance, which aims to have a wide variety of rules to cover more cases. From the point view of statistics, a diverse population is composed of a number of different small groups of individuals, which are significantly different from each other. Therefore, we propose an ensemble method called statistical selection learning (SSL). Suppose there is a population with different evolved individuals, and each individual i contains two terms, fit_i and s_i, where fit_i is the fitness value of the individual. Fitness value is calculated by finding the average of support and confidence. s_i is the status of the individual i, which indicates if it is selected or not. At the beginning, a set of individuals with small size (e.g., 3) is randomly selected from the Pareto front as the primary set, and then the same number of individuals are randomly selected from the whole population as the secondary set. We determine if the two sets are different by using t-test. The two sets are merged to form a new primary set if they are significantly different from each other at the 5 percent significance level. Then the above steps are repeated to compare the new primary set with another secondary set. On the other hand, the secondary set is reselected if it is not significantly different from the primary set. Once the termination conditions are satisfied, the final ensemble is constructed.

In summary, the flowchart of the whole framework is depicted in Fig. 9.

4 Experiments and Results

This section describes the experiment preparation and experiment results. In this study, a number of data mining techniques are applied to solve four financial fraud detection problems. The experiment preparation is described in Sect. 4.1. Section 4.2 shows the parameter setting for the proposed method and briefly introduces several

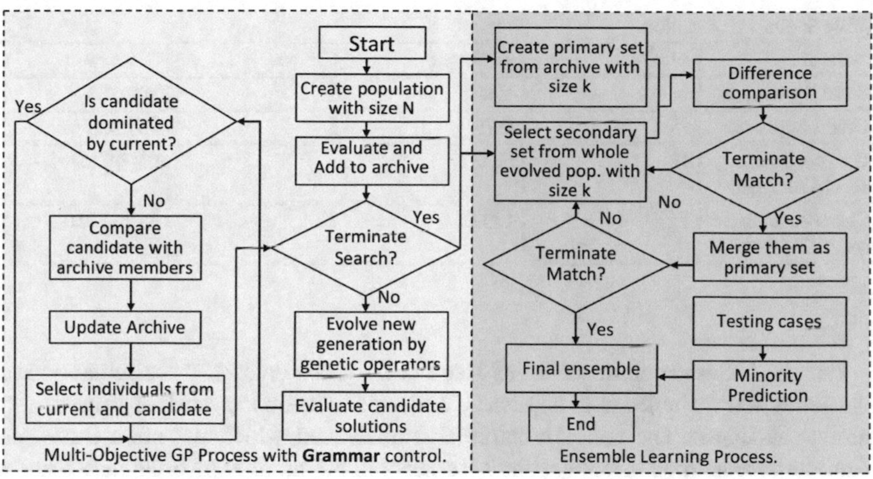

Fig. 9 The flowchart of the proposed method

variants of the proposed method. Section 4.3 presents the experiment results of all methods and discusses the results in detail.

4.1 Introduction to Experiment Preparation

In order to compare the performance of Grammar-based Multi-objective Genetic Programming (GBMGP) with Token Competition and the other well-known data mining techniques, we apply Waikato Environment for Knowledge Analysis (WEKA) [11] in the experiments. Logistic Regression (LR), Neural Networks (NNs), Support Vector Machine (SVM), Bayesian Networks (BNs), Decision Trees (DTs), AdaBoost, Bagging, LogitBoost, variants of GBGP, and variants of GBMGP (including the proposed method) are evaluated in the study.

4.1.1 Data Description

Four financial fraud detection problems are considered. Two of them have been taken from the UCI machine learning repository [2] and the other two are real-life financial fraud problems. The description of datasets is shown in Table 3.

Australian credit and Credit approval are similar, but the latter has one more attribute, which may affect the results. However, they are often used together as benchmark problems in many data mining studies. Their class distributions are balanced.

Table 3 Data description

Dataset	Attributes	Instances	Classes	Class ratio
Australian credit	14	690	2	307:383
Credit approval	15	690	2	307:383
U.S. corporate securities fraud	41	68332	2	63: 68269
China corporate securities fraud	17	18373	2	855:17518

For the U.S. corporate securities fraud (U.S. CSF) dataset, the original dataset has about 200 variables with duplicated and useless attributes, such as firm identity number and name. The dataset is extremely imbalanced, which may affect the results if models are learned directly from it. In general, the number of fraudulent firms is much smaller than the number of non-fraudulent firms. Therefore, it is better to maintain all fraudulent instances. Otherwise, it is difficult to learn the fraudulent information based on a small number of instances. If the fraudulent firms have too many missing values (e.g., more than 40% missing values) in some attributes, we remove those attributes directly. On the other hand, if the fraudulent firms have few missing values in some attributes, we replace them based on the data distributions of those attributes (e.g., take the mean of the variable as the value for the missing data). For non-fraudulent firms, we remove the instances with many missing values. For the attributes with few missing values, we also replace them based on the data distributions of those attributes.

The China corporate securities fraud (CCSF) dataset contains records of corporations with their firm, financial, governance, and trade characteristics. The original dataset has 21,396 instances with 24 attributes for all listed firms from 1998 to 2011. Each instance with more than 20 missing values in these 24 attributes is directly removed. Moreover, seven attributes about trade characteristics are removed since more than two-thirds of firms have not this trade information. The final dataset contains 18,373 records with 17 attributes. This dataset is also highly imbalanced with 4.7% fraudulent and 95.3% non-fraudulent examples.

4.1.2 Model Evaluation Criteria

As discussed in Sect. 3.1, accurate rates are the most important criteria of a model when solving FFD problems. Each problem has two classes. The first class is regarded as *positive* (i.e., fraudulent) and the other is *negative* (i.e., non-fraudulent). Table 4 shows the possible outcomes for binary classification.

The accurate rate of positive class is called *true positive rate (TPR)*, which is calculated by Eq. (7).

$$TPR = TP/(TP + FN) \tag{7}$$

Table 4 Contingency table with four outcomes of binary classification

	Classified as true	Classified as false
Actual is true	True positive (TP)	False negative (FN)
Actual is false	False positive (FP)	True negative (TN)

where *TP* is the number of positive examples that are correctly classified. *TP* + *FN* is the total number of positive examples including the number of correctly classified positive examples (i.e., TP) and the number of positive examples incorrectly classified as negative (i.e., FN). The accuracy rate for the negative class is called *true negative rate (TNR)*, which is calculated by Eq. (8).

$$TNR = TN/(TN + FP) \tag{8}$$

where *TN* is the number of negative examples that are correctly classified. *TN* + *FP* is the number of total negative examples including the number of correctly classified negative examples (i.e., TN) and the number of negative examples that are incorrectly classified as positive (i.e., FP). It is easy to observe the performance of each model for each class by using TPR and TNR as evaluation criteria.

4.2 Parameter Settings

Table 5 shows the parameter setting for the Grammar-based Multi-objectives Genetic Programming (GBMGP) with Token Competition. In addition to the proposed GBMGP, several GBGP variants and GBMGP variants are developed for model comparisons. GBGP variants include GBGP(s,c), GBGP(s,c) with majority voting and GBGP(s,c) with weighted voting, where *s* and *c* indicate support and confidence, respectively. In majority voting, the rules matching the testing case will make their own predictions about the class of the case and the final prediction is determined by the votes. On the other hand, each rule has a weight, which is the average of its support and confidence, for weighted voting and the final prediction for the testing case is determined by the weighted votes. GBMGP variants include GBMGP(s,c), GBMGP(s,c) with majority voting and GBMGP(s,c) with weighted voting. In GBGP variants, support and confidence are combined in a linear equation. On the other hand, support and confidence are the two objectives in GBMGP variants. GBGP(s) and GBGP(c) are special variants of GBGP that only consider support and confidence, respectively, in their fitness functions.

GBGP variants use *elitism* to select the best individual(s) of the current population for the next generation directly. The *elitism* operator always selects the individual with the highest fitness value for the next generation directly without using any genetic operators. GBMGP variants do not use *elitism*, since the non-dominated

Table 5 Parameters and values for the proposed method

GBMGP with token competition	
Parameter	Value
Population size	200
Max. no. of generation	500
Use elitism	No
Selection scheme	Tournament
Tournament size	2
Crossover rate	0.8
Mutation rate	0.2
Ensemble method	Statistical selection
Max. ensemble size	0.6

solutions in the current population are already considered in the evolutionary process automatically. Other experiments settings are the same as shown in Table 5. Moreover, *tournament* selection is used. It randomly selects a number of solutions with tournament size k, and chooses the best (i.e., winner) for genetic operation (e.g., crossover or mutation). The default tournament size is 2. The last two parameters are only used in the proposed method for ensemble learning. It applies the proposed statistical selection with the maximum of 60 percents of the whole population. For example, if the population size is 100 and ensemble size setting is 0.6 (i.e., 60%), then at most 60 individuals will be selected to form an ensemble. On the other hand, majority voting and weighted voting used all evolved rules (i.e., ensemble size is 100%).

The parameter settings of other data mining methods are shown in Table 6 and separated by a double line.

4.3 Results and Analysis

Table 7 summarizes the average accuracies for each class on the four financial datasets. The name of each method is shown in the first column. Four datasets are evaluated by nine methods in this experiment. Each dataset has two classes: positive and negative, and the corresponding accuracies are indicated by TPR and TNR, respectively, which are shown in the second row of Table 7. The Standard Deviation (S.D.) of each method is also given below the corresponding accuracy result. For example, Logistic Regression obtains 81% accurate rate in classifying positive class on the Australian dataset, and its S.D. is 5.8%.

For Australian credit and credit approval datasets, all the approaches are promising with regard to their TPRs and TNRs. For the two real-life datasets (U.S.CSF and CCSF), the performance values are not stable using different methods. Some meth-

Table 6 Parameter settings for the compared approaches

Method	Parameter	Value
Logistic Regression (LR)	Ridge	1.0E-8
	Max. iterations	−1
Neural Networks (NNs)	Learning rate	0.3
	Momentum value	0.2
	No. hidden layers	1
	Weight update	Back-propagation
	Training epochs	500
	Random seed	0
Support Vector Machine (SVM)	Kernel function	Polykernel
	Complexity	1
	Tolerance rate	0.001
	Exponent value	1
Bayesian Networks (BNs)	Estimator	Simple estimator
	Search algorithm	Hill climbing
Decision Trees (DTs)	Min number of nodes	2
	No pruning	False
	Number of folds	3
	Min variance probability	0.001
AdaBoost	Classifier	Decision stump
	Number of iterations	20
	Seed	1
	Use resampling	False
	Weight threshold	100
Bagging	Classifier	REPTree
	No. iterations	20
	Bag size percent	100
LogitBoost	Classifier	Decision stump
	No. iterations	20
	Use resampling	False
	Seed	1
	Weight threshold	100
	Likelihood threshold	−1.798

Table 7 Accuracies of data mining techniques and the proposed method

	Australia credit		Credit approval		U.S.CSF		CCSF		
Methods	TPR	TNR	TPR	TNR	TPR	TNR	TPR	TNR	w/t/l
LR	0.81^{++}	0.81	0.85^{++}	**0.84**$^{-}$	0.41^{++}	0.90^{--}	0.41^{++}	0.47^{++}	6/0/2
	0.058	0.063	0.046	0.042	0.222	0.013	0.045	0.017	
NNs	0.80^{++}	0.83	0.80^{++}	0.83	0.47^{++}	0.94^{--}	0.31^{++}	0.83^{--}	5/0/3
	0.089	0.064	0.047	0.048	0.191	0.014	0.067	0.053	
SVM	0.87	0.79^{++}	0.80^{++}	0.79	0.51^{+}	0.94^{--}	0.41^{++}	0.73^{--}	6/0/2
	0.027	0.069	0.043	0.051	0.183	0.006	0.037	0.019	
BNs	0.80^{++}	**0.88**	0.80^{++}	0.81	0.52^{+}	0.91^{--}	0.28^{++}	**0.94**$^{--}$	4/0/4
	0.059	0.017	0.046	0.029	0.185	0.008	0.022	0.006	
DTs	0.82^{++}	0.84	0.82^{++}	**0.84**	0.10^{++}	**1.00**$^{--}$	0.24^{++}	0.93^{--}	5/0/3
	0.079	0.105	0.078	0.103	0.071	0.001	0.029	0.009	
AdaBoost	0.81^{++}	0.84	0.77^{++}	0.81	0.51^{+}	0.88^{--}	0.47^{++}	0.71^{-}	5/0/3
	0.053	0.062	0.046	0.069	0.174	0.016	0.046	0.031	
Bagging	0.79^{++}	0.82	0.78^{++}	0.80	0.11^{++}	**1.00**$^{--}$	0.23^{++}	0.89^{--}	5/1/2
	0.068	0.067	0.057	0.058	0.076	0.001	0.025	0.008	
LogitBoost	0.81^{++}	0.82	0.83^{++}	0.82	0.50^{+}	0.90^{--}	0.44^{++}	0.80^{--}	5/0/3
	0.038	0.069	0.032	0.047	0.211	0.013	0.046	0.028	
GBMGP $(s,c,S)_i$	**0.89**	0.85	**0.89**	0.80	**0.64**	0.76	**0.66**	0.67	n/a
	0.063	0.069	0.058	0.055	0.118	0.133	0.074	0.069	

1. $^{++}$ Using paired t-test, the average accuracy is significantly worse than that of GBMGP$(s,c,S)_i$ at the 0.05 level
2. $^{+}$ Using paired t-test, the average accuracy is significantly worse than that of GBMGP$(s,c,S)_i$ at the 0.1 level
3. $^{-}$ Using paired t-test, the average accuracy is significantly better than that of GBMGP$(s,c,S)_i$ at the 0.1 level
4. $^{--}$Using paired t-test, the average accuracy is significantly better than that of GBMGP$(s,c,S)_i$ at the 0.05 level

ods such as Decision Trees and Bagging generate extremely biased results with very low TPRs and very high TNRs. Logistic Regression obtains about 41% in regard to classifying fraudulent firms in both real-life datasets. SVM, Bayesian Networks, AdaBoost, and LogitBoost obtain about 50% in regard to classifying fraudulent firms for the U.S.CSF dataset only, and worse TPR for the CCSF dataset. The proposed method with minority prediction, which is located in the last row of Table 7 can achieve better TPR results than all of the other techniques, but its TNR values for each dataset are relatively lower at the same time. According to the characteristics of financial datasets, especially for real-life FFD problems, the detection of positive class (i.e., fraudulent) is much more important than the detection of negative class (i.e., non-fraudulent). For example, if a firm is fraudulent, and it is incorrectly classified as non-fraudulent, then the loss to interested people (e.g., shareholders) may be destructive. However, if a firm is non-fraudulent, and it is incorrectly classified as fraudulent, it may need to be investigated by the Securities and Exchange Commission (SEC) or the China Securities Regulatory Commission (CSRC) at relatively much lower cost (i.e., investigation fees) compared to the destructive consequence caused by fraud without any investigations. Therefore, it is more important to classify fraudulent firms correctly than non-fraudulent firms.

In order to have a more comprehensive comparison for the proposed method, a number of GBGP variants and GBMGP variants are developed and the corresponding comparison results are shown in Table 8. The name of each method is located in the first column, and the meanings of notations are indicated in Table 9. For example, the first method is GBGP(s,c)$_a$, which is the original GBGP. The symbol "a" means that the first method used m**a**jority prediction. As another example, the last method is GBMGP(s,c,S)$_i$, which is the proposed method. It is a multi-objective GBGP (i.e., GBMGP), and the abbreviation "M" indicates that the system has the **m**ulti-objective component. Therefore, it uses support (i.e., s) and confidence (i.e., c) as the two objectives. The symbol "S" indicates that **s**tatistical selection learning is used as the ensemble learning method. The symbol "i" means that the last method uses m**i**nority prediction.

For the Australia credit and Credit approval datasets, the original GBGP can obtain about 85% accuracy for both TPRs and TNRs. Different ensemble learning techniques (i.e., majority voting and weighted voting) cannot improve the original GBGP, no matter whether majority prediction or minority prediction is used. In addition, variants with multi-objective (i.e., GBMGP) have slightly poorer performance in regard to TPRs and slightly better performance in regard to TNRs than the original GBGP. However, variants with multi-objective (i.e., GBMGP) and ensemble techniques perform similarly and even obtain better TPRs and TNRs than the original GBGP except for the proposed method. The proposed method obtains the highest TPRs and has slightly poorer performance in regard to TNRs.

For the U.S.CSF dataset, all methods using majority prediction have good performance in regard to TNRs. However the corresponding TPRs are very low, with only the GBMGP(s,c,S)$_a$ obtaining a result that is more than 50% for TPR. The TPRs are relatively improved by using minority prediction, but still less than 50%. The proposed method achieves 64%, which is the highest TPR value among all variants.

For the CCSF dataset, GBMGP without using any ensemble learning methods cannot improve the results over the original GBGP. The GBGP with majority voting even produces poorer TPR results. However, compared to the original GBGP, the TPR of GBGP with minority prediction has about 22.9% improvements. Except for the proposed method, the GBGP with majority voting and minority prediction obtains the second highest TPR, but the corresponding TNR is greatly reduced. The minority prediction performs well in this dataset, especially for GBGP(s,c,M)$_i$. Finally, the proposed method produces the highest TPR and relatively higher TNR compared to GBGP(s,c,M)$_i$.

The following empirical and statistical tests focus on the comparison between GBMGP with Token Competition and the other approaches.

Table 8 Classification accuracies of the proposed method and its variants

With tokens	Australia credit		Credit approval		U.S.CSF		CCSF		
Methods	TPR	TNR	TPR	TNR	TPR	TNR	TPR	TNR	w/t/l
GBGP(s,c)$_a$	0.84^{+}	0.86	0.84^{++}	0.85^{--}	0.33^{++}	0.84^{--}	0.48^{++}	0.78^{--}	4/0/4
	0.063	0.095	0.027	0.045	0.201	0.049	0.059	0.025	
GBGP(c)$_a$	0.68^{++}	0.79^{++}	0.66^{++}	0.75^{+}	0.29^{++}	0.83^{-}	0.33^{++}	0.65	7/0/1
	0.049	0.024	0.045	0.060	0.046	0.054	0.062	0.045	
GBGP(s)$_a$	0.09^{++}	0.97^{--}	0.09^{++}	0.97^{--}	0.02^{++}	0.93^{--}	0.07^{++}	0.94^{--}	4/0/4
	0.048	0.027	0.015	0.018	0.009	0.031	0.033	0.034	
GBGP(s,c, M)$_a$	0.82^{++}	0.79^{++}	0.82^{++}	0.79	0.39^{++}	0.85^{--}	0.25^{++}	**0.92**$^{--}$	6/0/2
	0.034	0.075	0.032	0.031	0.254	0.021	0.034	0.021	
GBGP(s,c, W)$_a$	0.82^{++}	0.80^{+}	0.84^{++}	0.79	0.30^{++}	**0.89**$^{--}$	0.54^{++}	0.67	6/1/1
	0.046	0.053	0.031	0.041	0.120	0.046	0.055	0.051	
GBMGP(s,c)$_a$	0.80^{++}	**0.91**$^{--}$	0.80^{++}	0.84	0.44^{++}	0.86^{--}	0.48^{++}	0.73^{--}	4/0/4
	0.054	0.027	0.074	0.066	0.229	0.034	0.058	0.054	
GBMGP (s,c,W)$_a$	0.85	0.90^{--}	0.85^{+}	**0.89**$^{--}$	0.32^{++}	0.83^{-}	0.38^{++}	0.88^{--}	4/0/4
	0.061	0.027	0.053	0.036	0.165	0.046	0.108	0.052	
GBMGP (s,c, M)$_a$	0.84^{++}	**0.91**$^{--}$	0.84^{++}	**0.89**$^{--}$	0.45^{++}	0.83	0.48^{++}	0.87^{--}	4/0/4
	0.040	0.047	0.034	0.049	0.235	0.084	0.127	0.061	
GBMGP(s,c,S)$_a$	0.82^{++}	0.90	0.86	0.87^{--}	0.54^{+}	**0.89**$^{--}$	0.53^{++}	0.81^{--}	4/0/4
	0.081	0.079	0.056	0.049	0.122	0.089	0.081	0.084	
GBGP(s,c)$_i$	0.82^{++}	0.78^{++}	0.83^{++}	0.80	0.43^{++}	0.81	0.59^{+}	0.59^{++}	6/1/1
	0.043	0.072	0.038	0.059	0.179	0.047	0.092	0.096	
GBGP(c)$_i$	0.75^{++}	0.69^{++}	0.73^{++}	0.68^{++}	0.34^{++}	0.58^{++}	0.43^{++}	0.49^{++}	8/0/0
	0.061	0.050	0.075	0.058	0.155	0.137	0.050	0.047	
GBGP(s)$_i$	0.13^{++}	0.99^{--}	0.14^{++}	0.97^{--}	0.08^{++}	0.97^{--}	0.10^{++}	0.96^{--}	4/0/4
	0.031	0.018	0.035	0.027	0.036	0.027	0.047	0.038	
GBGP(s,c, M)$_i$	0.84^{++}	0.78^{++}	0.85^{++}	0.79	0.47^{++}	0.86^{--}	0.61^{+}	0.60^{++}	7/0/1
	0.040	0.067	0.025	0.042	0.189	0.033	0.058	0.057	
GBGP(s,c, W)$_i$	0.85^{++}	0.79^{++}	0.86	0.79	0.44^{++}	0.76	0.60^{++}	0.58^{++}	7/1/0
	0.038	0.036	0.037	0.062	0.116	0.065	0.062	0.040	
GBMGP(s,c)$_i$	0.82^{++}	0.88	0.84^{++}	0.86^{--}	0.45^{++}	0.73	0.48^{++}	0.58^{++}	6/0/2
	0.037	0.070	0.045	0.048	0.258	0.143	0.052	0.055	
GBMGP (s,c,W)$_i$	0.86	0.89	0.85^{+}	0.87^{--}	0.35^{++}	0.81	0.55^{++}	0.68	4/0/4
	0.067	0.050	0.055	0.036	0.113	0.063	0.053	0.031	
GBMGP (s,c, M)$_i$	0.85	0.86	0.82^{++}	0.84^{-}	0.44^{++}	0.75	0.59^{++}	0.64	6/0/2
	0.064	0.087	0.047	0.022	0.186	0.154	0.042	0.075	
GBMGP(s,c,S)$_i$	**0.89**	0.85	**0.89**	0.80	**0.64**	0.76	**0.66**	0.67	n/a
	0.063	0.069	0.058	0.055	0.118	0.133	0.074	0.069	

1. $^{++}$ Using paired t-test, the average accuracy is significantly worse than that of GBMGP(s,c,S)$_i$ at the 0.05 level
2. $^{+}$ Using paired t-test, the average accuracy is significantly worse than that of GBMGP(s,c,S)$_i$ at the 0.1 level
3. $^{-}$ Using paired t-test, the average accuracy is significantly better than that of GBMGP(s,c,S)$_i$ at the 0.1 level
4. $^{--}$Using paired t-test, the average accuracy is significantly better than that of GBMGP(s,c,S)$_i$ at the 0.05 level

Table 9 Abbreviations of all the approaches

Abbreviation	Description
s	Objective: **s**upport
c	Objective: **c**onfidence
W	Ensemble: **W**eighted voting
M	Ensemble: **M**ajority voting
S	Ensemble: **S**tatistical selection
a	M**a**jority prediction
i	M**i**nority prediction

4.3.1 Empirical Analysis

Each problem has two results for TPR and TNR, respectively. We set each one as a competition, and therefore eight competitions are performed for the four datasets. The empirical *w/t/l* (i.e., win, tie, and lose) test results are given in the last column of Tables 7 and 8, where *w* means that GBMGP with Token Competition and minority prediction outperforms the compared approach, *t* means that GBMGP with Token Competition and minority prediction has the same results, and *l* means that GBMGP with Token Competition and minority prediction is worse than the compared approach.

In Table 7, compared with Bayesian Networks (BNs), the proposed method wins all competitions in regard to TPR, but also loses all of them with regard to TNR. BNs can be regarded as a generic method for solving FFD problems. Although it obtains the highest TNR for Australia and the second highest TPR for U.S.CSF, it also gets bad TPR results in the CCSF dataset. In this study, the real-life datasets are more important than the benchmark datasets. Compared with Logistic Regression (LR), the proposed method wins six competitions and loses two, which are TNRs from credit approval and U.S.CSF. Especially for U.S.CSF, LR has a biased result for TNR. However, logistic regression is still a competitive method compared with other approaches. Compared with Neural Networks (NNs), Support Vector Machine (SVM) and Decision Trees (DTs), the proposed method, respectively, wins 5, 6, and 5 competitions. On the other hand, it, respectively, loses in 3, 2, and 3 competitions, which are also related to TNRs. SVM has similar performance to that of BNs, but DTs have extremely biased results with regard TNRs. Therefore, DTs may not be an appropriate method for imbalanced financial datasets. Compared with AdaBoost, Bagging, and LogitBoost, the proposed method, respectively, wins 5, 5, and 5 competitions, and, respectively, ties in 0,1, and 0 competitions. Therefore, the proposed method is able to outperform other ensemble learning techniques even in the TNR competitions. Bagging generates extremely biased TNR results on the two real-life financial fraud datasets, and thus it may not be an appropriate method for imbalanced financial datasets. In addition, comparing the ensemble learning techniques with the

LR for U.S.CSF and CCSF datasets, the overall TPR results of ensemble learning techniques (except for Bagging) improve.

According to Table 8, compared with GBGP(s,c)$_a$, GBMGP(s,c,W)$_i$, GBMGP(s,c)$_a$, GBMGP(s,c,W)$_a$, GBMGP(s,c,M)$_a$ and GBMGP(s,c,S)$_a$, the proposed method wins all of the competitions for TPR, but also loses all of them for TNR. Compared with GBGP(s,c,M)$_a$, GBGP(s,c)$_i$, GBGP(s,c,W)$_a$, GBGP(s,c)$_i$, GBGP(s,c,M)$_i$, and GBGP(s,c,W)$_i$, the proposed method, respectively, wins 6, 6, 6, 6, 7, and 7 competitions, and, respectively, ties in 0, 0, 1, 1, 0, and 1. This indicates that the proposed method outperforms other GBGP variants no matter whether majority or minority prediction is applied. Moreover, it also indicates that the multi-objective and the new statistical selection learning technique together can improve the results for most TPRs and TNRs.

Finally, compared with GBMGP(s,c,M)$_i$ and GBMGP(s,c)$_i$, the proposed method wins 6 and loses in 2 competitions. The two lost competitions are for TNRs from the Australian credit and credit approval datasets. This indicates that the GBMGP(s,c,M)$_i$ and GBMGP(s,c)$_i$ with minority prediction maybe more suitable for balanced datasets. Comparing GBMGP(s,c,M)$_i$ and GBMGP(s,c)$_i$, it can be found that majority voting can improve the TPR and TNR results for the CCSF dataset.

4.3.2 Statistical Analysis

Pairwise t-test is applied to demonstrate the statistical significance of the experiments. The performance of the proposed method and other approaches is compared to calculate statistical significance. The results of the t-test are shown in Tables 7 and 8, which use the symbol "++" to indicate that the proposed method is significantly better than the compared method at the 5% level and apply the symbol "+" to represent that the proposed method is significantly better than the compared method at the 10% level. On the other hand, the symbols "− −" and "−" are used if the proposed method is significantly worse than the compared method at the 5% level and 10% level, respectively. For example in Table 7, compared with LR, the proposed method significantly outperforms it in 5 of the 8 metrics at the 5% level. However, the proposed method is significantly worse than LR for TNRs on credit approval and U.S.CSF at the 10% level and 5% level, respectively. In addition, the proposed method is significantly better than NNs in 4 of the 8 metrics at the 5% level. Compared with SVM, it is significantly superior in 3 of the 8 metrics at the 5% level, and 1 metric at the 10% level. Compared with BNs, it is significantly superior in 3 of the 8 metrics at the 5% level, and 1 metric at the 10% level. It is also significantly superior to DTs in 4 metrics at the 5% level. Moreover, it also outperforms AdaBoost and LogitBoost in 3 of the 8 metrics at the 5% level, and 1 metric at the 10% level. It outperforms Bagging in 4 of the 8 metrics at the 5% level.

For the benchmark problems in Table 7, the proposed method significantly outperformed all the data mining techniques for TPRs, except for SVM on Australia credit. Moreover, it also significantly outperforms SVR for TNRs on Australia credit at the 5% level but is significantly worse than the LR for TNRs on Credit approval at the 10% level.

For the U.S.CSF dataset, the proposed method significantly outperforms LR, NNs, DTs, and Bagging for TPRs at the 5% level, and it also significantly outperforms SVM, BNs, AdaBoost, and LogitBoost for TPRs at the 10% level. However, the TNRs of using the proposed method are significantly worse than all the data mining methods at the 10% level.

For the CCSF dataset, the proposed method significantly outperforms all the data mining methods for TPRs at the 5% level but is also significantly worse than all the data mining methods at the 10% level, except for the LR. It is easy to obtain very good results on the majority (i.e., negative) class by applying the compared approaches. As discussed before, the detection of fraudulent firms is much more important than the correct classification of non-fraudulent firms. The proposed method seems to reduce by a few percent of the accurate rate of classifying non-fraudulent firms, but it significantly increases the performance in identifying fraudulent firms. Thus the proposed method is promising for FFD problems.

5 Conclusion

5.1 Contributions

There are two major contributions of this study. Firstly, we have proposed a new method called Grammar-based Multi-objective Genetic Programming with Token Competition (GBMGP) that can take advantages of Grammar-based Genetic Programming (GBGP), Token Competition, Multi-Objective Evolutionary Algorithms, and ensemble learning. The new method applies token competition to maintain diversity among individuals. In general, NSGA-II can use the objective values of each individual to maintain the diversity of the population. Nevertheless, in this study, each individual is a classification rule and it is possible that some individuals are located in a crowded region, since they have close objective values. However, they are not truly similar to each other if they have different meanings. Token competition is used to adjust the objective values of each rule, and the rules with similar objective values but different meanings are separated. Moreover, a new ensemble technique called Statistical Selection Learning (SSL) has been developed. SSL can outperform majority voting and weighted voting in classifying fraudulent firms from the two real-life FFD datasets.

Secondly, we have performed a number of comprehensive comparisons between different data mining approaches in solving the FFD problem. These approaches include Logistic Regression, Neural Networks, Support Vector Machine, Bayesian Networks, Decision Trees, AdaBoost, Bagging, and LogitBoost. Moreover, we have also developed a number of GBGP variants and GBMGP variants with different ensemble methods for comparison.

5.2 Directions for Future Research

It will be interesting to incorporate other objectives such as risk and return into the proposed method and evaluate it on different real-life financial datasets to determine whether some useful and interesting rules can be discovered.

The proposed method can be used to solve other business problems such as direct marketing problems [5]. In direct marketing, the evaluation of a method is usually based on response rate and total profit. Response rate is the ratio of the number of respondents to the total number of customers in the dataset. Total profit is the sum of the profits generated from all respondents. A high value of response rate may not produce high total profit. Therefore, it is necessary to find respondents who can contribute high profits. In this problem, the minority class is the high-profit customers and the majority class contains the low-profit customers and the non-respondents. The learned rules can identify the high-profit customers, low-profit customers, and non-respondents if the objectives of the proposed method are changed to response rate and total profit.

Acknowledgements This research is supported by the LEO Dr. David P. Chan Institute of Data Science and the General Research Fund LU310111 from the Research Grant Council of the Hong Kong Special Administrative Region.

References

1. Agrawal, R., Imieliński, T., Swami, A.: Mining association rules between sets of items in large databases. In: ACM SIGMOD Record, pp. 207–216. ACM (1993)
2. Asuncion, A., Newman, D.: UCI machine learning repository (2007)
3. Bhattacharyya, S., Jha, S., Tharakunnel, K., Westland, J.C.: Data mining for credit card fraud: a comparative study. Decis. Support Syst. **50**(3), 602–613 (2011)
4. Coello, C.C., Lamont, G.B., Van Veldhuizen, D.A.: Evolutionary algorithms for solving multi-objective problems. Springer (2007)
5. Cui, G., Wong, M.L., Wan, X.: Cost-sensitive learning via priority sampling to improve the return on marketing and CRM investment. J. Manag. Inf. Syst. **29**(1), 341–373 (2012)
6. Cumming, D., Hou, W., Lee, E.: The role of financial analysts in deterring corporate fraud in China. SSRN Electron. J. (2011)
7. Deb, K., Pratap, A., Agarwal, S., Meyarivan, T.: A fast and elitist multiobjective genetic algorithm: Nsga-ii. IEEE Trans. Evol. Comput. **6**(2), 182–197 (2002)
8. Eskandari, H., Geiger, C.D.: A fast pareto genetic algorithm approach for solving expensive multiobjective optimization problems. J. Heuristics **14**(3), 203–241 (2008)
9. Goldberg, D., Richardson, J.: Genetic algorithms with sharing for multi-modal function optimization. In: Proceedings of the Second International Conference on Genetic Algorithms, pp. 41–49 (1987)
10. Goldberg, D.E.: Genetic Algorithms in Search, Optimization, and Machine Learning. Addison-Wesley, Reading, MA, USA (1989)
11. Hall, M., Frank, E., Holmes, G., Pfahringer, B., Reutemann, P., Witten, I.H.: The WEKA data mining software: an update. ACM SIGKDD Explor. Newsl. **11**(1), 10–18 (2009)
12. Holland, J.H.: Adaptation in Natural and Artificial Systems. The University of Michigan Press, Ann Arbor, MI, USA (1975)

13. Hopcroft, J.E.: Introduction to Automata Theory, Languages, and Computation, p. 3/E. Pearson Education India (2008)
14. Keane, M.A., Streeter, M.J., Mydlowec, W., Lanza, G., Yu, J.: Genetic Programming IV: Routine Human-competitive Machine Intelligence, vol. 5. Springer (2006)
15. Konak, A., Coit, D.W., Smith, A.E.: Multi-objective optimization using genetic algorithms: a tutorial. Reliab. Eng. Syst. Saf. **91**(9), 992–1007 (2006)
16. Kou, Y., Lu, C.-T., Sirwongwattana, S., Huang, Y.-P.: Survey of fraud detection techniques. In: Proceedings of 2004 IEEE International Conference on Networking, Sensing and Control, vol. 2. IEEE, pp. 749–754 (2004)
17. Koza, J.R.: Genetic Programming: On the Programming of Computers by Means of Natural Selection. MIT press (1992)
18. Ngai, E.W.T., Hu, Y., Wong, Y.H., Chen, Y., Sun, X.: The application of data mining techniques in financial fraud detection: a classification framework and an academic review of literature. Decis. Support Syst. **50**(3), 559–569 (2011)
19. Poli, R., Langdon, W., McPhee, N.F.: A Field Guide to Genetic Programming. LuLu Enterprises (2008)
20. Ponsich, A., Jaimes, A.L., Coello, C.A.C.: A survey on multiobjective evolutionary algorithms for the solution of the portfolio optimization problem and other finance and economics applications. IEEE Trans. Evol. Comput. **17**(3), 321–344 (2013)
21. Ravisankar, P., Ravi, V., Raghava Rao, G., Bose, I.: Detection of financial statement fraud and feature selection using data mining techniques. Decis. Support Syst. **50**(2), 491–500 (2011)
22. Syeda, M., Zhang, Y.-Q., Pan, Y.: Parallel granular neural networks for fast credit card fraud detection. In: Proceedings of the 2002 IEEE International Conference on Fuzzy Systems, vol. 1. IEEE, pp. 572–577 (2002)
23. Wong, M.L.: A flexible knowledge discovery system using genetic programming and logic grammars. Decis. Support Syst. **31**(4), 405–428 (2001)
24. Wong, M.L., Leung, K.S.: Evolutionary program induction directed by logic grammars. Evol. Comput. **5**(2), 143–180 (1997)
25. Wong, M.L., Leung, K.S.: Data Mining Using Grammar Based Genetic Programming and Applications. Kluwer Academic Publisher (2000)

Phenotypic Niching Using Quality Diversity Algorithms

Alexander Hagg

An animal's behaviour tends to maximize the survival of the genes "for" that behaviour, whether or not those genes happen to be in the body of the particular animal performing it.
- Richard Dawkins [13]

There is a power and utility to regarding the gene as the unit of selection, but equally there is value to seeing the organism as the unit of niche construction.
- Kevin Laland [46]

Abstract Here we describe quality diversity algorithms, a recent and powerful class of evolutionary algorithms that produces a diverse set of high-performing solutions. The optimization paradigm emphasizes phenotypic niching and egalitarian treatment of quality and diversity. We ground quality diversity in ecology, describe the historical development, and give an intuition and formalization of the algorithms. We present a practical example that we refer to for engineers and laymen readers to understand how and why quality diversity can be used. The main insights from research of quality diversity, performance metrics, and benchmarks are discussed. Finally, the open challenges are presented.

1 Introduction

One key feature of natural evolution is quality, or optimality, driven by the naturally occurring scarcity of nourishment. Only creatures that are fit enough will survive to create offspring. The second key feature is the diversity that arises from the driving forces of random variation and protection by *niching*. Mutations in genomes of the offspring of successful creatures cause the population to diverge and explore possible

---

A. Hagg (✉)
Leiden Institute of Advanced Computer Science, Leiden, The Netherlands

Bonn-Rhein-Sieg University of Applied Sciences, Sankt Augustin, Germany
e-mail: alex@haggdesign.de

M. Preuss et al. (eds.), *Metaheuristics for Finding Multiple Solutions*,
Natural Computing Series,
https://doi.org/10.1007/978-3-030-79553-5_12

genomes. Evolutionary niches protect species from having to battle for nourishment with all other species. Each species naturally converges to being fit enough within a niche, becoming specialized in surviving in that niche.

In evolutionary computation, diversity measures like niching [69] have classically been used to prevent a population-based optimization algorithm from getting stuck in local optima. These optima, which do not represent the best solution overall, can trap an algorithm inside a sub-optimal region of the solution space. The population within the algorithm is artificially spread out by the diversity measure, subdividing the population between multiple attracting optima. Niching is often tightly connected to the concept of speciation. Only individuals that are located in or close to the same niche interact directly. Only individuals that are genetically compatible are able to produce offspring together. In evolutionary computation, speciation is often seen as a core property of artificial niching methods. "Niching in evolutionary algorithms (EA) is a two-step procedure that (a) concurrently or subsequently distributes individuals onto distinct basins of attraction and (b) facilitates approximation of the corresponding (local) optimizers" [62]. Basins of attraction represent high fitness regions in the objective function out of which there is no escape path that does not go through a low fitness region. A species will therefore "nest" itself inside of a basin of attraction.

Let us now look at the concept of natural fitness as the simple metric of the amount of grass in kilograms eaten per day and compare species in their niches. The amount of grass a cow eats is many times larger than what a cutworm is able to consume. Both species are specialized in their own niche, both have reached a local optimum, but when using the proposed fitness metric, their absolute fitness is vastly different. From natural evolutionary effects, a highly diverse set of solutions emerges. Neither solution is preferred, as there is no invisible hand selecting between those niches.[1]

When observing this example from nature, a niche can certainly be seen as a basin of attraction in which only certain individual species can survive. A species that is effective at surviving in a niche is less likely to adapt toward another basin, as it would first have to cross unfit genetic states, although this is not impossible and is often seen in bird species that migrate to islands and become fit with respect to another niche, quite possibly losing their flight abilities due to genetic drift [84].

In computational evolution, a niche is often reduced to a basin of attraction in the objective function over the genome, or encoding, of a solution to a problem, especially in benchmark comparisons between algorithms. But this is not the full story. A natural niche can be filled by different genomes. It is not the cutworm alone that can eat grass. Cows can too. Both can digest grass and therefore have similarities in their phenotype, the expressed morphology (the ability to metabolize grass), and behavioral expression (searching grass for nourishment) of a genome. Describing a niche by looking at the expression of the genome seems to make more sense than only taking into account a species' genome. The phenotype, like the number of

[1] Popular literature about computational evolution often use the active tense when talking about evolution, as though it is a driving force rather than an emergent property of life. Even natural selection could be seen as a misnomer, as selection implies an entity acting upon the world, whereas selection seems to be an emergent property of complex interactions in nature.

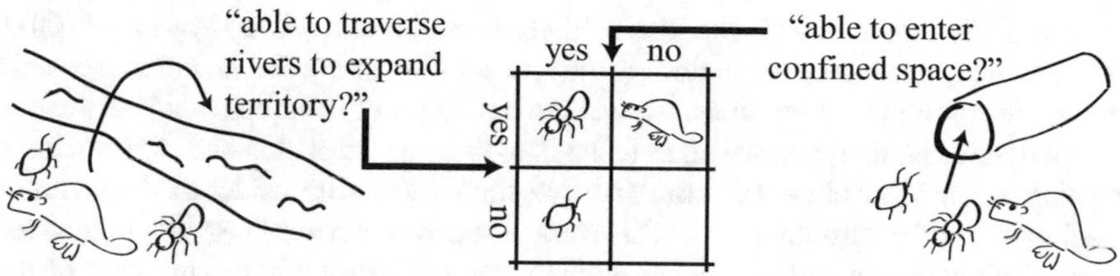

Fig. 1 Some phenotypic aspects, like the ability to traverse a river, put the fire ant and the beaver into the same phenotypic niche (rows). Aspects like body size allow separation between the two species into separate niches (columns). Stink bugs however can enter similar confined spaces as ants do, but they cannot swim. Both the morphology of the body as well as the behavior of a species determine what niche it can occupy

jumps performed by a robot biped controller (a behavior), or the amount of curvature in an aerodynamic shape (shape morphology), surely is a closer description of the individual in terms of what niche it can occupy, how successful it can be inside the niche, and how it achieves this success.

Even a genome that does not provide the ability to properly swim, like a fire ant, can swim across rivers through the expression of that genome—through cooperation with other fire ants by self-assembling into a raft [52]. A beaver and a fire ant, both genomes that occupy different basins of attraction in genetic space, can occupy the same niche according to a particular description of the phenotype (see Fig. 1). Now surely a beaver does not swim the same way a colony of fire ants does. The beaver's body makes swimming motions whereas the fire ants need to cooperate. But on some simplified descriptive level, the behavior is the same. On the other hand, let us take body size as a description. In that case, large body size will prevent the beaver from entering places fire ants can reach. Fire ants and beavers can both be described and compared, in terms of the two (extended) phenotypic diversity metrics. In one they fall into the same niche, but in the other, they do not.

It is behavior, the interaction between a creature and its environment, not the genome, that ultimately defines a natural niche. Dawkins poses in [13] that behavior could be subsumed into the *extended phenotype*. Furthermore, non-behavioral traits influence the conditions under which behavior can take place. A large animal will not be able to occupy niches that need a small body. The morphology of the species can limit or enable certain behaviors. So, certain aspects of the extended phenotype, be it behavioral or morphological, are important for fitness and some help to determine in what niche an individual can flourish.[2]

In optimization, we can perform niching on behavioral or other phenotypic aspects, not on the genome alone. A non-linear mapping between genome and phenotype is inherent to evolution and the interaction between a creature and its environ-

[2]Laland takes Dawkins' idea one step further and introduces the idea of *niche construction*. By influencing its environment, a creature can create its own niche. This introduces the idea of evolution being a causally cyclical process. A creature creates a niche by acting on its environment, its genome adapts to the niche, which in itself causes changes in the environment, and so forth [46].

ment is a key aspect. This is the idea behind computational *quality diversity*[3] (QD). We are not always interested in finding the best solution to a problem, nor in a genetic variety of solutions. Often times, we want to have a set of solutions that is diverse in their extended phenotype, as well as being fit to perform a task. Instead of performing niching on the basis of genetic similarity, we might want to consider the behavior of a solution in its environment, like the average speed of a neural network controlled robot, the amount of turbulence caused by a car, or the amount of curvature of the surface of a complex water drain system, as the basis for niching. This might enable us to find unexpectedly good solutions or make the optimization process more robust with respect to requirement changes. If one solution does not work well, we could take this new knowledge into account when selecting an alternative solution.

The rest of this chapter is organized as follows: a historical overview of the search for diversity and its connection to classical niching methods will be given in Sect. 2. The two main inspirations for the young field of QD will be introduced in Sect. 3, whereby we will also develop a general description of QD algorithms and their main components, followed by an overview of some of the most impactful example applications and extensions of QD. Please refer to the practical example given in Sect. 3.3 to understand how and why one would want to use QD. In Sect. 4, this is followed by a discussion of key insights in QD research. Performance metrics and benchmarks are given in Sect. 5. We conclude with an overview of current and future challenges in Sect. 6.

2 The Search for Diversity

There is a myriad of arguments for diversity in optimization (and beyond). In multimodal optimization, the concept of diversity is itself considered to be the goal [4] for a number of reasons: generating alternatives for engineers to choose from, finding trade-offs between features, learning properties of the decision space (*innovization* [18]), increasing the robustness of optimization, allowing solutions to adapt to a changing fitness landscape [7], performing model-based diagnostics, or crossing the reality gap between simulation and reality [45]. By understanding the diversity of high-performing solutions, we can hope to find more unexpected solutions that provide practitioners with options. This allows engineers to better understand the problem and improve the decision making process. We can also increase solution robustness by finding a set of options or by understanding more about the decision landscape's structure.

Nature provides us with ample evidence that speaks in favor of diversity. We will now examine the evidence that is provided from computational evolution itself.

---

[3] As in diversity of qualities. Both quality diversity as well as *illumination* are used in the field, although one could view the deeper concept to be *phenotypic niching*.

2.1 Genetic Diversity

Computer science research up to the 1970s had already developed interest in population-based evolutionary approaches to optimization, but in that decade specifically a number of approaches that increase genetic diversity were introduced. De Jong analyzed the problem of premature convergence to local optima in genetic space in his thesis in 1975. He recognized the need to integrate the concept of *niches* into the evolutionary optimization method. He introduced *crowding*, where child solutions only replace the most similar parent in the population, to improve performance on multimodal surfaces [14].

Alternative methods for *diversity management* were introduced, like *sharing* [16, 27, 39] or *clearing* [61], where the population is subdivided into subpopulations according to their similarity in genetic or phenotypic space, although the latter was done on a very low level basis. At that time, genes were strings and the "phenotype" constituted the decoding of that string, which represented a function value in a p-dimensional function space. Nowadays, this decoding would be seen as an intermediate representation of a solution in a complex real-world problem. This already gives us some first evidence that niching should take place on a higher phenotypic level.

Research on niching had been established for many decades at the start of the new millennium but it was mostly used to diversify the searching population, not to diversify the optimization result itself. One exception is multi-objective optimization (MOO), where diversity is defined in objective space. The result of a MOO algorithm is constituted by a Pareto front of non-dominated solutions, showing the trade-off between multiple optimization criteria. One expects the front to consist of a diverse set of solutions, all with different qualities in terms of objective functions. Since then, genetic diversity has been established as an important factor in the success of MOO algorithms [70, 76]. As Ulrich, Bader, and Zitzler notice: "*sometimes it might be more important to find a structurally diverse set of close-to-optimal solutions than to identify a set of optimal but structurally similar solutions*" [77].

Finding a diverse set of solutions is one of the main goals in multimodal optimization (MMO). The field of MMO is still quite young, with early work on restricted tournament selection to produce multiple solutions [37] and later being more established by using methods like restarted local search and nearest-better clustering [63, 64, 83].

Research toward producing multiple solutions in optimization was therefore introduced in MOO and the more novel field of MMO.

2.2 Phenotypic Diversity

The idea of measuring similarity on the phenotype level is not new. Pétrowski mentions the possibility [61]. The omni-optimizer by Deb takes a more narrow view on phenotypic space, describing it in terms of objectives and constraint values [19], though one can argue that a fitness function should not be seen as an inherent property of the extended phenotype but rather a metric on the (implicit) goal of a creature or solution's ability to survive.

The first optimization paradigm that embraced behavior-based niching is *novelty search* (NS). In this new paradigm, Lehman and Stanley choose to ignore fitness, which can be deceptive and lead a search toward a local optimum. Instead, they propose to use behavioral similarity as an objective function instead and show that this can circumvent deceptive fitness functions, outperforming fitness-based optimization. They pose that many points in the search space collapse to a small set of simple behaviors, making novelty-based search feasible. NS introduces an archive of past, novel solutions that serve as a basis for measuring novelty [48, 49].

One of the domains the approach was evaluated on was that of a medium-level controller for a biped robot in a three-dimensional simulation. The controller was able to use motor primitives of the six degree of freedom robot to assign poses, based on the inputs of two ground contact sensors only. The authors introduce a novelty metric based on a fixed-horizon sequence of changes in the horizontal coordinates of the robot's center of gravity. To compare the two controllers, the sum of squared distances of both sequences is calculated. This way, robots that fall down end up in the same niche (due to the fixed horizon). NS therefore ignores new solutions that fall down in the beginning and instead concentrates on finding new solutions that follow different trajectories. The results show that NS finds more biped controllers that do not fall down and finds better controllers than fitness-based search alone. This shows that NS is not just an exhaustive search method, but is able to ignore easy-to-reach local optima in a very hard search space.

The authors pose that the problem of deceptiveness resides in the fitness function, based upon the fact that the underlying neuroevolution algorithm they use provides (genetic) diversity management and still is not able to circumvent deceptiveness. The authors do acknowledge that removing the fitness function entirely is a problem in optimization, as it cannot be determined when solutions are fit or whether solutions are functional at all.

Phenotypic diversity allows us to judge different solutions in terms of how they behave in their environment. This can help us ignoring solutions that show similar behavior and concentrate the search on novel behavior.

3 Quality Diversity

The first two quality diversity algorithms built on the ideas of diversity or novelty being similarly important as qualare ity, as well as maintaining that diversity on a phenotypic level. In this section, we will see how the insights from the last section led to the first two major QD algorithms, then provide a general description of QD, discuss practical issues when using QD to solve a task, and finally provide a number of success stories that show how QD has been used so far.

3.1 First Algorithms

Inspired by Mouret's work on the reconciliation of NS with fitness [53], Lehman and Stanley extended NS by adding a competitive factor. They introduced novelty search with local competition (NSLC) [50]. NSLC can be seen as the first true QD algorithm. It formulates fitness and diversity as a bi-objective problem, treating both objectives as equally important and is open-ended.

The algorithm is introduced on the problem of evolving virtual creatures [71], using the height, mass, and the number of active joints as phenotypic aspects to perform niching. While searching for new creature morphologies, NSLC allows those that belong to a known phenotypic niche to compete locally. Reintroducing fitness-based competition, they make sure that the resulting creatures are functional.

NSLC keeps a growing archive of diverse solutions next to its population to measure phenotypic diversity. Solutions are only added to the archive if their morphology is novel enough. For any new candidate solution, its novelty score is calculated by the average Euclidean distance to a fixed number of k nearest individuals in the archive, where k is determined experimentally. Then, the candidate receives a local competition score, represented by the sum of morphologically nearest individuals it outperforms. Both the novelty and local competition scores are then used in an adapted version of the non-dominated sorting genetic algorithm (NSGA-II) [15]. The Pareto front that is produced consists of solutions that show the trade-off between novelty and local competition. Parents for a new generation are selected from the individuals in the front to produce new solutions that are different from those saved in the archive and high-performing when compared to similar solutions.

Although NSLC is shown to better exploit morphological niches than fitness or novelty-based optimization, a number of problems exist with the approach. The archive keeps growing which makes the comparison of new candidates ever more expensive. The reliance on NSGA-II also makes it difficult to scale up to multiple diversity metrics. These problems are addressed by Mouret et al. A year after the introduction of NSLC, Mouret formulates multi-objective landscape exploration, which works in a similar way, but emphasizes the aspect of exploring the landscape of phenotypic traits with respect to their fitness [54]. Phenotypic niching is performed by collecting elites mapped by behavioral traits. Combining this with the idea of building behavior repertoires, formulated one year later [9, 11], lead Mouret and his group to formulate a simplification of the multimodal QD algorithm, multidimensional archive of phenotypic elites (MAP-Elites) [55]. It *illuminates* the fitness landscape through the lens of phenotypic aspects.

MAP-Elites uses a fixed, discretized n-dimensional grid (Fig. 2), with each dimension representing one phenotypic aspect. This prevents comparisons from becoming more expensive as the algorithm progresses and emphasizes the phenotypic space. The number of aspects, which serve as diversity measures, is limited by the fact that the niching space becomes exponentially larger when adding dimensions. Contrary to NSLC, in MAP-Elites, parents are selected randomly from the archive. There is no distinction between the population and the archive, so the archive is updated in place.

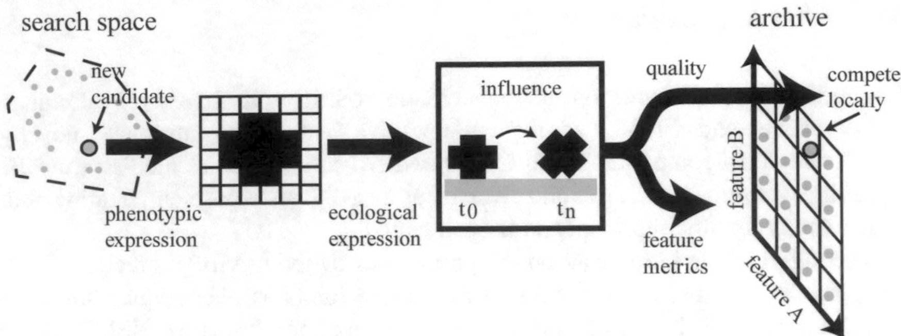

Fig. 2 Candidate genomes are expressed into their phenotypes and, extendedly, into their behavior, through ecological expression. In QD, phenotypic aspects determine similarity and niche assignment. Candidate solutions only compete within the phenotypic niches they occupy

MAP-Elites solve some of the issues in NSLC in such an elegant way that we feel it should be seen as the base QD algorithm, the first in which the archive becomes the central evolvable object. The fixed-grid archive can be seen as a lattice, whose points move through genetic space as elites are replaced over time, turning QD into a meta-version of evolutionary algorithms.

3.2 General Description

QD searches in genetic space for solutions to fill phenotypic niches (Fig. 2). Solutions for tasks that are to be optimized, usually encoded in high-dimensional genomes, are expressed into their phenotypes. A low-dimensional phenotype space is defined by selecting a number of phenotypic aspects that can be based on behavioral, morphological, or other aspects. The aspect values of the phenotypes, *phenotypic descriptors*, are then calculated to determine in which niche they belong. Solutions that occupy the same niche compete to be positioned inside an archive of phenotypic niches. The archive is used to compare new solutions and represents the final outcome of the QD algorithm.

QD algorithms follow the pattern described in Algorithm 1. After initializing a population of solutions and setting an evaluation budget, parents are selected based on a scoring scheme. An explanation of various scoring strategies is elaborated on shortly. In NSLC, parents are selected from a separate population, whereas in MAP-Elites, they are selected from the archive itself. Offspring is created, usually using mutation only. The performance of the offspring is evaluated and their phenotypic descriptors are calculated. The individuals can be added to existing niches through local competition, or new niches can be created, depending on the archive structure that is used. Finally, the selection scores are updated.

Algorithm 1 Quality Diversity

Define and formalize phenotypic descriptors
Initialize population
Initialize archive $\mathcal{A}$
for iter $= 1 \rightarrow$ generations budget **do**
  Select parents to form new offspring population-based on scoring scheme
  Evaluate performance and phenotypic descriptors of offspring
  Add individuals (potentially) to niches in archive $\mathcal{A}$
  Update selection scores
end for

Cully and Demiris propose a modular framework in which the main features of a QD algorithm are described [7]: the type of container or archive and the procedure for selecting the parents of the next generation. We will now discuss the main choices the algorithm designer has to make when building QD algorithms.

3.2.1 Phenotypic Aspects

The phenotypic aspects used for niching can be based on solutions' behavior [6], morphology [25], or other properties of the extended phenotype. For example, with QD, we can evolve an intuition of well-performing neural controllers for a hexapod robot. By expressing each robot controller, certain phenotypic traits can be measured, like the amount of time each leg touches the ground, along which we perform phenotypic niching. Each niche accepts only those controllers whose legs touch the ground in certain temporal patterns. Within that niche, the best solutions compete to survive. The resulting archive of locally well-performing behaviors provided the authors with many different walking gait strategies, which allows us to quickly switch strategies when necessary [6] (see Fig. 6 and accompanying description for further details).

Selecting phenotypic aspects is very much a domain-dependent issue. Some work has been done on unsupervised and adaptive learning of aspects. A major breakthrough in unsupervised generation of aspects is *innovation engines*. The technique, introduced, applied on image generation in [58, 59], uses a deep neural network to learn phenotypic aspects that are "interesting" at an abstract level. By training the network on 1000 image classes, its output confidence can be used to determine whether an image generated by a QD algorithm is a good representative of this class. Similarly, in [5, 8, 34, 35], the aspects used in the archive are found using an autoencoder in an unsupervised manner. Another idea is to measure how well a network is able to compress an image. The better it is able to do this, the less interesting or novel an image is. In [26], the authors evolve indirectly encoded images. They periodically train an autoencoder on all images in the archive and thereby allowing themselves to measure a kind of novelty score represented by the image reconstruction error. The higher the error, the more novel an image is. If the image is novel enough, it gets saved in the QD archive. The diversity metric is therefore not only unsupervised, but also adaptive.

In some cases, unsupervised descriptors might help us to find phenotypic aspects that are hard to define in a particular or high-dimensional domain. This does reintroduce the issue of *explainability*, because there is no direct way to show what the aspects actually represent. We can only perform a posteriori analysis on the deep neural networks, often only by using simple metrics, like a confidence level on the classification output of the network. Using deep neural networks to find aspects is not something that can be done in every domain. It is no coincidence that innovation engines were introduced for images, a domain in which a large amount of data generally is available.

Finally, it is important to take into account the cost of calculating the aspect descriptors, which has to be done for each candidate solution. In this respect, an aspect that is based on, e.g., morphology might be less expensive than one that is based on behavior in a simulation.

We might not always want to define phenotypic aspects that are as influential as possible on diversity, but instead help us to better understand or control solutions in a certain domain. The use of unsupervised aspects can help us discover aspects in more complex domains, although issues around cost and explainability might prevent this in many cases.

3.2.2 Archive

The archive is the central object in QD. It is constructed based on phenotypic aspects and represents our current intuition of the diversity of high-quality solutions to a problem. The archives used in the first QD algorithms were either be fixed, as is done in MAP-Elites, or unbounded as is done in NSLC (Fig. 3). Quite a lot of interesting work has been published since, allowing QD archives to be more efficient, effective, compact, and capable of long-term optimization.

Tackling the problem with the original definition of the grid is the growth of the number of niches, which is exponential in the number of aspect dimensions, the ability of the MAP-Elites archive to deal with high-dimensional niching spaces is

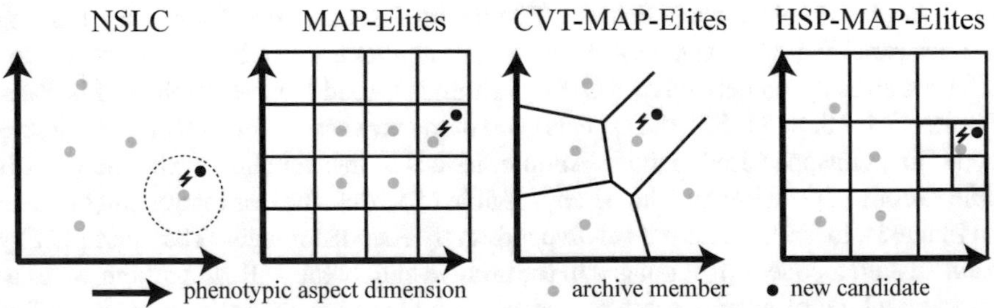

Fig. 3 The first two QD algorithms introduced an unbounded (NSLC) and fixed-grid bounded (MAP-Elites) version of the phenotypic archive. The fixed grid in the latter can be replaced with a Voronoi or a hierarchical subdivision of the phenotype space

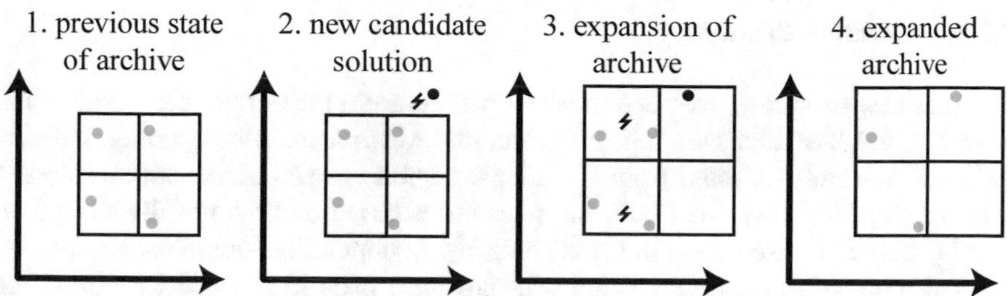

Fig. 4 Expansive MAP-Elites. The bounds of the MAP-Elites archive can be expanded over time in case the bounds cannot be predetermined [80]

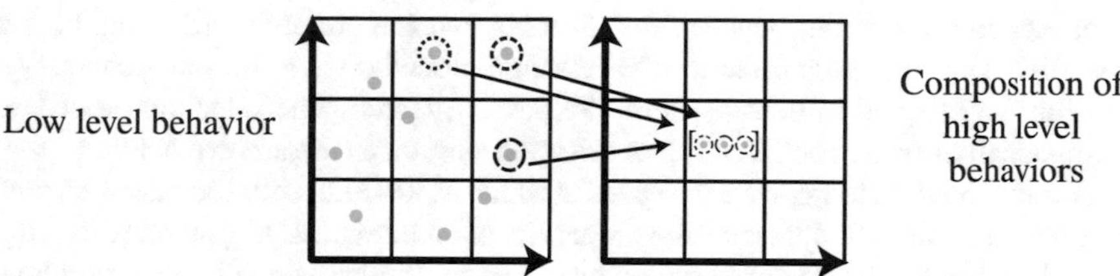

Fig. 5 Hierarchical behavior repertoire. By building up a hierarchy of stacked archives, each archive can be filled using compositions of primitives from another layer

improved. The equidistant grid is replaced with a fixed number of niches, independent of the dimensionality, using the Voronoi tessellation (CVT) [81] (Fig. 3). Others use hierarchical spatial partitioning (HSP), and the speed of exploring the niching space was shown to be increased [73].

In the other work, the fixed bound of the MAP-Elites archive can be expanded during the QD process [80] (Fig. 4), which does lead to changing niche composition and possible removal of solutions. The now unbounded archives perform similarly to the bounded versions, without having to manually predefine the bounds of the niching space. Another unbounded archive is introduced in [80], which bases on the idea of using clustering to dynamically determine the phenotypic niches. Over time, the number of niches can be increased to increase the resolution of diversity.

The idea of using multiple archives was first introduced in [66] to give QD more options to sidestep local optima that can still appear in QD. Taking this idea one step further, a hierarchical, stacked archive, where each layer uses the layer beneath as a set of primitives can decompose the problem of generating complex behaviors [8]. Finally, the archive can be decentralized and used in swarm robotics [38], which allows to simultaneously evolve multiple behaviors across multiple robots (Fig. 5).

Archives can be seen as building blocks for knowledge containers and we can certainly look forward to a myriad of extensions, applications, and novel containers over the next few years. Phenotypic aspects allow us to create effective knowledge representations.

3.2.3 Selection Procedure

To form new offspring, we need to select parents from the current set of solutions. Selection in MAP-Elites was originally introduced as an unbiased operator. Individuals are randomly selected from the archive, whereby each niche can be selected with equal probability. In NSLC, the selection is based on both novelty as well as local quality and takes place in the accompanying population, not in the archive.

A number of alternative scoring schemes have been introduced as a basis for selection. In [67], the probability of selecting a parent in MAP-Elites is proportionate to the novelty of the solution within the niche of the parent. The authors analyze when this selection method makes sense by changing the alignment, which basically translates to correlation, between the diversity and the quality metric. The higher this alignment, the more effective the selection procedure is to fill the archive. QD can thus be adapted to behave more like novelty search. The selection procedure automatically reverts back to random selection as soon as the archive is filled.

Another method to improve the speed at which MAP-Elites fills the map is *curiosity*-based selection. It represents the propensity of an individual to generate offspring that gets successfully added to the archive. The metric is defined by counting how often a niche produces offspring that is accepted into the archive because of novelty or better performance. The niches that produce more offspring, are more novel, or higher performing are selected with a higher probability. The method [7] not only is able to fill the archive more quickly than random selection, but also it often finds better solutions in the niches.

The selection method can have a significant impact on QD performance, although random selection as a default performs well.

3.3 A Practical Example

Let us look at an example domain to see how we would apply QD to it, with a clear separation between genome and phenotype. We intend to find neural network controllers to be used in an autonomous driving setup. The controllers' representation will not further be described other than that we can evolve it and takes the form of a neural network. In the end, we are not interested in what its topology looks like but rather how it behaves in a vehicle in the real world, which is not possible to extrapolate other than simulating the controller or actually running experiments in the real world. Let us assume we have at our disposal simulation software, a variety of typical situational descriptions (like driving on a highway) and the ability to insert a controller into the simulation and measure its behavior.

The fitness function will be to get from a position on the map to a given goal position as fast as possible. The diversity metrics can be things like the energy usage per mile, the average distance to moving objects, or the number of lane changes on the highway. These are things we can measure in simulation, although we will have to find ways to make this measurement as computationally inexpensive as possible.

In the scenario described here, quality and diversity are not always aligned. The average distance to moving objects might confine the ways a controller moves through traffic, but this does not necessarily cause longer travel times. The energy usage per mile is more aligned to quality, as faster speed will cause higher energy usage. Let us use MAP-Elites with a fixed three-dimensional archive consisting of the phenotypic aspects; energy usage per mile, the average distance to moving objects, and the number of lane changes. The archive can be bounded, as was done in MAP-Elites, because the number of aspects is low. This way we can produce an archive of controller behaviors.

There is a number of reasons to not just evolve a single controller and instead look at the space of controller behaviors. If we build up an archive of controllers, along a number of interesting phenotypic/behavioral dimensions, we get a good intuition on what kind of controllers there are that fulfill our optimization criteria. The fastest controllers might cause unforeseen difficulties, for example, produce unsafe driving behavior. We can use these controllers in an adaptive strategy by being able to switch controllers if, for example, a certain controller does not show the same behavior in the real world as in simulation due to damage or unforeseen environmental circumstances, as was done in [6]. For example, if the traffic density increases, we can switch to a controller that performs less lane changes, although it might reach the goal less often. This however might give us a safer driving strategy for the given environment. We can also use the diversity along the axis of the distance to other moving objects to select which controller fits which vehicle model or given safety laws best.

Furthermore, we can use the archive to analyze the stability of controller classes. We can determine the size of the controller classes in genetic space. If the class is too small, the controller might be too sensitive to external influences. Or, if the class is present in many phenotypic niches, we could induce that it might work well under different circumstances. We can also use the insights gained from the QD result and observe unexpected behavior in controllers that might not have been discovered using single-solution optimization strategies.

Another strategy could be to use the hierarchical archives that were introduced in [8] to find higher-level controllers that can be used on lower-level hardware-specific primitives. In this case, we might train the archives once with a random selection procedure but then speed up the process with a curiosity-based selection when exchanging the lower-level archive for a different vehicle model. Finally, we could use QD to test the simulation itself, find diverse controllers, test those in the real world, and then evaluate any discrepancies between it and the simulation.

The expectation is that not only will we find more and possibly better solutions to a problem but also find more insights into the structure of the solution space. We will be able to make more informed decisions on the stability of solutions and what typical behaviors appear and possibly find more and more interesting complex solutions to a problem that is often seen as having a single target, more so than in multi-objective optimization.

3.4 Success Stories

Here is a short overview of some of the most impactful uses of QD in various fields. Quite early on, Nguyen, Yosinski, and Clune [57] show that QD can help to find high confidence false positive images for deep neural networks, by evolving image representations that lead to misclassification for 1000 image classes simultaneously.

QD has mostly been applied in the field of robots. Cully, Clune, Tarapore, and Mouret [6] use MAP-Elites in an adaptive robotics setting, combining it with a trial and error technique based on a preliminary intuition on the behavior of a hexapod robot, the same example that is described in Sect. 3.2.

Figure 6 shows how at first an archive is created as a first intuition to how well various controllers would perform in the domain. The initial intuition (lower left) is then updated after the damage is detected, using an intelligent trial and error technique. QD can also be used to learn diverse robot throwing movements [44]. Jegorova, Doncieux, and Hospedales [43] use generative policy networks that allow generating a vast amount of different robot behaviors to increase the robustness of a robot with respect to changing environments. Multiple authors apply the idea of finding a primitive behavior repertoire that is used to compose complex robot behaviors [21, 29].

QD was applied to reinforcement learning very successfully. It can be used to intelligently explore hard reinforcement learning problems. It has been used to solve Montezuma's Revenge in a never before seen quality [22] and beat human players at the game of StarCraft II, using QD to maintain a diverse population of solutions [2].

The application of QD in design engineering has been taken up by a number of authors. An efficient, surrogate-assisted, variant of QD was applied in an aerodynamic shape design optimization domain, decreasing the number of necessary fluid dynamics simulations by up to three orders of magnitude, while creating hundreds of designs [24, 25, 36]. The approaches allow scaling up the map without the need for more simulations. QD can produce a large array of solutions, which can be impeding to an engineer's capabilities of making a design decision. In order to integrate it

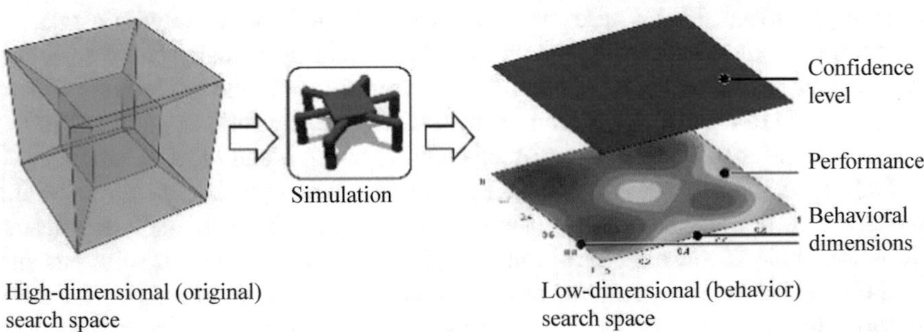

Fig. 6 After finding an archive of walking gaits for a hexapod robot, this initial intuition can be adapted when the robot gets damaged [6]

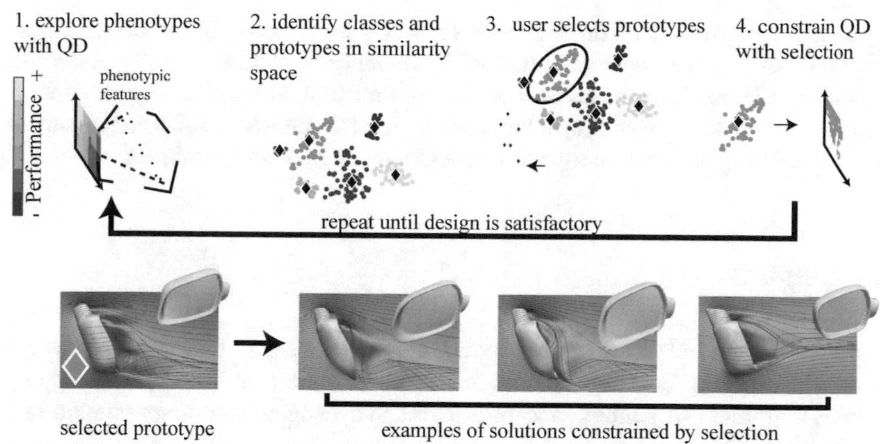

Fig. 7 In this interactive approach, QD explores the space of solutions (1), after which a prototypical representation is extracted from the solution classes (species) found by QD result (2). This provides the user with a more succinct representation of the resulting archive and allows them to select prototypes (3). QD can then be guided to start the search only using the selected solution classes (4) [32]

into real-world design processes, an iterative computer-aided ideation procedure is introduced in [32] (Fig. 7). In this work, user interaction is established by selecting preferred car side mirror classes. The authors use a similarity space that allows comparing high-dimensional optimization solutions and influencing QD with the user's choice. As such, QD is shown to be able to work as a divergent counterpart to convergent human selection.

Phenotypical aspects are usually manually defined. Although this is not generally a big issue per se, as diversifying aspects can be extracted from domain knowledge, automated discovery of these aspects would be even more empowering to break out of the confines of user experience. This is the idea behind innovation engines, which were introduced by Nguyen, Yosinski, and Clune [58, 59]. They combine QD with a deep neural network that is able to evaluate whether behaviors are interesting enough to be used. The work was accompanied by the widely acclaimed work on fooling deep neural networks [57]. Cully and Demiris used innovation engines in their work on hierarchical behavior archives to discover behavioral descriptors with an unsupervised neural network [8]. Building upon this idea, a combinatorial multi-objective evolutionary algorithm evolves multiple QD archives simultaneously, one for every subtask and combination thereof in a multi-task robotics problem. The archives allow the decomposition of more complex tasks and the complexification of solutions to simpler tasks [41].

Finally, taking advantage of the multitude of solutions, a large repertoire of solutions for a workforce scheduling and routing problem can be generated, providing communities with choice, solving many instances of a problem at once [78].

In the short period of time since it has been introduced, QD is becoming an important cog in the wheels of artificial intelligence. Due to the effectiveness and efficiency of its archive, it can be used for optimization, to build more general intuitions for engineers, to build up a knowledge base that increases solution robustness and adaptability, and can be used as a tool to challenge other techniques.

4 Insights

Here we will provide a survey containing the major insights that were discovered about and by using QD: alignment of quality and diversity, the creation of stepping stones, alignment of genome and phenotype, and insights about exploitation and exploration.

4.1 Alignment of Quality and Diversity

A major insight is that where NS performs well when quality and diversity are aligned, QD performs much better when they are orthogonal, which was shown in [66, 67]. The authors introduce a number of diversity metrics, with different degrees of alignment to quality. They show that NS and QD perform very similarly when alignment exists, but QD algorithms generally perform better than NS alone when the niching dimensions are uncorrelated to fitness. One could make a case that the niches allow “sidestepping” deceptive local optima in an unaligned niching space, but this is only speculation and should be evaluated in a dedicated article. However, there is a connection to stepping stones, intermediate goals, which we will now describe.

4.2 Stepping Stones

Diversity is often used to maintain possible stepping stones that can lead to more complex, hard-to-reach solutions [48]. This is also observed when using phenotypic diversity. Evolutionary paths often pass through several basins of attraction before ending up in the target basin. In [59], MAP-Elites is used to generate images for a particular image class. It is shown that the generated images often traverse through other image classes. Decomposing a problem into subtasks is often necessary to reach complicated tasks, as was done in animal training [72]. Called *stepping stones*, subtasks can be used as intermediate goals that allow a system or organism to be guided to a certain complex goal in a multimodal domain. Huizinga, Clune, and others [40, 41, 51] clarify how this effect occurs in QD algorithms. The order of the subtasks (simple to complex) is clearly of importance to the performance of an optimization algorithm. QD algorithms allow avoiding ordering altogether by allowing all subtask combinations to be explored using the QD archive.

The performance gain of the curiosity selection procedure that was mentioned in Sect. 3.2.3 is probably successful due to stepping stones. Selecting niches that produce the most successful offspring are the most *evolvable* and can be translated to actively selecting stepping stones for reproduction. Conclusive work has not been published on this subject.

Diversity measures that are lateral to fitness, like the phenotypic diversity used in QD, seem to improve the stepping stones that can be used. The ability to save phenotypic basins of attraction in the fitness function allows QD to decompose a problem into simpler, easier-to-reach problems.

4.3 Alignment of Genome and Phenotype

As we have seen so far, the two main features of QD are its egalitarian treatment of quality and diversity, and the use of phenotypic niching.[4] We can ask ourselves what effects can occur due to effects caused by the separation between genetic and phenotypic space.

In QD, niching takes place in the space that is spanned by some predefined phe notypic aspects. Keep in mind that the phenotype itself can almost never be fully described. As an example, depending on which view on evolution you take, you might or might not include intra-individual influence in the extended phenotype [13]. This however means that it is computationally and phenomenologically impossible to fully describe the phenotype, as there might not be clear boundaries between what does and does not belong to it. Even a more narrow view of the extended phenotype poses a problem, e.g., when only taking into account the behavior of a neural robot controller in a maze. The observed trajectory really only is a sample of the full behavior. When starting conditions change, due to external or internal noise or when the environment is changed, the exposed behavior of the robot will vary.[5]

When a non-linear and/or adaptive phenotype encoding is present, phenotypic niches are often not aligned to genetic space. The phenotypic variation can be much

---

[4] The field of evolutionary algorithms does not always make a distinction between genetic or phenotypic niching. Confusingly, after Deb mentions the biological definition of species and niches, he defines niches as being artificial subpopulations and niching as a method to force population diversity [17]. There is a case to be made to use more rigor in the definition of niches and species. Niches are basins of attraction in the objective function in phenotypic space. Species are phenotypic solutions that tend to fill certain niches. The genomes of species tend to be similar or at least compatible. One could then argue that, especially when the phenotype is indirectly encoded in the genome, niching has to take place based on phenotypic characteristics. Speciation still takes place on a genetic level, to ensure compatibility between genes. Placing speciation on this level is compatible with Dawkins' understanding of genes as a primary evolutionary unit [12].

[5] One can also argue that the behavior of a neural robot controller in a particular maze is the extended phenotype's embedding in its environment. When a controller was evolved to act in a specific environment, it is not of interest to describe its behavior in other environments.

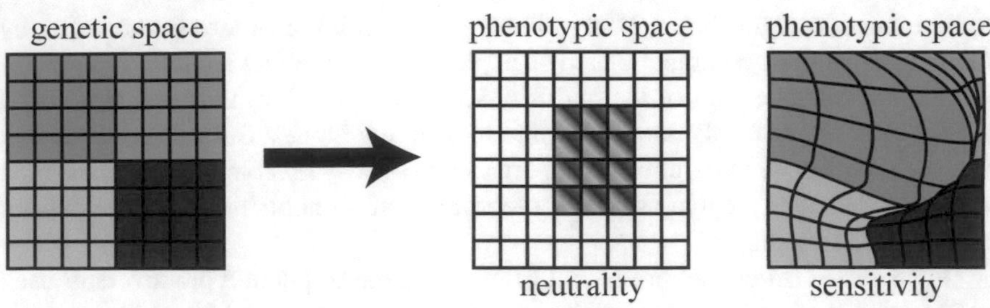

Fig. 8 Species and phenotypes are not necessarily aligned. Due to genetic neutrality, different genomes can expose similar behavior and be put into the same niche. Genetic sensitivity can cause members of the same species to occupy different niches through a non-linear phenotypic mapping

higher than that of the genome. This can cause some species to occupy multiple niches and some niches to be occupied by multiple species. Something similar happens in nature, where more than one biological species can occupy the same niche. Both cows and sheep can graze grass from the same pasture, as long as there is enough nourishment. On the other hand, one species can occupy multiple niches as well, e.g., humans can live as rice farmers as well as hunter-gatherers. We will now look at three phenomena that illuminate the effects of non-alignment between species and niches.

In evolutionary computation, the encoding can cause *genetic neutrality* (Fig. 8), a non-injective genome to phenotype mapping. Individuals with different genomes can therefore show the same functional behavior.[6] We can therefore expect to find different genomes in the same phenotypic niche.

A second effect, *ecological neutrality*, describes functional equivalence between (phenotypic) species [1]. It is a phenomenon that allows multiple species to occupy the same ecological space. There is evidence that this phenomenon explains the co-existence of species in species-rich biological environments [82]. How much this effect can be observed in QD is unknown and certainly an interesting subject of research.

In QD, the genetic diversity of individuals within the same niche is expected to be large because of the effects of genetic and ecological neutrality. Since niches are only described by a small number of aspects, we can expect a high variety of solutions in every niche. This might have an adverse effect when comparing solutions in genetic space, which impacts any method that needs these comparisons.

In [74], the effect of indirect encoding on QD is analyzed. An example of an indirect encoding is HyperNEAT, in which the weights in large neural networks are determined by another network. The authors show that indirect encodings are adversarial to the performance of QD algorithms and point to the fact that locality in the encoding is the main cause here. A small change in the producing network can

[6] Igel and Toussaint analyzed the effects of neutral encodings on computational evolution and showed they are necessary for self-adaptation while only marginally increasing the number of necessary function evaluations [42].

lead to large changes in the resulting network [75]. This effect will be coined *genetic sensitivity* (Fig. 8), the third effect of the alignment between genome and phenotype. The more sensitive a representation is, the larger phenotypic changes can be when applying small changes to the underlying genome.

The observation that (genetic) species sometimes fill more than one phenotypic niche is not restricted to such indirect encodings such as HyperNEAT, but can happen in any mapping between genome and phenotype. Some evidence that species discovered in QD are not homogeneous in terms of their phenotypic aspects is provided in [32, 33], in which genomes discovered by QD are clustered in a low-dimensional space, which allows visual inspection and robust clustering. The resulting solution *classes*[7] are not always mapped to a single phenotypic niche. Vassiliades and Mouret [79] use the insight that in nature, genomes between (phenotypic) species are often quite alike, although they might not occupy the same niche. They introduce the *elite hypervolume*, the part of the genetic space that contains the possible phenotypic elites.

The alignment of genome and phenotype can have a major impact on the performance of QD algorithms. Three effects, namely that of genetic and ecological neutrality and genetic sensitivity, have been described in this chapter. More research should be done on archive-level effects like ecological neutrality and the interaction between genetic sensitivity, neutrality, and QD performance.

4.4 Exploitation and Exploration

In classical optimization, we usually speak of *convergence* of an algorithm when it finds a (local) optimum from which it cannot escape. QD is a class of *divergent* algorithms that prevent getting stuck when a fitness function is *deceptive*, i.e., contains many local optima that hide the true optimum and are easier to reach. In evolutionary optimization, there is a trade-off between *exploitation*, improving what we know is good to become better, and *exploration*, discovering novel solutions. In this subsection, we will look at the influence on exploitation and exploration of some of QD's internals.

Mutation. Vassiliades and Mouret [79] adapt the mutation operator to become biased to move solutions between elites, along the direction of correlation between the two, instead of using an isotropic mutation. The introduced method resembles a mix of a mutation and a crossover operation, moving solutions toward other elites with added isotropic "salt" to allow some exploration to take place. The authors therefore provide an elegant control of the amount of exploitation and exploration that takes place. However, more research should be done to find out how well this method works for highly non-linear mappings. One could suspect that every hypervolume could be treated separately.

---

[7] Classes translate to species as defined in [3].

Nordmoen, Samuelsen, Ellefsen, and Glette [60] evaluate a number of dynamic mutation schemes for MAP-Elites, allowing autotuning of exploitative and explorative behavior. They find some evidence that these schemes outperform fixed-mutation MAP-Elites in terms of the overall fitness and archive coverage.

Selection. The use of random uniform selection in algorithms like MAP-Elites causes a loss of selection pressure which is inversely proportional to the size of the archive [7]. To deal with this loss of selection pressure, score proportionate selection can be applied. Cully and Demiris [7], for example, introduce the *curiosity score* on the archive. The score is higher for individuals that produce offspring that is accepted in the next archive. This can indeed lead to a better filled, higher performing archive. A question that has not yet been answered is whether using score proportionate selection not only leads to a higher phenotypic diversity but also to higher genetic diversity. Lehmann and Miikkulainen [47] find evidence that unbiased mass extinction events implicitly increase pressure for evolvability. Individuals that are highly evolvable, i.e., can quickly generate more different phenotypic traits, have a bigger chance of surviving such events.

Erosion. Cully and Mouret [10] found that archive borders in unbounded QD, e.g., NSLC, tend to erode. When a new but less novel individual is found near the border which outperforms the closest individual, it tends to be selected, hence discouraging exploration and increasing the density of solutions in high-performance regions. MAP-Elites due to their fixed phenotypic grid-like archive do not suffer from this effect.

Increasing Exploration. Pugh, Soros, and Stanley [65] use multiple non-parallel archives to circumvent deceptive regions in the fitness function, possibly allowing a more consistent explorative behavior of QD.

Gravina, Liapis, and Yannakakis [30] introduce *surprise* as an alternative to novelty search. Surprise is defined as the deviation of a measured response to an expected response. It is hypothesized to be orthogonal to novelty, as it measures the change in novelty with respect to the expectation of novelty. They introduce a simple model of the expected behavior of a generation of individuals based on past generations' behaviors. Individuals that do not behave as expected are preferred to be selected. The authors compare NSLC with different combinations with surprise search. There is some evidence that the variant using a linear combination of novelty and surprise performs better than NSLC alone [31].

Thoughts on Generalization of Exploration and Exploitation. In QD, exploration is implemented by making use of phenotypic niching, and exploitation by the local competition that takes place within a niche. Yet, the QD archive itself can be considered to move through genetic space. Due to the non-linear mapping between genetic and phenotype space, and genetic and ecological neutrality (Sect. 4.3), this movement is not bounded. It consists of local improvements but also of large jumps of points in the archive's lattice. But every niche in the archive can contain only points of a subspace of genetic space, those that adhere to certain values of phenotypic aspects [33]. Surely, we can again distinguish between diversity in genetic and phenotypic space. We therefore could generalize exploitation and exploration to the archive's movement through genetic space.

We have shown some of the work that has been done on improving exploration and exploration in QD: on mutation, a crossover-like operation, a score proportionate selection mechanism, archive erosion, and a number of ways to diversify the notion of diversity.

5 Comparing Performance

The current issues in reproducibility of scientific results and lack of understanding of neural network techniques remind us how important well-designed benchmarks and performance metrics are. A QD algorithm's primary goal is to find a diverse set of well-performing solutions, without a bias toward any of the two. In some of the cases that were reviewed, authors are more interested in quality and neglect diversity. In this section, an overview is given on the metrics available for QD performance and a benchmark that might be more appropriate than classical optimization benchmarks is introduced.

5.1 Performance Metrics

In order to measure the performance of QD algorithms, a number of metrics have been introduced. Some of the most commonly used are described in this section. The *total quality* [67], the sum of all fitness values in all niches, is the most straightforward measure for QD, as it can increase when either filling new niches or by increasing the performance in a niche. To understand the performance in more detail, other metrics are used as well. The *collection size*, which can be described as an absolute or relative value, measures the coverage of the phenotypic space. *Maximal quality* (in any of the niches) allows comparing QD to classical optimization algorithms. Because of the way NSLC and MAP-Elites are framed, as an unbounded and bounded version of QD, metrics specific to the boundedness exist. The *total novelty* is specific to the unbounded version, as it measures whether solutions are distributed evenly or more concentrated in phenotypic space. In MAP-Elites, this metric does not make sense, as the niches are evenly distributed by design. Instead, Mouret and Clune introduce *global reliability*, which compares the average performance per niche to the best performance of any (other) algorithm [55] in that niche. Finally, to compare unbounded and bounded variants, the solutions of an unbounded result can be projected into a bounded archive for comparison, as was done in [67].

5.2 Benchmarks

The design of a benchmark for QD is not as straightforward as taking a typical benchmark function like the multimodal Rastrigin function [68]. QD does not perform niching in the same space it uses to search for solutions and there is quite some evidence that this is one of the core insights that allows a divergent search. We therefore need at least a solution representation that entails a mapping between genome and phenotype, which is a common feature of evolutionary optimization, but neglected in purely analytical benchmark functions.

The most commonly used kind of domain in evaluation of QD algorithms are neurally controlled robots acting in a simulated maze environment [20, 28, 33, 41, 67] and robot arm control [7, 8, 79].

Here we explain the maze benchmark used in [33] to show how such a benchmark can look like. The maze can be configured to have multiple rings that all contain a number of exits (three in the example in Fig. 9). A solution's phenotype is defined as a path through the maze. These paths represent multiple basins of attraction. In the neutral configuration, there are multiple paths through the maze that are symmetric. The maze could be configured to contain blocked paths to make traversing the maze more difficult or to make certain paths shorter than others. The phenotypic niches can be based on the final position of the robots, which would align the diversity metric to the maze itself to allow easy visualization of the resulting niches and to allow easy qualitative and quantitative analysis.

In Sect. 4.1, it was discussed that QD algorithms perform better than novelty search alone when quality and diversity are not aligned. To allow comparison between aligned and non-aligned mappings between genome and phenotype (at least), two tasks can be defined in the benchmark (Fig. 10). The first task is a path planning problem in which the genome to phenotype mapping is very simple. A path planning

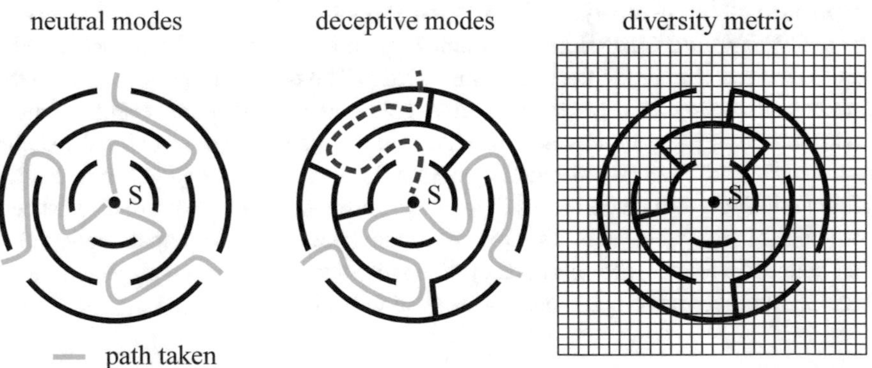

Fig. 9 Neutral and deceptive configurations of a multimodal maze domain with a phenotypic diversity metric that is aligned to the world coordinates for easy visualization and analysis. S is the starting position

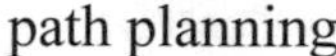

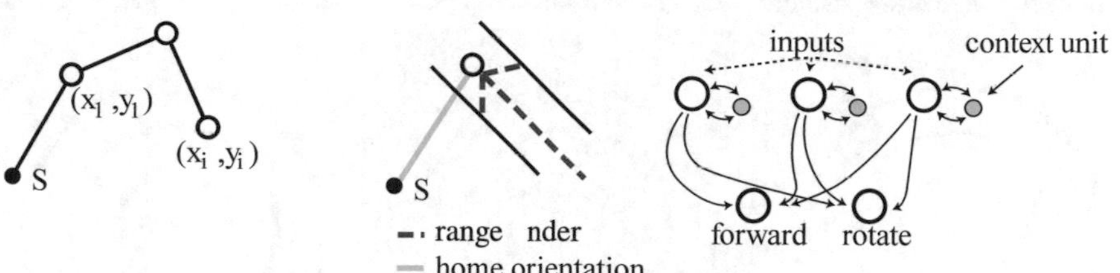

Fig. 10 Two different tasks to allow comparing a direct (path planning) and indirect (robot control) genome to phenotype mapping. The robots are controlled by a simple recurrent Elman network in this example, as was done in [33], whereby the context units are used as a local memory

solution is encoded by a sequence of a number of (x, y) nodes that, when connected, form a path through the maze. The encoding ensures that solutions showing the same phenotypic behavior, e.g., take similar paths, are also similar in terms of their genomes.

In the second task of finding neural robot controllers that traverse the maze, small changes can lead to large changes in the phenotype due to the non-linearities in and reactivity of the neural controller. We recommend using the simulation that was created in [56]. A robot is equipped with three laser sensors that are able to detect the distance to the nearest walls and a home beacon that detects the quadrant in which the direction to the start position of the robot lies. The robots can be controlled by a recurrent Elman [23] network which controls forward/backward and rotational movement, but alternative strategies can be used as well. In both tasks, the representation can be scaled or complexified.

We can use the benchmark to compare QD algorithms using the total quality metric Fig. 11. The fitness function is defined as the shortest path to a niche (Fig. 11). Quality is specifically not defined as "getting out of the maze", as we want to prevent the alignment of diversity and quality. When comparing QD with novelty-based search, we can apply the same diversity measure. To compare QD with classical single objective optimization methods, a (or multiple) goal(s) G can be defined.

The environment and quality and diversity measures can be used for many evaluation scenarios and degrees of complexity in the representation and problem space. A benchmark for QD should fix the quality and diversity measures and most of all the environment with which a phenotype interacts. The proposed benchmark is a suggestion that was used in [33].

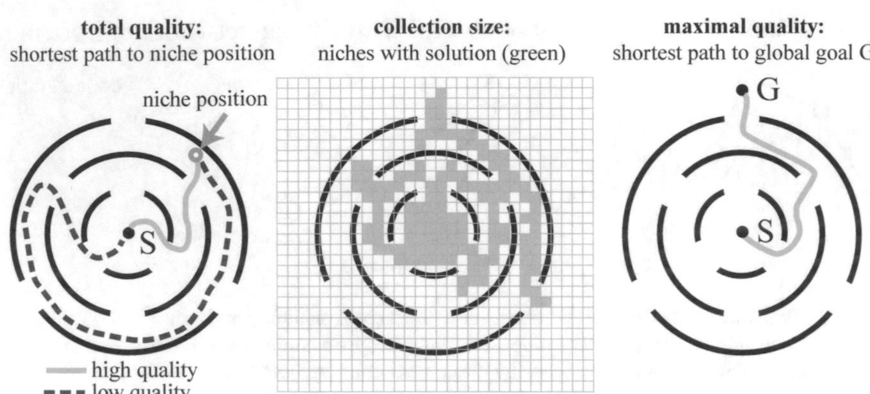

Fig. 11 Quality metrics to compare QD to other QD algorithms using the sum of all quality values over all niches (left), to novelty search using the size of the collection as projected onto a MAP-Elites archive (center), and to classical global optimization using a global goal at one of the exits (right) in the multimodal maze domain

6 Conclusions and Open Challenges

Quality diversity algorithms are a class of algorithms that treat quality and phenotypic diversity in an egalitarian way. The extended phenotype provides a broader interpretation of phenotypic aspects and allows comparison between individuals *in their environment*. QD can serve as a knowledge base for the discovery of many diverse, robust, and stable solutions and can be used as an analytical tool. In this chapter, we have seen the research paths that lead to the development of QD, a generalized description of QD algorithms, some impactful applications, and key extensions. You were provided with the key insights about QD, its embedding in ideas from ecology, a way to evaluate QD algorithms, and a practical example on how and why to apply QD to your optimization problem.

There are some issues and effects that still have to be analyzed, as we already noticed throughout this chapter. In Sect. 4.1, the alignment of quality and diversity was discussed. There is some evidence that unaligned niching space allows sidestepping around local optima (see Sect. 4.2). One could generalize local optima for single solutions to local optima of an entire archive. When thought of as a surface that moves through genetic space, the archive could get stuck on easier-to-reach species or classes within the elite hypervolume that was mentioned in Sect. 4.3. Curiosity score-based selection might actually suffer from falling into these archive-wide local optima.

Efficient QD, mentioned in Sect. 3.4, has already shown that QD can be used on expensive domains. Possible extensions would be the surrogate modeling of phenotypic aspects. Some aspects, especially those that are derived from a solution's behavior, need a full simulation of an individual to be determined. Instead, surrogate aspect models could be built.

Integrating multiple criteria with QD through multi-objective optimization has up to this date not been performed. Also, integrating cooperative aspects in, e.g., multi-robot optimization would be an interesting research track. Finally, the integration into user-driven design processes, solution, and landscape analysis, as was discussed in Sect. 3.4, has just started.

We can imagine that by now the reader might have had the thought that QD seems to be a more generalized interpretation of evolutionary algorithms, cementing in the concept of phenotypic niches and diversity on a supra-population level. Natural evolution is more than genes swimming in an objective landscape—it is the morphological expression of genes and the behavior that phenotype allows to happen that determines how and how well a task is performed. QD takes into account the evolutionary effects that arise from the interaction between the environment and the gene pool, driven by the selfish gene and its extended phenotype.

The move to comparing solutions in all aspects that inform an engineer's, designer's, or user's intuition in terms of the real-world domain has already helped understanding black boxes like deep neural networks and will continue to provide insight in a time where machine learning is becoming more and more powerful and intrusive. This seems to be the time where the computational science community is becoming more and more aware that quality without diversity does not provide us with choice or understanding. QD can unveil some of the intricacies of models, systems, products and make algorithms and solutions more robust with respect to the unknowns we know of and those we have not discovered yet.

Acknowledgements I would like to thank Alexander Asteroth, Adam Gaier, and Jörg Stork for their feedback. This work received funding from the German Federal Ministry of Education and Research and the Ministry for Culture and Science of the state of North Rhine-Westfalia (research grants 03FH012PX5 and 13FH156IN6).

References

1. Alonso, D., Etienne, R.S., McKane, A.J.: The merits of neutral theory. Trends Ecol. Evol. **21**(8), 451–457 (2006)
2. Arulkumaran, K., Cully, A., Togelius, J.: Alphastar: an evolutionary computation perspective. In: Proceedings of the Genetic and Evolutionary Computation Conference Companion, pp. 314–315 (2019)
3. Basto-Fernandes, V., Yevseyeva, I., Deutz, A., Emmerich, M.: A survey of diversity oriented optimization: problems, indicators, and algorithms. In: EVOLVE–A Bridge Between Probability, Set Oriented Numerics and Evolutionary Computation, vol. 7, pp. 3–23. Springer (2017)
4. Basto-Fernandes, V., Yevseyeva, I., Emmerich, M.: A survey of diversity-oriented optimization. EVOLVE 2013-A Bridge Probab. Set Oriented Numer. Evol. Comput. **1**(2013), 101–109 (2013)
5. Cully, A.: Autonomous skill discovery with quality-diversity and unsupervised descriptors. In: Proceedings of the Genetic and Evolutionary Computation Conference, pp. 81–89 (2019)
6. Cully, A., Clune, J., Tarapore, D., Mouret, J.-B.: Robots that can adapt like animals. Nature **521**(7553), 503–507 (2015)
7. Cully, A., Demiris, Y.: Quality and diversity optimization: a unifying modular framework. In: IEEE Transactions on Evolutionary Computation, pp. 1–15 (2017)

8. Cully, A., Demiris, Y.: Hierarchical behavioral repertoires with unsupervised descriptors. Presented at the (2018)
9. Cully, A., Mouret, J.-B.: Learning to walk in every direction. Evol. Comput. **24**(1), 59–88 (2013)
10. Cully, A., Mouret, J.-B.: Evolving a behavioral repertoire for a walking robot. Evol. Comput. **24**(1), 59–88 (2016)
11. Cully, A., Pierre, U., Mouret, J.-B.: Behavioral repertoire learning in robotics. In: Proceeding of the Fifteenth Annual Conference on Genetic and Evolutionary Computation Conference - GECCO '13, pp. 175–182 (2013)
12. Dawkins, R.: The Selfish Gene. Oxford University Press, Oxford (1976)
13. Dawkins, R.: The Extended Phenotype, vol. 8. Oxford University Press, Oxford (1982)
14. De Jong, K.A.: An analysis of the behavior of a class of genetic adaptive systems (1975)
15. Deb, K., Agrawal, S., Pratap, A., Meyarivan, T.: A fast elitist non-dominated sorting genetic algorithm for multi-objective optimization: Nsga-ii, pp. 849–858. Springer (2000)
16. Deb, K., Goldberg, D.E.: An Investigation of Niche and Species Formation in Genetic Function Optimization (1989)
17. Deb, K., Spears, W.: Speciation methods. Evol. Comput. **2**, 93–100 (2010)
18. Deb, K., Srinivasan, A.: Innovization: innovating design principles through optimization. Presented at the (2006)
19. Deb, K., Tiwari, S.: Omni-optimizer: a generic evolutionary algorithm for single and multi-objective optimization. Eur. J. Oper. Res. **185**(3), 1062–1087 (2008)
20. Doncieux, S., Coninx, A.: Open-ended evolution with multi-containers QD. In: Proceedings of the Genetic and Evolutionary Computation Conference Companion, pp. 107–108 (2018)
21. Duarte, M., Gomes, J., Oliveira, S.M., Christensen, A.L.: Evorbc: evolutionary repertoire-based control for robots with arbitrary locomotion complexity. In: Proceedings of the Genetic and Evolutionary Computation Conference 2016, pp. 93–100 (2016)
22. Ecoffet, A., Huizinga, J., Lehman, J., Stanley, K.O., Clune, J.: Go-explore: a new approach for hard-exploration problems (2019). arXiv:1901.10995
23. Elman, J.L.: Finding structure in time. Cogn. Sci. **14**(1990), 179–211 (1990)
24. Gaier, A., Asteroth, A., Mouret, J.-B.: Aerodynamic design exploration through surrogate-assisted illumination. In: 18th AIAA/ISSMO Multidisciplinary Analysis and Optimization Conference, AIAA AVIATION Forum, (AIAA 2017-3330) (2017)
25. Gaier, A., Asteroth, A., Mouret, J.-B.: Data-efficient exploration, optimization, and modeling of diverse designs through surrogate-assisted illumination. In: Proceedings of the Genetic and Evolutionary Computation Conference, pp. 9–106 (2017)
26. Gaier, A., Asteroth, A., Mouret, J.-B.: Are quality diversity algorithms better at generating stepping stones than objective-based search? In: Proceedings of the Genetic and Evolutionary Computation Conference Companion, pp. 115–116 (2019)
27. Goldberg, D.E., Richardson, J.: In: Genetic algorithms with sharing for multimodal function optimization, pp. 41–49. Lawrence Erlbaum, Hillsdale, NJ (1987)
28. Gomes, J., Lyhne Christensen, A.: Comparing Approaches for Evolving High-level Robot Control based on Behaviour Repertoires. In: 2018 IEEE Congress on Evolutionary Computation (CEC), pp. 1–6 (2018)
29. Gomes, J., Oliveira, S.M., Christensen, A.L.: An approach to evolve and exploit repertoires of general robot behaviours. In: Swarm and Evolutionary Computation, pp. 265–283 (2018)
30. Gravina, D., Liapis, A., Yannakakis, G.N.: Surprise search for evolutionary divergence (2017). arXiv:1706.02556
31. Gravina, D., Liapis, A., Yannakakis, G.N.: Quality diversity through surprise. IEEE Trans. Evol. Comput. PP(c):1 (2018)
32. Hagg, A., Asteroth, A., Bäck, T.: In: Prototype Discovery Using Quality-diversity, pp. 500–511. Springer, Berlin (2018)
33. Hagg, A., Asteroth, A., Bäck, T.: Modeling user selection in quality diversity. In: Proceedings of the 2019 on Genetic and Evolutionary Computation Conference - GECCO 2019 (2019)

34. Hagg, A., Asteroth, A., Thomas, B.: A deep dive into exploring the preference hypervolume. In: ICCC (2020)
35. Hagg, A., Preuss, M., Asteroth, A., Bäck, T.: An analysis of phenotypic diversity in multi-solution optimization. In: BIOMA 2020 (2020)
36. Hagg, A., Wilde, D., Asteroth, A., Bäck, T.: Designing air flow with surrogate-assisted phenotypic niching (2020)
37. Harik, G.R.: Finding multimodal solutions using restricted tournament selection. In: ICGA, pp. 24–31 (1995)
38. Hart, E., Steyven, A.S.W., Paechter, B.: Evolution of a functionally diverse swarm via a novel decentralised quality-diversity algorithm. In: Proceedings of the Genetic and Evolutionary Computation Conference, pp. 101–108 (2018)
39. Holland, J.H.: Adaptation in Natural and Artificial Systems. MIT press (1975)
40. Howard, D., Eiben, A.E., Kennedy, D.F., Mouret, J.-B., Valencia, P., Winkler, D.: Evolving embodied intelligence from materials to machines. Nat. Mach. Intell. **1**(1), 12–19 (2019)
41. Huizinga, J., Clune, J.: Evolving multimodal robot behavior via many stepping stones with the combinatorial multi-objective evolutionary algorithm (2018). arXiv:1807.03392
42. Igel, C., Toussaint, M.: Neutrality and self-adaptation. Nat. Comput. **2**(2), 117–132 (2003)
43. Jegorova, M., Doncieux, S., Hospedales, T.: Generative adversarial policy networks for behavioural repertoire (2018). arXiv:1811.02945
44. Kim, S., Doncieux, S.: Learning highly diverse robot throwing movements through quality diversity search. In: Proceedings of the Genetic and Evolutionary Computation Conference Companion, pp. 1177–1178 (2017)
45. Koos, S., Mouret, J.-B., Doncieux, S.: The transferability approach: crossing the reality gap in evolutionary robotics. In: IEEE Transactions on Evolutionary Computation, pp. 1–25 (2012)
46. Laland, K.N.: Extending the extended phenotype. Biol. Philos. **19**(3), 313–325 (2004)
47. Lehman, J., Miikkulainen, R.: Enhancing divergent search through extinction events. Presented at the (2015)
48. Lehman, J., Stanley, K.O.: Exploiting open-endedness to solve problems through the search for novelty. In: Alife, pp. 329–336 (2008)
49. Lehman, J., Stanley, K.O.: Abandoning objectives: evolution through the search for novelty alone. Evol. Comput. **19**(2), 189–222 (2011)
50. Lehman, J., Stanley, K.O.: Evolving a diversity of virtual creatures through novelty search and local competition. In: Proceedings of the 13th Annual Conference on Genetic and Evolutionary Computation, pp. 211–218 (2011)
51. Meyerson, E., Miikkulainen, R.: Discovering evolutionary stepping stones through behavior domination. In: Proceedings of the Genetic and Evolutionary Computation Conference, pp. 139–146 (2017)
52. Mlot, N.J., Tovey, C.A., David, L.H.: Fire ants self-assemble into waterproof rafts to survive floods. Proc. Natl. Acad. Sci. **108**(19), 7669–7673 (2011)
53. Mouret, J.-B.: Novelty-based multiobjectivization. In: New Horizons in Evolutionary Robotics, pp. 139–154. Springer (2011)
54. Mouret, J.-B., Clune, J.: An algorithm to create phenotype-fitness maps. Proc. Artif. Life Conf. **375**(2012), 593–594 (2012)
55. Mouret, J.-B., Clune, J.: Illuminating search spaces by mapping elites (2015). arXiv:1504.04909
56. Mouret, J.-B., Doncieux, S.: Encouraging behavioral diversity in evolutionary robotics: an empirical study. Evol. Comput. **20**(1), 91–133 (2012)
57. Nguyen, A., Yosinski, J., Clune, J.: Deep neural networks are easily fooled: high confidence predictions for unrecognizable images. In: Proceedings of the IEEE Conference on Computer Vision and Pattern Recognition, pp. 427–436 (2015)
58. Nguyen, A., Yosinski, J., Clune, J.: Innovation engines: automated creativity and improved stochastic optimization via deep learning. In: Proceedings of the 2015 on Genetic and Evolutionary Computation Conference - GECCO '15, pp. 959–966 (2015)

59. Nguyen, A., Yosinski, J., Clune, J.: Understanding innovation engines: automated creativity and improved stochastic optimization via deep learning. Evol. Comput. **24**(3), 545–572 (2016)
60. Nordmoen, J., Samuelsen, E., Ellefsen, K.O., Glette, K.: Dynamic Mutation in Map-Elites for Robotic Repertoire Generation, pp. 598–605. MIT Press (2018)
61. Pétrowski, A.: In: A clearing procedure as a niching method for genetic algorithms, pp. 798–803. IEEE (1996)
62. Preuss, M.: Niching prospects. In: Proceedings of Bioinspired Optimization Methods and their Applications (BIOMA 2006), pp. 25–34 (2006)
63. Preuss, M.: Niching the cma-es via nearest-better clustering. In: Proceedings of the 12th Annual Conference Companion on Genetic and Evolutionary Computation, pp. 1711–1718 (2010)
64. Preuss, M.: Multimodal Optimization by Means of Evolutionary Algorithms. Springer (2015)
65. Pugh, J.K., Soros, L.B., Stanley, K.O.: An extended study of quality diversity algorithms. In: Proceedings of the 2016 on Genetic and Evolutionary Computation Conference Companion, pp. 19–20 (2016)
66. Pugh, J.K., Soros, L.B., Stanley, K.O.: Searching for quality diversity when diversity is unaligned with quality. In: Lecture Notes in Computer Science (including subseries Lecture Notes in Artificial Intelligence and Lecture Notes in Bioinformatics), 9921 LNCS:880–889 (2016)
67. Pugh, J.K., Soros, L.B., Szerlip, P.A., Stanley, K.O.: Confronting the challenge of quality diversity. In: Proceedings of the 2015 Annual Conference on Genetic and Evolutionary Computation, pp. 967–974 (2015)
68. Rastrigin, L.A.: Systems of extremal control. Nauka (1974)
69. Schaaf, L.J., John Odling-Smee, F., Laland, K.N., Feldman, M.W.: Niche Construction: The Neglected Process in Evolution. Princeton University Press, Princeton (2003)
70. Shir, O.M., Preuss, M., Naujoks, B., Emmerich, M.: Enhancing decision space diversity in evolutionary multiobjective algorithms. In: Lecture Notes in Computer Science (including subseries Lecture Notes in Artificial Intelligence and Lecture Notes in Bioinformatics), 5467 LNCS:95–109 (2009)
71. Sims, K.: Evolution of Virtual Creatures. In: Proceedings of the 21st Annual Conference on Computer Graphics and Interactive Techniques (1994)
72. Skinner, B.F.: Reinforcement today. Am. Psychol. **13**(3), 94 (1958)
73. Smith, D., Tokarchuk, L., Wiggins, G.: Rapid phenotypic landscape exploration through hierarchical spatial partitioning. In: Lecture Notes in Computer Science (including subseries Lecture Notes in Artificial Intelligence and Lecture Notes in Bioinformatics), 9921 LNCS:911–920 (2016)
74. Stanley, K.O., D'Ambrosio, D.B., Gauci, J.: A hypercube-based encoding for evolving large-scale neural networks. Artif. Life **15**(2), 185–212 (2009)
75. Tarapore, D., Clune, J., Cully, A., Mouret, J.-B.: How do different encodings influence the performance of the map-elites algorithm? In: Proceedings of the Genetic and Evolutionary Computation Conference 2016, pp. 173–180 (2016)
76. Toffolo, A., Benini, E.: Genetic diversity as an objective in multi-objective evolutionary algorithms. Evol. Comput. **11**(2), 151–167 (2003)
77. Ulrich, T.: Integrating decision space diversity into hypervolume-based multiobjective search categories and subject descriptors. In: GECCO 2010, pp. 455–462 (2010)
78. Urquhart, N., Hart, E.: Optimisation and illumination of a real-world workforce scheduling and routing application (WSRP) via map-elites. In: Lecture Notes in Computer Science (including subseries Lecture Notes in Artificial Intelligence and Lecture Notes in Bioinformatics), 11101 LNCS:488–499 (2018)
79. Vassiliades, V., Mouret, J.-B.: Discovering the elite hypervolume by leveraging interspecies correlation. In: Proceedings of the Genetic and Evolutionary Computation Conference, pp. 149–156 (2018)
80. Vassiliades, V., Chatzilygeroudis, K., Mouret, J.-B.: A comparison of illumination algorithms in unbounded spaces. In: Proceedings of the Genetic and Evolutionary Computation Conference Companion, pp. 1578–1581 (2017)

81. Vassiliades, V., Chatzilygeroudis, K., Mouret, J.-B.: Using centroidal voronoi tessellations to scale up the multidimensional archive of phenotypic elites algorithm. IEEE Trans. Evol. Comput. **22**(4), 623–630 (2017)
82. Vergnon, R., Dulvy, N.K., Freckleton, R.P.: Niches versus neutrality: uncovering the drivers of diversity in a species-rich community. Ecol. Lett. **12**(10), 1079–1090 (2009)
83. Wessing, S.: Two-stage methods for multimodal optimization. PhD thesis, Universitätsbibliothek Dortmund (2015)
84. Wright, N.A., Steadman, D.W., Witt, C.C.: Predictable evolution toward flightlessness in volant island birds. Proc. Natl. Acad. Sci. **113**(17), 4765–4770 (2016)

·inted in the United States
Baker & Taylor Publisher Services